IVR X

EQUALITY
INTERNATIONAL AND

Papers of the
World Congress on Philosophy of Law
and Social Philosophy
St. Louis, 24-29 August 1975

Supported in part by grants from
National Endowment for the Humanities
American Bar Endowment

Recognized as an Official Function of the
American Revolution Bicentennial

IVR X

& FREEDOM:
COMPARATIVE JURISPRUDENCE

VOLUME I

Edited by authorization of
Internationale Vereinigung für Rechts- und
Sozialphilosophie (IVR) by

GRAY DORSEY

1977

OCEANA PUBLICATIONS, INC. DOBBS FERRY, NEW YORK
A.W. SIJTHOFF, LEIDEN

Library of Congress Cataloging in Publication Data

World Congress on Philosophy of Law and Social Philosophy, St. Louis, 1975.

    Equality and freedom, international comparative jurisprudence.

    At head of title: IVR X.
    English, French, German, or Spanish.
    Includes index.
    1. Civil rights—Congresses. 2. Equality before the law—Congresses. 3. Liberty—Congresses. 4. Law—Congresses.
I. Dorsey, Gray L. II. Internationale Vereinigung für
Rechts - und Sozialphilosophie. III. Title.
K3239.6.W67   1975   342.08'5   77-76800
ISBN 0-379-00658-8 (v. 2)
ISBN 90 286 0517 7 (Sijthoff)

© Copyright 1977 by Oceana Publications, Inc.

All rights reserved. No part of this publication may be reproduced or transmitted in any form or by any means, electronic or mechanical, including photocopy, recording, xerography, or any information storage and retrieval system, without permission in writing from the publisher.

Manufactured in the United States of America

Table of Contents

Table of Contents

Bibliographic Note   xix

Introduction, Gray Dorsey   xxvii

I. PERSONS. Equal protection of basic aspects of human existence and identity: Life, health, liberty, work, dignity, privacy, and freedom to choose one's own pattern of life. Freedom from destructive or intrusive acts of government, groups, or persons: Freedom from oppression, discrimination, exploitation, experimentation. Norms for the future viewed in the light of historical development of the law of persons and concepts inherent in current liberation movements.   1

DIGNITY, LIBERTY AND EQUALITY
    Luis Recaséns-Siches, Mexico City   3

PERSONS AS MORAL BEINGS
    John Plamenatz (deceased), England   27

HISTORICAL DEVELOPMENT OF THE PRINCIPLES OF EQUALITY AND FREEDOM AND THE CONCEPTION OF MAN, Maria Borucka-Arctowa, Krakow   51

A GLIMPSE OF PERSONS IN THE MODERN LEGALISTIC WORLD, Mitsukuni Yasaki, Osaka   75

LES DEUX FIGURES DE LA LIBERTE AU XVIIIème SIECLE, Simone Goyard-Fabre, Caen   91

THE LEGAL STATUS OF THE INDIVIDUAL: PRESENT AND FUTURE, A. F. Shebanov, Moscow   103

PERSONA Y ESTADO
    Agustín Basave Fernández del Valle, Monterrey   109

THE HUMAN PERSON AND THE LEGAL PERSON
    Iredell Jenkins, University, Alabama   117

LAW, PERSON AND PRIVACY
Lincoln Reis, Brooklyn ... 125

PERSONS, MALE AND FEMALE
Virginia Held, New York City ... 135

LIBERTAD, IGUALDAD Y DIGNIDAD DE LA PERSONA
Marcelino Rodríguez M. ... 143

ÜBER PERSÖNLICHE FREIHEIT UND SOZIALISTISCHES
RECHT, Karl A. Mollnau, East Berlin ... 151

EL HOMBRE, LA SOCIEDAD Y LA LIBERTAD
Domingo A. Labarca P., Maracaibo ... 159

EQUAL WORTH OF PERSONS
Robert Ginsberg, Media ... 169

THE EXHAUSTION OF THE IDEALS OF FREEDOM AND
EQUALITY IN THE UNITED STATES, Lester J.
Mazor, Amherst ... 175

BROWN v. BOARD OF EDUCATION AND EDUCATIONAL
EQUALITY, Klaus H. Heberle, Richmond ... 183

LOS DERECHOS HUMANOS Y EL BIEN COMUN
Jorge Iván Hübner G., Santiago ... 191

SOME QUESTIONS CONCERNING THE CORRELATION
BETWEEN PERSONALITY AND LAW, Vilmos Peschka,
Budapest ... 199

LA RECEPTION ET LE DEVELOPPEMENT DES IDEES
OCCIDENTALES DE LA LIBERTE ET DE L'EGALITE
AU JAPON MODERN, Yukio Uehara, Tokyo ... 205

EQUALITY OF BASIC RIGHTS
Gene G. James, Memphis ... 211

PERSONS AND PUNISHMENT
Wade L. Robison, Kalamazoo ... 219

HUMAN NATURE AND THE STATE
Edward Sankowski, Evanston ... 227

THE PARADOXES OF EQUALITY
Alejo de Cervera, Rio Piedras ... 237

THE FULL DEVELOPMENT OF PERSONALITY IN
SOCIALIST SOCIETIES, Peter Popoff, Sofia ... 245

TABLE OF CONTENTS   ix

II. PARTICIPATION. Democratic ideals, their
    historical development and institutional forms:
    What is the nature of individual participation
    in the new massive groups, governmental and
    non-governmental? What place is there for
    persons participating as functioning parts of
    an organism with definite common objectives?
    For persons participating as equal parties to
    a contractual enterprise? For face-to-face
    and participatory democracy?

    In what ways can new forms of participation be
    developed or old ones refined to broaden lib-
    erties and secure greater equality in
    participation?   253

    THE CONCEPT OF PARTICIPATION
       Kazimierz Opalek, Krakow   255

    PARTICIPATION: A DISCUSSION BASED ON
       SCANDINAVIAN EXPERIENCES, Britt-Mari
       Blegvad, Copenhagen   277

    INDIVIDUAL FREEDOM AND SOCIAL ACTIVITY
       D. A. Kerimov, Moscow   297

    PARTICIPATORY EQUALITY--AN EMERGING MODEL
       Elizabeth Flower, Philadelphia   309

    DERECHO Y PARTICIPACION
       José Luis Curiel, Mexico City   315

    ON A THEORY OF PARTICIPATION
       Joachim K. H. W. Schmidt, Cologne   323

    CLOGS ON PERSONAL PARTICIPATION IN A DEMOCRACY
       B. Sivaramayya, Delhi   333

    PEOPLE'S MASSES PARTICIPATION IN THE MANAGE-
       MENT OF SOCIETY IN THE SOCIALIST DEMOCRACY,
       Ioan Ceterchi, Bucharest   343

    PARTICIPATION: AN OVERVALUED, IMPRACTICAL IDEAL
       Michael D. Bayles, Lexington   351

    PRINCIPLE OF COOPERATION IN THE RELATIONSHIP
       OF THE CITIZEN AND THE SOCIALIST STATE,
       V. S. Shevtsov, Moscow   359

    LA UTOPIA ROUSSEAUNIANA: DEMOCRACIA Y
       PARTICIPACION, Andrés Ollero, Granada   367

PARTICIPATION FROM SOME ASPECTS OF SOCIALIST LEGAL THEORY, Mihaly Samu, Budapest   379

BÜRGER-MITBESTIMMUNG AN DEUTSCHEN GERICHTEN: DEFIZITE UND CHANCEN IN DER LAIEN-AUSWAHL UND DER EINSTELLUNG DER BERUFSRICHTER, Ekkehard Klausa, West Berlin   387

III.A. ANTICIPATION: Environment and Natural Resources. In the light of the increased rapidity of scientific and technological developments and the problems of readjustment they entail, what new methods of social control and new forms of social organization are desirable?

For instance, what should be the forms of legal and social control on newly available natural resources such as those thought to be in the seabed and subsoil of the deep ocean; what mechanisms are available for preventing abuse of the environment and for dealing with over-population?   399

PERSONAL FREEDOM AND ENVIRONMENTAL ETHICS: THE MORAL INEQUALITY OF SPECIES, S. I. Benn, Canberra   401

PERSONAL FREEDOM AND ENVIRONMENTAL ETHICS: THE MORAL INEQUALITY OF SPECIES, Bart Landheer, Gröningen   425

ECOLOGICAL PROBLEMS AND PERSUASION John Passmore, Canberra   431

EQUALITY, FREEDOM, AND GENERAL WELFARE IN ECOLOGICAL PERSPECTIVE, Stephen W. White, Johnson City   443

THE CODIFYING CONFERENCE AS AN INSTRUMENT OF INTERNATIONAL LAW-MAKING: FROM THE "OLD" LAW OF THE SEA TO THE "NEW", Edward McWhinney, Burnaby   453

III.B. ANTICIPATION: Information Science and Human Relationships. In the light of the increased rapidity of scientific and technological developments and the problems of readjustment they entail, what new methods of social control and new forms of social organization are desirable?

In what way can a greater spread of benefits

be achieved? Problems of displaced skills,
ownership of new resources, control of new
modes of power. The effective use of new
technologies of communication and the pro-
tection of privacy.   475

IMPLICATIONS FOR EQUALITY AND FREEDOM WITH
  REGARD TO THE SOCIAL AND LEGAL PROBLEMS
  RESULTING FROM LATEST DEVELOPMENTS IN THE
  FIELDS OF COMPUTER TECHNOLOGY AND INFOR-
  MATION SCIENCE, Wolfgang Kilian, Hannover   477

SOCIALIST DEMOCRACY AND THE SCIENTIFIC AND
  TECHNOLOGICAL REVOLUTION, B. N. Topornin,
  Moscow   499

ANTICIPATION CONCERNING ISSUES ARISING FROM
  THE USE OF COMPUTERS, Claes-Göran Källner,
  Stockholm   507

COMPUTERIZATION FROM THE VIEWPOINT OF AN
  INTERPRETATION OF HISTORICAL MATERIALISM,
  Leszek Nowak, Poznań   513

LA ANTICIPACION COMO TECNICA JURIDICA
  CONTRADICTORIA, Jesús López Medel, Madrid   523

POTENTIALITIES AND PERSPECTIVES OF FORMAL
  LEGAL REASONING, Ilmar Tammelo, Salzburg   531

IV. TENSIONS BETWEEN THE GOALS OF EQUALITY
    AND FREEDOM. The 19th century saw numerous
    attempts to pit liberty against equality as
    the assault of the masses upon freedom. What
    are the lessons of human experience in its
    institutional experiments on the possibility
    of reconciling both and developing unified
    ideals?

    TENSIONS BETWEEN THE GOALS OF EQUALITY AND
      FREEDOM, D. D. Raphael, London   543

    BASIC PROBLEMS IN THE PURSUIT OF FREEDOM
      AND EQUALITY, Berislav Perić, Zagreb   559

    DAS SPANNUNGSVERHÄLTNIS DER SOZIALEN ZIELE
      FREIHEIT UND GLEICHHEIT, Robert Walter,
      Vienna   583

    ISSUES REGARDING TENSIONS BETWEEN GOALS OF
      EQUALITY AND FREEDOM, Stuart S. Nagel,
      Urbana   603

NOTE SUR LES ASPECTS JURIDIQUES DES RAPPORTS
ENTRE L'EGALITE ET LA LIBERTE, Pierre
Goyard, Caen   611

SOME COMMENTS ON THE TENSIONS BETWEEN THE
GOALS OF FREEDOM AND EQUALITY, Shlomo
Avineri, Jerusalem   623

BEMERKUNGEN ZU DEN IDEEN DER GLEICHHEIT UND
FREIHEIT, Vladimir Kubeš, Brno   627

EQUALITY, FREEDOM, AND SECURITY: SOME
INTERRELATIONSHIPS, Lyman Tower Sargent,
St. Louis   635

GLEICHHEIT UND FREIHEIT: KOMPLEMENTÄRE
ODER WIDERSTREITENDE IDEALE, Ota Weinberger,
Graz   641

DIE MAXIMIERUNG DES HANDLUNGSSPIELRAUMS ALS
DEFINITION DER FREIHEIT, Rupert Schreiber,
Cologne   655

EQUALITY'S DEPENDENCE ON LIBERTY
Tibor Machan, Fredonia   663

FREIHEIT ALS ZENTRALWERT EINER
IDEOLOGIEFREIEN RECHTSORDNUNG, Peter Paul
Müller-Schmid, Fribourg   671

LIBERTE ET EGALITE (PASSE, PRESENT, AVENIR)
K. Stoyanovitch, Paris   681

INEQUALITY AS A CONDITION FOR PEACE: SOME
REMARKS ON CHAPTERS 13-17 OF HOBBES'
LEVIATHAN, J. P. van Twist, Nijmegen   689

FREEDOM, EQUALITY, JURIDICAL RELATIONSHIPS
Beniamino Scucces-Muccio, Modica   693

V. THE SCIENTIFIC MANIPULATION OF BEHAVIOR AND
LEGAL PROTECTION OF FREEDOM. New scientific
techniques, such as psychosurgery, genetic
engineering, and operant conditioning, mingle
danger with promise and raise serious moral
and legal issues. Can the law protect freedom in such a way as to maximize human benefits
of the new knowledge? And can the new knowledge help lighten the burden which law has
been carrying as an instrument of social
control?   697

SCIENTIFIC MANIPULATION OF BEHAVIOR AND THE
LEGAL PROTECTION OF FREEDOM, S. I. Shuman,
Detroit ... 699

THE SCIENTIFIC MANIPULATION OF BEHAVIOR AND
THE LEGAL PROTECTION OF FREEDOM,
Shigemitsu Dando, Tokyo ... 727

FREEDOM AS AN OPERATIONAL CONCEPT
Gyan S. Sharma, Jaipur ... 745

ON THE NOTION OF "MANIPULATION"
Zygmunt Ziembiński, Poznan ... 757

A VIEW ON THE BASIC ISSUES
H. Ph. Visser't Hooft, Leiden ... 765

REFLECTIONS ON INFORMED CONSENT
W. T. Blackstone, Athens ... 775

TREATMENT OF PATIENTS AND PROTECTION OF
FREEDOM, B. O. Osuntokun, Ibadan ... 783

LA ULTIMA INDIGNIDAD DEL PROCESO DEMOCRATICO
Alberto E. Serrano, Maracaibo ... 793

BEHAVIOR TECHNOLOGY AND HUMAN FREEDOM
William J. Winslade, Los Angeles ... 801

VI. NEW LEGAL INSTITUTIONS FOR NEW SOCIAL
RELATIONS. A focus on constructive attempts
to assess emerging legal institutions that
attempt to cope with new social problems
(both international and intra-national),
and suggestions for possible lines of fresh
construction. (E.g., UN agencies, legal
instruments in international communication
developments, ombudsman, new uses of insurance, fresh possibilities arising from
computer developments, etc.) ... 807

PERDIDA Y DESPERDICIO DE LOS INTELECTUALES
EN LOS PAISES SUBDESARROLLADOS, Carlos
Cossio, Buenos Aires ... 809

COERCION, RESOURCES AND THE RIGHT TO PARTICI-
PATE IN DECISIONS: SOME TRENDS IN THE LEGAL
DEVELOPMENT, Vilhelm Aubert, Oslo ... 843

BASIC INSTITUTIONS OF PARTICIPATORY DEMOCRACY
Mihailo Marković, Belgrade ... 865

A STUDY OF NYAYA PANCHAYAT OF SULLIA VILLAGE,
KARNATAKA STATE, INDIA, M. G. Narasimha
Swamy and Gurushri Swamy, Bangalore   887

TAKING PURPOSE SERIOUSLY
Philippe Nonet, Berkeley   905

THE IMPACT OF SOME NEW SOCIAL CHANGES ON
LEGAL INSTITUTIONS, Moh. H. Ghanem, Cairo 925

ENFORCING EQUALITY: TWO STATUTORY ATTEMPTS
Geoffrey Marshall, Oxford   933

L'EVOLUTION DE LA TECHNIQUE JURIDIQUE DANS LES
SOCIETES LIBERALES AVANCEES, Paul Amselek,
Strasbourg   941

EQUALITY AND SOCIAL PROGRESS: LEGAL ASPECTS
A. M. Vassilyev, Moscow   953

A COMMENTARY ON CARLOS COSSIO'S LOSING AND
WASTING OF SCHOLARS IN THE UNDERDEVELOPED
COUNTRIES, J. Jorge Klor de Alva, San José 959

THE LEGAL SETTLEMENT OF THALIDOMIDE CASES IN
JAPAN, Yoichiro Yamakawa, Tokyo   965

VII. EQUALITY AMONG NATIONS. What does the ideal
of the equality of nations, traditionally tied
to national sovereignty, call for in a world
in which there is a resurgence of nationalism
in the face of increasing global economic
penetration? How can equality be strengthened
among nations that differ vastly in size,
degree of industrial development, wealth and
power?   971

A NORMATIVE FRAMEWORK FOR REDUCING ECONOMIC
INEQUALITIES, Oscar Schachter, New York City 973

EQUALITY AMONG NATIONS AND THE CONTRADICTIONS
OF TECHNOLOGICAL CHANGE, Ali A. Mazrui,
Nairobi   993

SOVEREIGN EQUALITY AND THE SETTLEMENT OF
INTERNATIONAL DISPUTES, Cornelius F. Murphy,
Jr., Pittsburgh   1013

TOPICAL REFLECTIONS ON THE MAXIM FIAT
JUSTITIA ET PEREAT MUNDUS, Mieczyslaw
Maneli, Flushing   1023

TABLE OF CONTENTS

L'EGALITE ET LA LIBERTE. AUJOURD'HUI ET
DEMAIN, Zygmunt Jedryka, Paris 1033

FORMAL EQUALITY AND SUBSTANTIVE EQUALITY
Fouad Abdel-Moneim Riad, Cairo 1041

INEQUALITY AMONG NATIONS
William L. Blizek, Omaha 1049

A THEORETICAL FRAME FOR EQUALITY AMONG
NATIONS, James King, DeKalb 1059

VIII. EQUALITY AND FREEDOM: PAST, PRESENT AND
FUTURE. What are the significant implications of Equality and Freedom as goals for legal and social organization and action, in historical experience, in immediate problems, and in hopes for the future. 1067

SOCIALISM AND DEMOCRACY
Zdenek Krystufek, Boulder 1069

EQUALITY OF OPPORTUNITY: SOME AMBIGUITIES
IN THE IDEAL, A. M. Macleod, Kingston 1077

FREEDOM AND EQUALITY IN CONSTITUTIONAL AND
CIVIL LAW, H. J. van Eikema Hommes,
Amsterdam 1085

THE PROSPECTS OF MAN'S EQUALITY AND FREEDOM
Adam Lopatka, Warsaw 1095

LE DROIT AU CHOIX DU METIER DANS LES
SOCIETES INDUSTRIELLES: LIBERTE OU
EGALITE, Jeanne Parain-Vial, Dijon 1099

REFLEXIONES SOBRE LA IGUALDAD JURIDICA
Hermann Petzold-Pernía, Maracaibo 1109

MAX WEBER AND THE PROBLEM OF FREEDOM AND
EQUALITY, Gert Schmidt, Dortmund 1119

L'EGALITE, LA LIBERTE ET LE DROIT
Octavian Ionescu, Jassy 1127

LIBERTAD E IGUALDAD: BREVES CONSIDE-
RACIONES SOBRE EL ASPECTO HISTORICO
DEL TEMA, Nelson Saldanha, Recife 1135

RAPPORTS ENTRE LIBERTE ET RESPONSABILITE
DANS L'ECOLE DU DROIT NATUREL MODERNE,
Odile Zanelli, Orsay 1141

DES DROITS DE DIEU AUX DROITS DE L'HOMME
  EN DROIT MUSULMAN, Jean-Paul Charnay,
  Paris   1151

LIBERTE-DEFENSE ET LIBERTE-POUVOIR
  Monique et Roland Weyl, Paris   1161

MORAL CONTENT OF LAW
  Joseph Shatin, Melbourne   1169

LA LIBERTAD Y LA IGUALDAD, PASADO,
  PRESENTE Y FUTURO, Francisco X. González
  Díaz L., Mexico City   1175

PARTICIPANTS   1181

INDEX   1191

Bibliographic Note

Bibliographic Note

IVR is used around the world to refer to an international association of scholars interested in legal and social philosophy. The association conducts its activities in German, French and English. Its name, in those languages is, respectively, Internationale Vereinigung für Rechts- und Sozialphilosophie (IVR), Association Internationale de Philosophie du Droit et de Philosophie Sociale, and International Association for Philosophy of Law and Social Philosophy. It was founded in 1909 in Germany.

The quarterly journal known as ARSP (Archiv für Rechts- und Sozialphilosophie) which was started in 1907 in Germany by Josef Kohler and Fritz Berolzheimer is published under the sponsorship of IVR. Beginning in the 1950's IVR has encouraged the formation of national sections, has held a series of world congresses, and has encouraged conferences and symposia on legal and social philosophy at national and regional levels. Lectures presented and papers discussed at the world congresses, and at national or regional conferences and symposia have been published--usually, but not always, as supplementary, or beiheft, volumes to the journal ARSP.

The Executive Committee of IVR, in order to make the work of its members more readily available, has decided to initiate a series title for all publications resulting from world congresses or national or regional conferences and symposia. The series title for world congresses will consist of the acronym IVR together with a Roman numeral to indicate the place in the series. The name of the country or region will be added in the case of national or regional conferences and symposia.

The publications resulting from the plenary World Congress held at St. Louis, 24-29 August 1975, inaugurate the use of series titles. The public lectures and the papers prepared for discussion of the Property topic, are being published by Franz Steiner Verlag, GmbH, Wiesbaden, West Germany, under the series title, and title, IVR IX, EQUALITY AND FREEDOM: PAST, PRESENT AND FUTURE, edited by Carl Wellman. The present two volumes con-

taining papers discussed at all other sessions of the
St. Louis Congress, bear the series title, and title,
IVR X, EQUALITY AND FREEDOM: INTERNATIONAL AND COMPARA-
TIVE JURISPRUDENCE. It is hoped that librarians will
catalogue and shelve these and future IVR publications
in accordance with the series title, will make individual title/subject analytics for each congress, and to
the maximum extent possible will either recatalogue and
reshelve, or at least note the location of, earlier
volumes in the series, as indicated below.

WORLD CONGRESS PUBLICATIONS

Series title: IVR I   (Presently assigned; not printed on the volume).
Title: Published without an overall title.
Edited by: Not indicated.
Contents: Lectures, World Congress, Vienna, 1959.
Published as: ARSP Beiheft Nr. 38, Neue Folge 1 (1960).
Publisher: Hermann Luchterhand Verlag GmbH, Neuwied am Rhein and Berlin.

Series title: IVR II   (Presently assigned; not printed on the volume).
Title: PHILOSOPHY OF HUMAN RIGHTS AND THE RIGHTS OF THE CITIZEN.
Edited by: Not indicated.
Contents: Lectures, IVR World Congress, Istanbul, 1963.
Published as: ARSP Beiheft Nr. 40, Neue Folge 3 (1964).
Publisher: Hermann Luchterhand Verlag GmbH, publisher of ARSP.

Series title: IVR III   (Presently assigned; not printed on the volume).
Title: VALIDATION OF NEW FORMS OF SOCIAL ORGANIZATION.
Edited by: Gray L. Dorsey and Samuel I Shuman.
Contents: American Symposium, IVR World Congress, Gardone Riviera, 1967.
Published as: ARSP Beiheft Neue Folge Nr. 5 (1968).
Publisher: Franz Steiner Verlag GmbH, Wiesbaden, publisher of ARSP.

Series title: IVR IV   (Presently assigned; not printed on the volume).
Title: SEIN UND SOLLEN IM ERFAHRUNGSBERICH DES RECHTES.
Edited by: Peter Schneider.

BIBLIOGRAPHIC NOTE   xxi

Contents: Lectures, IVR World Congress, Gardone
                  Riviera, 1967.
Published as: ARSP Beiheft Neue Folge Nr. 6 (1970).
Publisher: Franz Steiner Verlag GmbH, Wiesbaden.

Series title: IVR V   (Presently assigned; not printed
                            on the volume).
Title: LE RAISONNEMENT JURIDIQUE.
Edited by: Hubert Hubien.
Contents: Papers, IVR World Congress, Brussels, 1971.
Published as: Book, 1971
Publisher: Établissements Émile Bruylant, Brussels.

Series title: IVR VI   (Presently assigned, not printed
                            on the volume).
Title: DIE JURISTISCHE ARGUMENTATION.
Edited by: Not indicated.
Contents: Lectures, IVR World Congress, Brussels,
                  1971.
Published as: ARSP Beiheft Neue Folge Nr. 7 (1972).
Publisher: Franz Steiner Verlag GmbH, Wiesbaden.

Series title: IVR VII   (Presently assigned; not printed
                            on the volume).
Title: DIE FUNKTIONEN DES RECHTS.
Edited by: Luis Legaz y Lacambra.
Contents: Lectures, IVR Extraordinary World
                  Congress, Madrid, 1973.
Published as: ARSP Beiheft Neue Folge Nr. 8 (1974).
Publisher: Franz Steiner Verlag GmbH, Wiesbaden.

Series title: IVR VIII (Presently assigned; not printed
                            on the volume).
Title: LAS FUNCIONES DEL DERECHO.
Edited by: Not indicated.
Contents: Papers, IVR Extraordinary World Congress,
                  Madrid, 1973.
Published as: ANUARIO DE FILOSOFIA DEL DERECHO, Tomo
                  XVII-1973-1974 (1973).
Publisher: Ministerio de Justicia y Consejo Superior
                  de Investigaciones Cientificas, Madrid.

Series title: IVR IX
Title: EQUALITY AND FREEDOM: PAST, PRESENT AND
                  FUTURE.
Edited by: Carl Wellman.
Contents: Lectures; Papers on topic, Property, IVR
                  World Congress, St. Louis, 1975.

Published as: ARSP Beiheft, IVR IX, forthcoming.
Publisher: Franz Steiner Verlag, GmbH, Wiesbaden.

Series title: IVR X
Title: EQUALITY AND FREEDOM: INTERNATIONAL AND
                COMPARATIVE JURISPRUDENCE.
Edited by: Gray Dorsey.
Contents: Papers, IVR World Congress, St. Louis,
                1975.
Published as: Book, in two volumes.
Publisher: Oceana Publications, Inc., Dobbs Ferry,
                New York and A. W. Sijthoff, Leiden.

NATIONAL OR REGIONAL PUBLICATIONS

AUSTRALIA

Series title: IVR-AUSTRAL.-I   (Presently assigned; not
                                  printed on the volume).
Title: AUSTRALIAN STUDIES IN LEGAL PHILOSOPHY.
Edited by: Ilmar Tammelo, Anthony Blackshield, Enid
                Campbell.
Contents: Symposium of the Australian Society of
                Legal Philosophy
Published as: ARSP Beiheft Nr. 39, Neue Folge Nr. 2
                (1963).
Publisher: Hermann Luchterhand Verlag, Neuwied am
                Rhein and Berlin.

FEDERAL REPUBLIC OF GERMANY

Series title: IVR-BRD-I   (Presently assigned; not
                                  printed on the volume).
Title: IDEOLOGIE UND RECHT.
Edited by: Werner Maihofer.
Contents: Oral presentations, Conference of Federal
                Republic of Germany Section of IVR in
                Köln, 1966.
Published as: Book, 1966.
Publisher: Verlag Vittorio Klostermann, Frankfurt/
                Main.

Series title: IVR-BRD-II   (Presently assigned).
Title: RECHT UND SPRACHE.
Edited by: Theodor Viehweg, Frank Rotter.
Contents: Lectures, Conference of Federal Republic
                of Germany Section of IVR in Mainz, 1974.
Published as: ARSP Beiheft Neue Folge Nr. 9 (1977).
Publisher: Franz Steiner Verlag GmbH, Wiesbaden.

BIBLIOGRAPHIC NOTE   xxiii

LATIN AMERICA

Series title: IVR-LATINAM-I   (Presently assigned).
Title: LATEINAMERIKANISCHE STUDIEN ZUR RECHTS-
                PHILOSOPHIE.
Edited by: Ernesto Garzón Valdés.
Contents: Latin American Symposium on Philosophy of
                Law.
Published as: ARSP Beiheft Nr. 41, Neue Folge 4 (1965).
Publisher: Hermann Luchterhand Verlag, Neuwied am
                Rhein and Berlin.

NORTH AMERICA (UNITED STATES & CANADA)

Series title: IVR-NORTHAM-I   (Presently assigned; not
                                    printed on the volume).
Title: HUMAN RIGHTS: AMINTAPHIL I.
Edited by: Ervin H. Pollack.
Contents: Papers, Second Plenary Conference of
                North American Section of IVR in Newark,
                New Jersey, 1970.
Published as: Book, 1971.
Publisher: William S. Hein, Buffalo, New York. (Jay
                Stewart Publications).

Much of the scholarly work done by members of IVR is published individually, in ARSP or elsewhere. IVR is beginning to make the study of legal and social philosophy more cooperative and collegiate. The purpose of adopting series titles for publications resulting from IVR regional or national conferences or symposia and world congresses is to encourage this trend and to make fully available different points of view so that synthesis can occur where it is warranted.

The present list, above, of IVR sponsored publications does not represent, by any means, a full range of points of view. It is to be hoped that IVR national sections will soon conduct conferences or symposia and that views not now represented will be available in publications in the IVR series.

Introduction

Introduction

The 105 papers in these two volumes were written for the plenary World Congress of the International Association for Philosophy of Law and Social Philosophy which convened at St. Louis, Missouri, 24-29 August 1975. A list of all participants, including the authors of these papers, appears at the end of the second volume. The International Association, founded in Germany in 1909 and generally known as IVR, the acronym of its name in German (Internationale Vereinigun für Rechts- und Sozialphilosophie), meets quadrennially in plenary World Congresses and occasionally in extraordinary World Congresses. (See the Bibliographic Note, supra, for publications resulting from previous congresses.)

Prior to 1975, IVR had never met outside of Europe. At the 1971 World Congress in Brussels, the American Section invited IVR to hold its next quadrennial World Congress in the United States as one of the events marking the Bicentennial of the American Revolution. The invitation was accepted and the Executive Committee selected the theme "Equality and Freedom: Past, Present and Future." The Organizing Committee of the World Congress,* with approval of the Executive Committee of IVR, developed a program which focused upon problem situations in which the implications of different meanings of Equality or Freedom can be tested and understood, rather than upon abstract analysis of the concepts of Equality and Freedom.

---

*The Organizing Committee was composed of Thomas A. Cowan, Rutgers University Law School; Eugene E. Dais, University of Calgary Department of Political Science; President, Gray Dorsey, Washington University School of Law; Secretary, James F. Doyle, Philosophy Department, University of Missouri at St. Louis; Abraham Edel, Philosophy Faculty, Graduate Center, City University of New York; Lon L. Fuller, Harvard Law School; Jerome Hall, Hastings Law School; Iredell Jenkins, Philosophy Department, University of Alabama; and Samuel I. Shuman, Law and Psychiatry Faculties, Wayne State University.

The topics selected for consideration by the Congress are indicated by the title of the first seven chapters of this book.* The Organizing Committee discussion about each of these topics was summarized by Abraham Edel in succinct notes (which appear on the title page of each chapter and under the chapter headings in the Table of Contents) which were sent to all prospective authors. The first two to four papers in each chapter are longer than those that follow, because the authors of these papers were asked to assume special responsibility for stating the issues raised by the chapter topic or for suggesting alternative approaches. Some participants chose to write on the general theme of the Congress, "Equality and Freedom: Past, Present and Future." These papers are collected in Chapter VIII.

Organizing a congress of legal and social philosophers around concepts entails a risk that discussion will become so abstract, so remote from the realities of concrete legal and social structures and processes that pertinence, and more seriously, perspective, are lost. On the other hand, a risk of organizing such a congress around problems is that discussion of the concrete becomes an end in itself and philosophical analysis is abandoned in favor of pursuit of practical solutions. Admittedly, some of these papers suffer from this characteristic defect. This fact will disappoint philosophers, but may make the book more interesting to the general reader.

In most cases, however, the problem focus of the St. Louis World Congress strengthened the philosophical discussion--because theoretical discussion proceeded on the basis of broad experience and knowledge of the social and legal structures and processes by which alternative meanings of equality and freedom come to have significance in people's lives, or because desire to solve a legal or social problem induced creative developments in philosophy. Papers in these volumes that immediately come to mind as exhibiting one or another of the virtues of such a symbiotic relationship between theory and experience include those by John Plamenatz, Iredell

---

*An additional topic, PROPERTY, was also selected. The papers prepared for discussion of that topic, and the public addresses delivered at the Congress are published in a separate volume: IVR IX, EQUALITY AND FREEDOM: PAST, PRESENT AND FUTURE, ed. Carl Wellman, Franz Steiner Verlag GmbH, Wiesbaden, 1977.

Jenkins, Kazimierz Opalek, Bart Landheer, Leszek Nowak, and Zdenek Krystufek. The reader may find other papers particularly cogent.

A further risk of encouraging consideration of experience as well as theory is that each participant's emotions and interests may be so strongly associated with the philosophical justifications underlying his own country's legal and social orders that meaningful discussion of radically different views is impossible. The St. Louis World Congress proved reassuring on this point. Association in a professional organization, such as IVR, apparently promotes an atmosphere of mutual respect and civility which permits candid discussion of radically different premises and presuppositions. Polemics and utopian claims were not eliminated but they were kept to a minimum.

Not surprisingly, individualism and socialism are the philosophical positions that figure most prominently in these volumes. In the form of paradigm statements these are polar opposites but any given author may refer to either in modified form. Summaries of the paradigm positions, as stated in these papers, are:

Individualism. The individual person is primary. The social order is formed by the voluntary assent of individuals. The purpose of law is to protect individuals as the makers of decisions controlling their own lives and as participants in decisions that affect the lives of others, who are free to so function and who are presumed to be equally capable of making such decisions.

Socialism. Objective conditions and laws that can be scientifically known form the basis for development of society and of personality. Development of society in accordance with these objective necessities will eliminate exploitation of man by man and provide the material basis for full development of personality. Participation by individuals in this process will result in the individual freedom of putting subjective interests in correspondence with the objective process of social and individual development. Socialist law is the means by which the state moulds, regulates, and protects the process of realizing the freedom of the individual. All persons are equal in having the right to such development of society and of personality.

Luis Recaséns-Siches forcefully states the position of individualism in his paper, "Dignity, Liberty and Equality," in the PERSONS Chapter. D. A. Kerimov forcefully states the position of socialism in his paper, "Individual Freedom and Social Activity," in the PARTICIPATION

Chapter. Lincoln Reis' paper, "Law, Person and Privacy," and Karl A. Mollnau's paper, "Über Persönliche Freiheit und Sozialistisches Recht," both in the PERSONS Chapter, are cogent shorter statements of the positions of individualism and socialism. The reader will find implications of one or the other of these positions running through many of the papers in these two volumes.

It is unfortunate that other positions are not more fully developed in these papers. Phenomenological premises and presuppositions as the basis for the solution of problems concerning the person, society, and law, appear in Carlos Cossio's paper, "Perdida y Desperdicio de los Intelectuales en los Paises Subdesarrollados," in Chapter VI, NEW LEGAL INSTITUTIONS FOR NEW SOCIAL RELATIONS, and in other papers by Latin American authors. The Indian tradition of freedom as a state of consciousness rather than a relationship between persons in society is introduced by Gyan Sharma in his paper, "Freedom as an Operational Concept," in Chapter V, SCIENTIFIC MANIPULATION OF BEHAVIOR AND LEGAL PROTECTION OF FREEDOM. However, the implications of the phenomenological and Indian non-materialist positions are not discussed in enough papers so that their full potentialities as alternatives to individualism or socialism are set forth.

The titles of the papers by Mollnau and Cossio, referred to above, call attention to the fact that 30 of the papers in this book are in languages other than English. Seven are in German, eleven in French and twelve in Spanish. However, a short English summary follows each. Many of the papers in English were written by persons for whom English is not a first language. These have been edited--in some instances extensively--with the intention of preserving meaning but improving clarity and felicity. The authors have approved the revisions. The index, at the end of the second volume, includes entries in all four languages, integrated alphabetically.

The World Congress could not have occurred without the financial support of the St. Louis legal and business communities and many individuals, and without the dedicated help of many persons, all of whom have been personally and individually thanked. I reiterate those expressions of gratitude, and add my thanks to the authors of the papers in these two volumes and to the other participants in the sessions of the St. Louis World Congress on Equality and Freedom

Gray Dorsey
St. Louis, Missouri
28 February 1977

Chapter One

Persons

Equal protection of basic aspects of human existence and identity: Life, health, liberty, work, dignity, privacy, and freedom to choose one's own pattern of life. Freedom from destructive or intrusive acts of government, groups, or persons: Freedom from oppression, discrimination, exploitation, experimentation. Norms for the future viewed in the light of historical development of the law of persons and concepts inherent in current liberation movements.

DIGNITY, LIBERTY AND EQUALITY

Luis Recaséns-Siches

I. DIGNITY OF THE HUMAN PERSON

The idea of human dignity consists in recognizing that man is a being that has ends proper to himself, his own ends, to be freely complied with by himself. Or putting it in other words, maybe clearer, man ought not to be treated as a mere means for ends which are not his own, which are strange or alien to him for ends which do not belong to him. Although this formulation evokes some words of Kant, it is not necessarily tied to his philosophy. In defining human dignity, Kant did not express an especial idea of his own system, but presented in a clear and concise way an idea generally admitted since many centuries ago, an idea which appears in the Bible, especially in the New Testament.

In the Old Testament we read that "man was created in God's image and likeness"[1] and that men ought to be treated equally, since they all are God's Children. In the New Testament, Jesus Christ appears as Redeemer of all men and as the origin of all of them, as well as their common destiny. For one who believes in Jesus Christ, there is no difference between Jews and Greeks, between slaves and free men, between men and women, because in the faith all are alike, identified with Christ.[2] The idea of human dignity is a characteristic of Christian culture, however not exclusive to it. Such an idea appeared also in the old Chinese thought which attributed to man the uppermost importance.

In the classic Greek philosophy[3] (Plato and Aristotle) we find the idea of human dignity, though frustrated as to its consequences of freedom for all men. In fact, ancient Greeks, as far as they underlined the primacy of reason, opened a way for humanistic ethics, though they did not succeed in drawing the consequences of such an ethical direction when they developed moral and legal philosophy. That way consisted in recognizing that man is not a thing submitted to blind causality, subordinated

to strange ends or powers. On the contrary, in virtue
of his reason, man can reach the goal of a good life.
Certainly man can degrade himself and submit to a mere
animal passion, and subordinate himself to matter. However, he can live also in a divine way, in so far as he
satisfies the requirements of his soul and acts according
to his reason. The chariot of his soul is driven by all
the forces of his nature. The triumphant man, however,
is capable of checking and restraining the bestial
forces. Thanks to his rational mind, man is capable of
succeeding in the knowledge of the highest truths. This
grants man his proper dignity and makes him superior to
all the other living beings on earth.

Notwithstanding these ideas, that lead to the recognition of human dignity, the greatest philosophers of ancient Greece did not formulate this principle with a
universal dimension, since they held that there were some
men with no rights at all--the slaves. Equal dignity
and equal rights were reserved only to free hellenic persons, especially of the male sex. According to Aristotle, the different and inferior treatment of slaves, women
and children was tentatively justified, because women
and children had a lesser participation in reason than
men, and because slaves had no participation in reason
at all.[4] In classic antiquity, only Stoic philosophers,
especially as to the Roman developments--Epictetus,
Seneca, and Marcus Aurelius--held a universal idea of
mankind, of essential equality of all human beings as to
the dignity that belongs to everyone.

Although it was Christianity that gave utmost importance to the idea of dignity of the human person, this
idea has its transcription in the philosophical field.
As a matter of fact, what Kant did, when he formulated
the idea of human dignity, was to give a philosophic
foundation and expression to Christian thought.[5] Modern
age thinking contributed paramount strength to this idea,
stressing that man is the center and the goal of all culture. Kant expressed that all things in this world have
a price--that is, a relative or instrumental value--except man, who has no price, because he has dignity, because he is an end in himself and for himself, because
he is the substratum for the realization of the supreme
values, namely of moral values.

It seems to me that through the ways of purely philosophic meditation it is possible to establish the principle of the dignity of the human individual. That is
what Kant achieved. This has been accomplished also by
other philosophers, for example, by Max Scheler and
Nicolai Hartmann, taking as the basis their theory of
high moral values which the individual complies with by

his free decision. Also I have attempted to contribute to the same purpose by means of the philosophic justification of humanism.[6]

According to humanism, the state, the government (and consequently the law) as well as all culture, will have meaning and justification as a means put at the service of individual human persons, as an instrument for the realization of their ends. This may be expressed, to paraphrase in relation to institutions some Biblical words relating to the Sabbath, "The State or Government was made for the sake of man and not vice versa." On the contrary, antihumanism, (for example totalitarianism) affirms that man embodies values only as he is part of the state or a vehicle for the objective products of culture; that is, that the individual man, as such, lacks any dignity of his own, and only acquires value when he serves as the effective means for some transpersonal end of the state (glory, power, conquest, etc.) According to antihumanism--also called transpersonalism--man is degraded to a pure mass of dough at the service of some supposed objective functions to be realized in state glory, in race, in proletarian class, that is, in transpersonal dimensions. I think that it is possible to justify humanism and to show the absolute error of antihumanism or transpersonalism, by means of several philosophic arguments.

Although philosophic idealism has been superseded by the philosophy of human life or existence, it preserved as a firm truth that my consciousness constitutes the center, support, and proof of all the other realities. The consciousness is inescapably and necessarily the born center of the universe; since its vision or conception is articulated in a perspective which converges in a necessary way on the mental pupil of my mind which contemplates it. I am not a thing among other things, for I am the witness of everything else. The perspective created by individuality is inescapable and necessary; it constitutes one of the components of reality.

The universe is my universe. The world appears as a correlative of the I, as my world. And if I disappear, with me there disappears also my world. It will perhaps be said, that the world will continue to exist for others after I have disappeared. This is correct. But it happens that this affirmation is a theory which I have fabricated, and, consequently a part of my world.

Human life (the individual life, my life) constitutes the primary and basic reality, as co-present and inescapable correlate between the I and my world, between the subject and the objects. To live is to be occupied with a world in which I find myself necessarily. We meet each other, forcibly joined, in inexorable company. Therefore, my life requires these two essential ingredients--the world and I, like inseparable twins. But the objects of the world, the same as I, occur solely in the reality of my life, which is the indubitable reality, and which is also the reality which sustains and conditions all the other realities.

Now, if all that is beyond me obtains expression only in my life, which is the individual life; if all the other things depend on me--although it is also true that I depend on them; if all the other things occur solely in the reality of my life, it is evident that the primacy in a conception of the universe belongs to my life. And from this it follows necessarily that the realization of values has meaning only in my life, which is the individual life.

So-called culture--religion, philosophy, science, art, morals, law, state, government, technology, economics, etc.--is an aggregate of things and works which man makes in his life. Consequently they have meaning only in his life and for his life. Culture consists of human acts and works which aspire to realize ideas of value. It is integrated by actions and products which try to embody truth, in the philosophic and scientific knowledge of the universe; to give sensible form to beauty, in art; to obtain the fulfillment of good in conduct, by morals; to obtain the reign of justice in society, by the law; to utilize nature and overcome its resistances, thanks to the technology; and so forth. Culture, then, as an intention of approaching the values of truth, goodness, justice, beauty, utility, power, etc., has meaning only for man, who does not possess such values in full measure, and who, nevertheless, feels the need of striving to achieve them. Therefore culture has no meaning for unconscious nature, nor for animals; nor has it either for God, who is by essence, absolute wisdom and truth, total good, supreme justice, complete beauty, infinite power. What need has God of science, if He knows everything in eternal actuality; of morals, if He is pure good; and of the law, if He is supreme justice; and of art, if He is perfect beauty, and of technology, if He is omnipotent?

On the other hand, culture seems to us filled with meaning, so far as we regard it as a human function and work. Because man does not know, but needs to know, science is constructed. Because man, who does not shelter within himself pure beauty, nevertheless desires it, art is created. Because man is a sinner, but longs for redemption, we have ethics and religion. Because society must be organized according to justice, law is elaborated. Because man is helpless, but desires to dominate his environment, technology is produced. Therefore, man is necessarily the born center of all culture and its point of final gravitation. And as the supreme values which can be referred to man are the ethical ones, hence the idea of personal dignity must rule always above all his other tasks.

Miguel de Unamuno, in true exaltation of the value of the human wrote: "Who are you?, you ask me, and with Obermann I answer: For the Universe, nothing. For me everything. I am the center of my Universe, the center of the Universe, and in my supreme anguishes, I cry with Michelet: my I, they will take it away from me! Egoism, do you say? There is nothing more universal than the individual, since what it is of each it is of all; and then it is of no use to sacrifice each one to all, but rather sacrifice all to each one. That which you call egoism is the principle of psychic gravity, the necessary postulate. I shall never deliver myself willingly and granting my confidence to any leader of peoples who had not learned that in leading a people he is leading men, men of flesh and blood, men who are born, suffer, and will die, although they do not wish to die; men who are ends in themselves, not only means; men who have to be what they are and not otherwise. It is inhuman, for example, to sacrifice one generation of men to the generation which follows, when there is no feeling of destiny for the sacrificed."[7]

If our life--the individual life of each one--is the basic reality; if values, although objective, occur in our life--as all the rest which there is in the universe--and have an intravital dimension; if the agent of the realization of values is man, alone capable of understanding them and devoting himself to their call, it follows that the realization of values has meaning only for man. Things in which values dwell--among them society, which is a mechanism, an instrument--constitute goods, only in the measure in which they represent apparatus serviceable for man, in the measure in which they are conditions for both his consciousness and conscience being able to embody the supreme values, which are those destined for the individual as such. I stress the fact that I do not deny that in the collectivity there may be and are values in so far as they constitute instruments or conditions for the realization of the values proper to the individual.

The great error committed by antihumanism or transpersonalism is the following: It does not take into account that the collectivity has no substantive reality, that it does not have a being for itself independent of the being of the individuals who compose it. On the other hand, the being of the individual consists of a being for itself, of an autonomous and independent being. On this account the collectivity ought to respect the individual, in the mode of being peculiar to him, in the values which are destined for him, and to recognize his autonomy.

The individual is not simply and purely a part of the whole. Although he must of course be a member of society, he is at one and the same time superior to it. He is superior to the collectivity, because he is a person in the fullest and most authentic sense of this idea, which society can never be. The collectivity would lack meaning if it did not affirm itself as a medium for the individuals.

What has been said must not be misinterpreted in the sense--which would be a great error--that it is denied that in the collectivity there can and ought to be embodied values; nor either in the sense--which also would be a mistake--that society is considered as something purely fortuitous. The individual is essentially social; even to the point that the isolated individual is not anything real, not even possible, but a pure abstraction. The individual exists only in society; he lives on its historic level, supporting himself on it and making use of the goods which he finds in it. Both the anchorite and Robinson Crusoe carry the collectivity within themselves and live upon its level. The individual rests upon the values realized through history and transmitted to him by the collectivity; and almost everything which he does rests upon those communal goods; and sometimes he succeeds at last in raising himself above the historic level of those goods which the collectivity has transmitted to him.

But although the social is something essential to man, the goods which are realized in the collectivity are goods only of an instrumental character. They are means for the realization of higher values, which belong only to the individual and which can and ought to be fulfilled or enjoyed only by the individual. Of course, without society there is no man. But man--understood as the individual man, who is the only one who constitutes a basic and substantive reality--is axiologically superior to society. For society is something made by him and for him. Man has both consciousness and conscience, which society can never have.

A good example of the projections which the idea of dignity has on juridical axiology are the maxims set up

by Rudolf Stammler[8] as auxiliary help for producing a
just law. Stammler developed a purely formalistic juridical axiology. He held that, in relation to the law,
the only value with an absolute dimension, that is, universal and necessary, is the idea of justice. According
to him, the idea of justice is a formalistic method to
order or organize the social ends and means. The idea
of justice consists in the idea of absolute harmony.
Stammler then established some maxims or principles to
help in the elaboration of just law. (In doing so, he
impliedly gave up his formalism, for he had recourse to
the idea of human dignity.) These maxims or principles
are four, and grouped into two classes, as follows:

(1.) Principles of reciprocal respect:

(a.) A person's ends and means ought never to be left
to the arbitrary will of other persons.

(b.) Any juridical requirement ought to treat the
obliged person as a person endowed with dignity, who is
an end in himself, and never a mere means for another
individual.

(2.) Principles of participation:

(c.) Never ought anybody to be excluded from a community or from a legal relationship by the arbitrary decision or the mere subjective whim of another person.

(d.) Any juridical power of disposal, granted to a person, of excluding another person ought to be exercised
in such a manner that the excluded one is treated as a
person with dignity, that is, as an individual with proper ends and never as a mere means for other people or a
mere object of a legal right of other persons.

Some ideas of Del Vecchio[9] help to clarify this theme.
According to Del Vecchio, the idea of justice related to
the dignity of the individual and to legal equality, implies the idea of reciprocity. This means that a person
in acting with respect to other persons ought to do so
only on condition that he recognize as legitimate the
same conduct of others in relation to him. This is a
philosophical expression of the Golden Rule.

II. INDIVIDUAL LIBERTY

The idea of dignity of the human person necessarily
implies the principle of individual liberty. If man is
a being with ends, his proper ones, if he is an end in
himself, and if such end can be fulfilled only by his
individual decision, it becomes clear that the human

person needs a field of liberty or freedom, inside of which the person can operate by himself. Because man has ends, which are his own, to be fulfilled by his decision, he needs respect for and guarantee of his freedom; he needs to be exempt from coercion by other individuals and by public authorities.

Moreover there is another argument to justify man's legal freedom. Juridical liberty is essential to the human being, because man's life is the utilization and development of a series of potential capabilities, of a series of creative possibilities, which cannot be actualized in any preestablished way. The person's development can succeed only by means of the latent creative energies of the human individual. Although society and authority are essentially necessary, they, as well as all institutions, are not creative. Man develops his own person only by means of his freedom. It is true, however, that man needs the help of society, government and law; but only the individual human being, who enjoys freedom, can develop his own creative forces.

From a social and legal viewpoint, freedom has various aspects. It has <u>negative</u> aspects, of a barrier, a stockade, hedge, wall, which defend the person's sanctity against other individuals' intrusion and against the interference by public power. Freedom has also <u>active</u> dimensions, among them political freedom, the democratic right to participate in the government of one's own country, and, to some extent, certain of the so-called social, economic and cultural rights, thanks to which men obtain the material and social conditions as well as the collective services, for the free development of their own possibilities.

At the moment, however, it is necessary to study and to underline the negative aspects, that is, juridical freedom as a set of barriers and defenses against the encroachments or obstacles or interferences of other individuals as well as against any unjust intrusion by the public powers. These negative aspects of the right of juridical liberty include two classes of defenses: (a.) Defense of the individual by the law against the government and its officials; and (b.) Defense of the individual by the law against the attacks of other individuals or against abusive pressures of social groups.

Juridical freedom, which consists in being liberated from undue coercions and intrusions, comprises various dimensions, among which it is convenient to emphasize the following:

(A.) Liberty consisting in being owner of one's own

destiny, that is, in not being slave or serf of anybody, either of another individual, or of any institution, or of the government.

(B.) Security of the person. Here "security" is not used in the legal sense of something deposited to assume performance of a promise, but in the sense of a well-founded conviction, free from doubt or apprehension, that one's moral and physical integrity will be respected, that one's rights of individual freedom will not be infringed, and that one's procedural rights will be honored.

(C.) Freedom of conscience, religion, thought and opinion, and freedom of expression.

(D.) The right to marry with a consenting person and to found a family, without any limitation based on race, nationality or religion.

(E.) The right to choose freely an occupation, profession, work or trade.

(F.) The right to freedom of movement and residence, national and international.

(G.) The right to the inviolability of his privacy, family, home, and correspondence.

(H.) The right to choose one's own pattern of life.

(I.) The right to freedom of peaceful assembly and association.

(J.) The right not to be compelled to attend an assembly or to belong to an association.

III. PERSONAL LIBERTY OR AUTONOMY

Personal liberty is usually declared in a negative form stating that no one shall be held in slavery or servitude. Slavery is the definite negation of man's dignity as well as of all his basic human rights. The notion of slavery, which should be negated and prohibited, includes:

(A.) The slavery of classic antiquity, established by the Roman law, which denied personality to the slave, who was deemed a mere thing owned by his master;

(B.) Slavery in the American continent of the imported Negroes and of their descendants;

(C.) Any form of serfdom which negates man's dignity

and autonomy;

(D.) Forced labor as existed for example in Nazi concentration camps and still exists in the Soviet Union and other communist countries.

(E.) Any other situation, regardless of its name, which is similar to slavery or servitude, that is, which negates or diminishes the dignity and freedom essential to the individual person.

It is not superfluous today to keep on proclaiming the absolute condemnation of slavery, because this abominable institution has not entirely disappeared. Under various forms it persists in some regions of the world. Thus, the United Nations has set up a special commission on slavery to promote an international treaty prohibiting any form of slavery. It so happens that there are still certain cases of slavery in some Arab countries and among some African and Asiatic peoples. And the forced labor camps in the USSR are tantamount to a severe type of slavery.

IV. FREEDOM OF THOUGHT, CONSCIENCE, RELIGION, OPINION AND EXPRESSION

Freedom of thought, and conscience, is an absolute right, because it does not need a special regulation. It simply consists in demanding of other individuals and the public authorities a total forbearance of any interference; it consists in a full respect for anybody's thinking.

Of what does freedom of thought consist? In fact nobody, no human person can impede a person's thinking. A tyrant can chain, torture, and even kill a person, whose thinking he dislikes; but he cannot prevent him from thinking what he thinks. What then does freedom of thought mean? It means that nobody shall be persecuted, prosecuted, punished, discriminated against or curbed in his rights because of his thinking.

The absolute right to freedom of thought is based on the persons' dignity. Certainly an individual's thinking can err. But the only appropriate response is to set out the reasons that demonstrate the error. Any violation of freedom of thought is both a crime and a stupidity. Socrates was condemned to death, yet the socratic philosophy became the basis of Western culture. Christians were sent to the lions; yet the Church kept on growing and expanding.

"Liberty of conscience" habitually means freedom of

thought in religious and moral matters. Actually liberty of conscience is a part of liberty of thought. If it usually is mentioned separately, this is due to the fact that historically freedom of religious conscience was the first to be achieved. Besides, liberty of religious conscience is the most important, because it refers to man's supreme destiny. And this destiny--whatever it be--can be attained only freely, by sincere adhesion, by intimate recognition and by personal decision. Man can go to God only by himself: He can not be forced by the police to seek God. Liberty of conscience of course includes also freedom of doubting; freedom to have no religion; freedom to change one's religion or belief; and freedom, either alone or in community with others, to manifest one's religion in teaching, practice, worship and observance.

Freedom of expression is a consequence of freedom of thought. However, it is not an absolute right. Freedom of thought is an absolute right, because it does not need any regulation and it is not submitted to any limitation. But the right of free expression needs regulation, and has some justified restrictions. First of all, it is necessary to differentiate between expression of thought on the one side, and incitement to action on the other side. A mere thought should never be classified as a penal offense. On the contrary, the incitement to act in a criminal way should be prohibited and punished. Expression of opinions about people is free, provided that it does not involve abusing, insulting, defaming, libelling, or slandering, which should be prohibited and penalized by the law. For reasons of keeping public peace and order, the authorities may prohibit a demonstration or parade which might provoke opposite reactions of disorderly conduct and of fights in the streets.

As indicated above, liberty of conscience includes freedom of cult and observance. However, the law must prohibit the practice of a cult which includes human sacrifices (such as the Aztec one); it must also prohibit rites consisting in the public practice of sexual intercourse (as in certain African tribes).

Freedom of expression includes the right to receive and impart information through any media and regardless of frontiers. This involves the illegitimacy of censorship of press, books, etc.

V. THE SECURITY IN LIBERTY

In the Anglo-Saxon countries the right to a fair legal procedure and judgment is called "due process of law." I think that a good summary of this subject are articles

5, 6, 8, 9, 10, and 11, of the Universal Declaration of Human Rights, proclaimed in Paris by the General Assembly of the United Nations on the 10th of December of 1948. Since the length of this paper is limited by the Congress organizing committee and I have still to deal with other freedoms and with equality, I shall in relation to the legal procedural guarantees only quote the mentioned articles of the Universal Declaration.

"<u>ARTICLE 5</u> No one shall be subjected to torture or to cruel, inhuman or degrading treatment or punishment.

"<u>ARTICLE 6</u> Everyone has the right to recognition everywhere as a person before the law.

"<u>ARTICLE 8</u> Everyone has the right to an effective remedy by the competent national tribunals for acts violating the fundamental rights granted him by the constitution or by law.

"<u>ARTICLE 9</u> No one shall be subjected to arbitrary arrest, detention or exile.

"<u>ARTICLE 10</u> Everyone is entitled in full equality to a fair and public hearing by an independent and impartial tribunal, in the determination of his rights and obligations, and of any criminal charge against him.

"<u>ARTICLE 11</u> (1) Everyone charged with a penal offence has the right to be presumed innocent until proved guilty according to law in a public trial at which he has had all the guarantees necessary for his defence.

(2) No one shall be held guilty of any penal offence on account of any act or omission which did not constitute a penal offence, under national or international law, at the time when it was committed. Nor shall a heavier penalty be imposed than the one that was applicable at the time the penal offence was committed."

VI. FREEDOM TO CHOOSE ONE'S OWN PATTERNS OF LIFE

This freedom is the projection of personal liberty or autonomy, and, of course, also of the dignity of the individual. Its scope is very large. It includes the freedom to remain single or to get married; the freedom to choose a place of residence or domicile; the freedom to travel anywhere; and the freedom to choose occupation, profession, business, employment, trade or calling.

The last one, freedom of occupation, profession, business, employment, trade, or calling, first of all consists in the right of not being prevented from working, and the right to look for the kind of work which the person deems best suited to him. Consequently, it also includes the right to choose freely one's own occupation. A person's occupation is a large part of his life. The negation of this right would be equivalent to the negation of his dignity and tantamount to a sort of serfdom.

There are, of course, some limitations for reasons of competence and safety; for the practice of some professions it is necessary to require proof that the candidate has fitting qualifications.

VII. EQUALITY

All human beings are, at one and the same time, alike and different, equal and unequal.[10] From an empirical viewpoint all human beings are similar:

(A.) In a set of biological characteristics.

(B.) In a set of psychological characteristics.

Men are alike also from the metaphysical standpoint of the functions which constitute human life. In every human life, in a greater or smaller proportion, we find a series of functions, or, saying it in a better way, a system of functions, such as knowledge of the world, and of fellow-men; a technological function of adapting oneself to nature and of dominating nature in order to satisfy human wants; the religious preoccupation about man's destiny after death; the artistic expression of emotions; the social organization, including the legal one; the economic activities; etc.

Nevertheless, at one and the same time, human beings are different from several viewpoints, among these are the following:

(A.) From the biological standpoint, in many aspects.

(B.) Human beings differ also as to psychological characteristics.

(C.) Human beings differ also as to the individual calling or vocation. Man is a born system of preferences and dislikes, something like an emotional battery of attractions and repulsions. Every individual, even before knowing the world around him, is emotionally thrown in a certain direction, towards certain values. Every individual is a singular project of life, to which he may be faithful, but he cannot change it, he cannot replace it by another one with the same degree of authenticity.

(D.) Human beings are different as to their conduct, that is to say, from a moral viewpoint. Thus, without precluding the universal analogies among all human beings and without precluding either the similarities, between many human beings, we must recognize that each human individual is different from all the other individuals, that he is unique. This uniqueness of each individual is precisely essential to human beings. This uniqueness has, among others, the following consequences: (1.) It has its correlative correspondence in a peculiar constellation of values which determines a personal calling, an individual vocation. (2.) The individual person represents a unique viewpoint on the world and on one's own task in life; and consequently embodies a singular perspective, both theoretical and practical.

Furthermore there is great variety as to the content which everyone gives to his life, as to his biography, as to what he does in his existence. José Ortega y Gasset[11] observes "Think for a minute all that man has been, that is to say, what he has done with his life, from the paleolithic savage to the surrealistic youth of Paris." Human beings, but how different!, were Saint Theresa of Avila, Madame de Pompadour, Helen Keller, Eleanor Roosevelt, Socrates, Alexander the Great, Saint Francis of Assisi, Genghis-Khan, Casanova, Rousseau, Charles Chaplin, Einstein and Trotsky.

All the preceding statements show, in the field of observable facts, on the one hand, the similarities of human beings, and, on the other hand, the differences among them from several viewpoints. It is convenient to keep in mind both the similarities and the differences, to deal now with juridical equality.

The principle of juridical equality, in its main postulate pertains to a level different from that of empirical facts. It is based on ethical grounds, and it is projected as juridical condition required by the idea of human person. From the moral and legal viewpoint it means above all--although not exclusively--equality as to dignity of the human person, and consequently equality as to fundamental human rights. It also means parity before the law; and moreover it means also, as a desideratum, the promotion of equality of opportunities.

What I have just affirmed--equality as to dignity and basic human rights, formal equality or parity before the law, and equality of opportunities--is not all that must be considered about the problems of equality in relation to the differences among human beings. Although it is true that we must assert those three dimensions of juridical equality (of human rights, before the law, and of opportunities) it is true also that in many legal relations justice requires taking into account many differences. It should be borne in mind that justice requires giving everyone his own, but not giving everyone the same. Because men, all of them, are alike as to dignity, are persons endowed with their own ends, persons who should never be degraded to the condition of mere means of somebody else, for this reason, all of them ought to be recognized as having equal dignity and equal human rights. But because human beings are different as to specific qualifications and capabilities, as to industriousness, as to conduct, as to the outcome of their activities, for these reasons they should be treated unequally in the relations in which such characteristics are relevant.

Generally speaking, physical differences (stature, muscular strength, etc.) and psychological differences should not be facts with legal relevance affecting human rights, or rights deriving from concrete juridical relations, such as purchaser, seller, lessor, lessee, etc. On the contrary, however, in appointing police officers, it is obvious that stature, muscular strength and agility are relevant. With respect to appointing employees whose activities will require special talent and cultural qualifications, appropriate mental differences should be given legal consequences.

Thus, human beings should be treated unequally as to the diversities that in justice should be taken into account. This second principle, that theoretically is obvious, gives rise in the practice to many problems. The main problem consists in investigating whether certain differences should, or should not, be relevant to the law, and consequently be sources of legal inequalities. It seems very clear that the working woman, when pregnant, should be granted two vacation periods, one before and another after childbirth. But as to the political suffrage for women there have been a variety of opinions, though at present the great majority are in favor of equal rights for women and men; thus the sexual difference being irrelevant to the legal right to vote.

The investigation of the problem whether certain actual inequalities, should, or should not, be relevant to the law, is a problem which cannot be solved in general terms that would be absolute and necessary. It is an imperative of applied juridical axiology that a person's possessory situation should be in harmony with his merits from the viewpoint of his work and his achievements. Not only the quantitative volume of his work, but also the qualitative dimension, and the human value of its results should be considered. For this purpose a correct axiological scale should be used. Undoubtedly the superior degree of this scale is represented by spiritual achievements--scientific, artistic, philosophic, technological, etc.--and by the leading tasks in government and also in great private institutions (social, industrial, commercial, etc.). There are other actual differences in status, which are legitimate sources of juridical inequalities--the diverse functions that the various human beings fulfill in society, for example, the functions of father, mother, child, public authority, etc.

Individual differences of all kinds are precisely the basis of social communities. Where there is diversity, community is possible. Without diversity there can exist unity, but not community. Community and society imply a reciprocal "give and take"; there are reciprocal interchange and complement. The one has in his individuality what the other does not have; the one needs what the other has, and vice versa.

The principle of essential equality among all human beings--as to dignity and human rights--does not exclude many legitimate differences as to concrete rights based on the following grounds: (A.) Diversity of conducts imputable to the individual (for example: legality or delinqency, industriousness or idleness, carefulness or carelessness, etc.); (B.) Diversity of individual capabilities and qualifications, which have a social bearing (physical and mental); and (C.) Diversity of social functions (as father, mother, child, husband, wife, private person, officer, etc.).

VIII. DISCRIMINATION

Juridical equality means equality without any discrimination. Discrimination is any action or omission which denies to individuals belonging to a certain social group equality of treatment.[12] Prejudice and discrimination exist when dislike or hatred is based on the fact that the disliked person belongs to a particular social circle, group or category.

It is necessary to distinguish between inter-individual and collective relations. Inter-individual relations are established between two or more persons on account of their individual characteristics, as such individuals, through the affinities of their peculiarly personal characteristics. In such a relationship each views the other as an individual. Ties of love, friendship and sympathy conform to this pattern. It is the special quality of the individual (as lover, friend, master or the like) that creates the bond. Needless to say hatred, enmity and antipathy also display similarly individualized forms. Prejudice, hostility, and conflict arising in inter-individual relations do not involve discrimination.

A second type of human relations is "social" or "collective" in the strict sense of the words--relations which are established between replaceable, interchangeable persons, as colleagues, co-partisans, co-religionists, comrades, citizens, etc. These relations are established on account of collective functions performed by categories of persons rather than on account of individual characteristics.

Such social or "collective" relations, and their respective attitudes are established not on account of the peculiar qualities of the real individuals involved, but on account of the particular social role which each plays in his capacity as a member of a special collective group or category, for example, as a national, an alien, a neighbor, a Negro, a Jew, a co-partisan, an Oriental, a professional, a manual worker, a woman, etc. These collective relations are mainly based on the fact that an individual is, or is not, included in a social group or category.

The idea of discrimination is not considered to include unjustified prejudice, hostile attitudes, or rejection, solely because of likes or dislikes based on strictly individual qualities. On the contrary, it deals with prejudice, dislike, enmity, or hatred and harmful conduct against a person, because that person belongs to a particular ethnic group, has a certain color of skin, is of the female sex, speaks a certain language, professes a particular religion, stands for a certain political opinion, maintains a certain philosophical or scientific theory, prefers a certain artistic style, is a foreigner, is a wealthy or is a poor person, does not belong to the nobility, is an illegitimate child, or is a soldier, or a lawyer, etc. Discrimination has legal relevance when it consists of acts or omissions which deny or violate legal rights of the person belonging to a disliked group, discriminated against.

There are other types of discrimination which present only social dimensions. For example nobody can force an individual to invite to a party a person whom he does not want to invite.

Discrimination with legal relevance can be sub-classified into two groups:

(A.) Discriminatory conduct by an official authority.

(B.) Discriminatory conduct by private people, which violated a legal right of the person discriminated against, for example, in working, in leasing of homes, etc.

Discrimination committed by public officials may be further classified as follows:

(1.) Unequal treatment in the form of disabilities.

(2.) Unequal treatment in the form of privileges, granted only to members of a favored group.

(3.) Unequal treatment in the form of odious obligations, for example: Imposition of compulsory labor; compelling members of a social group to wear special distinguishing marks, etc.

(4.) Persecution of the persons belonging to the group discriminated against.

All types of discrimination with juridical relevance should be legally prohibited and penalized. The law is a powerful instrument in society, but it is not omnipotent. Although law may not be capable of totally eradicating discrimination, many forms of discriminatory conduct may be suppressed by legal measures. The ways in which the law acts as a factor preventing or repressing discrimination are the following:

(A.) By its coercive dimension the law can help to prevent many discriminatory actions.

(B.) The law also can help repair the harm produced by unlawful conduct, so far as it can provide indemnities and reparation for the person wronged.

(C.) The law can foster the conviction that discrimination is wrong by fixing standards which are respected by the majority of people.

(D.) People who have little respect for the law are nevertheless afraid of the consequences of unlawful conduct; they therefore obey the law in order to avoid its penalties.

(E.) In both cases and whatever the motive, the resulting daily behavior tends to create social customs which are in harmony with the law: These customs constitute a powerful force.

Some forms of discrimination cannot be abolished by law, because they are beyond the reach of legal action;

but the law can and should suppress all discriminatory acts which imply a denial or violations of legal equality. The enjoyment of human rights without discrimination can be guaranteed to all persons within the jurisdiction of a particular state by the enactment and enforcement of legislation which (1.) abrogates all laws which permit or entail any discrimination, (2.) prohibits and penalizes discrimination by both official and private persons. Against merely social discrimination, it can be used only as an educational measure with a long-range, indirect result.

The main pretexts for discrimination are race, color, sex, religion, language, cultural circle or group, social class (including caste, origin, educational and economic status, etc.) and political or other opinion. The most famous anthropologists nowadays deny any scientific possibility of classifying mankind in races, because the differential characteristics (of color, form of head, nature and color of the hair, etc.) are not correlated; on the contrary they are intermingled. As a matter of fact, the most extreme dolichocephalic and brachycephalic types are found among blackskinned men; among white people there are many with blue eyes, but still more with dark eyes; and so forth--all differential characteristics are intermingled.[13] Moreover the best contemporary scientists in the fields of anthropology, psychology and sociology have shown that it is not possible to affirm that mental superiority of any kind can be correlated with any particular physical characteristics. The differences as to cultural achievements do not depend on racial characteristics, but on historical factors, conditions and environment.

In the great majority of cases, discrimination based on color is a particular manifestation of racial prejudice, which is favored and strengthened by the obvious fact of differences in color of the skin. The sexual difference is a real and obvious fact, which moreover has social and legal consequences--maternity with all its effects in the family, and labor law. But women and men, equal in dignity and human rights, deserve an equal treatment by the law. The justification of liberty of conscience refutes any attempt of discrimination on account of religion. Estimable and praiseworthy though it be, patriotism not infrequently becomes a cloak for discriminatory attitudes which are wholly alien to its nature. As to basic individual freedoms, there should be no difference between aliens and nationals, because these freedoms belong to the human being as such, without any specifications. On the contrary, differences between aliens and citizens as to democratic rights are fully justified.

Birth, lineage, origin, ancestry or extraction should be no reason for discrimination, because they do not depend on individual conduct. Nor should belonging to any particular cultural group be taken as ground for discrimination. In democratic societies, social class distinction does not usually engender discriminatory conduct denying or restricting human rights, but eventually only causes prejudices which give rise to forms of merely social discrimination.

IX. POLITICAL LIBERTY AND ITS LIMITATION--NO FREEDOM AGAINST FREEDOM

Freedom of thought and expression applied to political life acquire a special characteristic; they are in this aspect democratic rights. This paper is not intended to deal with human rights which belong to the group of democratic rights--in addition to political freedom, the rights of political assembly and association, the right of suffrage, the right of access to public service. Individual human rights of liberty and equality belong without distinction to all human beings as such. Political rights belong only to the citizens. Political freedom implies activity, participation. It is not studied in this paper, because another group is discussing "participation."

I should like, however, to draw attention to a very important principle concerning a fundamental limitation of political liberty, namely, "no freedom against freedom," that is, no political freedom to be used for the purpose of abolishing freedom and democracy.[14]

The politicians who, in the 20th century, set up totalitarian regimes took advantage of the democratic freedoms precisely to destroy democracy. People of the 18th and 19th centuries made the mistake of admitting that democratic freedom might be used in any direction, including activities aimed at the destruction of freedom and democracy. This limitation, however--no freedom to abolish freedom--is founded not only on that tragic experience, but has an intrinsic validity based on the very essence of political freedom.

People of the 18th and 19th centuries were much too naive and much too optimistic. They believed that once political freedoms were established, there was no place or motive for worrying. Political freedom was compared with the sun light; when political freedom shines, all shadows disappear. And it was said that political freedom itself would heal all the wounds that it might produce.

The exercise of political freedom and of democratic rights must rest on the condition of faithfully admitting the rules of liberty and democracy, that is to say, that all the parties involved in the discussion accept the rules of the play. Fair play is a necessary presupposition for the practice of political freedom. If such a principle does not rule, if certain people, for example people with totalitarian plans, take advantage of the democratic freedoms precisely for the abolition of democracy, they commit a disloyal, unfair and treacherous action. The notion of justice implies the idea of reciprocity. The instrument of government in liberal democracy does not impose a particular political content. On the contrary, that instrument has the possibility of being used in various directions, even the possibility of giving it the most diverse contents. The people will decide if the policy will develop towards the right, or the middle of the road, or the left. The people are free to make this decision, but they should not be free to destroy the democratic mechanism. We could compare democratic government to a ship. The people are free to decide which route the ship will take but they are not free to destroy the ship, namely the democratic mechanism.

Article 30 of the United Nations Universal Declaration of Human Rights has the same purpose against regimes denying freedom and democracy. It reads as follows: "Nothing in this Declaration may be interpreted as implying for any State, group or person any right to engage in any activity or to perform any act aimed at the destruction of any of the rights and freedoms set forth herein."

Several states in their constitutions or in other legal instruments outlaw totalitarian political parties and prohibit and penalize activities aiming at the establishment of a totalitarian regime. There should not be freedom of expression for advocating the suppression of freedom of expression.

In the first decade of this century, two Spanish congressmen held a conversation illustrative of the ideas here defended: Professor Gumersindo de Azcárate, advocate of a very democratic program, and Count Rodríguez de San Pedro, leader of the traditionalist party, most reactionary. Count Rodríguez de San Pedro told his colleague, Professor Azcárate: "In comparison with you I enjoy an advantageous situation; if you some day would rule Spain, according to your ideas and program, you would be bound to respect me entirely; whereas if I should succeed and govern Spain, I, according to my own ideas, I would let you be burnt alive."

FOOTNOTES

1. First Book of Moses: The Genesis, ch. 1, 26 and 27.

2. New Testament, First Epistle of Saint Paul to the Corinthians, VIII, 6; Epistle of Saint Paul to the Romans, XI, 36; Epistle of Saint Paul to the Colossians, I, 16; and principally, Epistle of Saint Paul to the Galatians, III, 38.

3. Recaséns-Siches, Luis, TRATADO GENERAL DE FILOSOFÍA DEL DERECHO 548-551, 4th edition, Editorial Porrúa, Mexico, 1965; the Anthology by Marias, Julian, "El Tema del Hombre," Revista de Occidente, Madrid, 1943; Bienenfeld, F. R., REDISCOVERY OF JUSTICE, Allen & Unwin, London, 1947; Davenport, Russell W., THE DIGNITY OF MAN, Harper, New York, 1955.

4. Recaséns-Siches, Luis, *op. cit.*

5. *Idem.*

6. See: Recaséns-Siches, Luis, HUMAN LIFE, SOCIETY AND LAW: FUNDAMENTALS OF THE PHILOSOPHY OF THE LAW, Translated by Prof. Gordon Ireland, in LATIN-AMERICAN LEGAL PHILOSOPHY BY LUIS RECASÉNS-SICHES AND OTHERS, Harvard University Press, Cambridge, Massachusetts, 1948; TRATADO GENERAL DE FILOSOFÍA DEL DERECHO, 4th edition, Editorial Porrúa, Mexico, 1965.

7. Unamuno, Miguel de, DEL SENTIMIENTO TRÁGICO DE LA VIDA EN LOS HOMBRES Y EN LOS PUEBLOS, Colección Austral, Espasa-Calpe Argentina, Buenos Aires, Mexico, Cuarta Edición, 1941.

8. Stammler, Rudolf, LEHRBUCH DER RECHTSPHILOSOPHIE 199-216, 3rd edition, Walter de Gruyter, Berlin und Leipzig, 1928.

9. Del Vecchio, Giorgio, LA GIUSTIZIA, 4th edition, Editrice Studium, Roma, 1951.

10. Recaséns-Siches, Luis, TRATADO GENERAL DE SOCIOLOGÍA 137-145, 12th edition, Editorial Porrúa, Mexico, 1972; TRATADO GENERAL DE FILOSOFÍA DEL DERECHO 587-594.

11. Ortega y Gasset, José, HISTORIA COMO SISTEMA, 1941, in OBRAS COMPLETAS, Vol. VI, pp. 34 and f., Revista de Occidente, Madrid, 1947.

12. This part concerning "discrimination" has been taken from The Main Types and Causes of Discrimination (Memorandum Submitted by the Secretary-General) United

Nations, 1949. I feel entitled to do it, because I was
the author who drafted that "Memorandum," when I was an
officer of the United Nations Secretariat. I must state
that for that job I had the most valuable assistance and
cooperation of my colleague in the Division of Human
Rights, Mr. Edward Lawson.

13. Recaséns-Siches, Luis, TRATADO GENERAL DE SOCIOLOGÍA
189-218, 12th edition, Porrúa, Mexico, 1972.

14. Recaséns-Siches, Luis, "El Art. 30 de la Declaración Universal de Derechos del Hombre," in Revista
Jurídica de la Universidad de Puerto Rico, Vol. XXX,
N° 3-4, 1961; Recaséns-Siches, Luis, TRATADO GENERAL DE
FILOSOFÍA DEL DERECHO, 597-600.

PERSONS AS MORAL BEINGS

John Plamenatz

What is a person?

Philosophers have put two questions, not unconnected and yet distinct. They have asked, In what does the identity of a person, or human being, consist? They have asked also, What is it about human beings that makes their behavior different in kind from that of other animals, and so makes persons of them, having claims upon and obligations to one another? Some philosophers - as, for example, Descartes - have given to the first question an answer quite different from the answer they would give to the question, In what does the identity of a dog consist? Others - whether they are crude materialists or empiricists or "phenomenalists" or (in France) "sensualists" - would give what is in principle the same answer to the question, whether it is put about a human being or an animal of another species. Hobbes, for example, would do so, or Hume or Condillac. Yet they too would agree that distinctively human behavior differs in fundamental respects from the behavior of other animals, so that men are uniquely rational and moral beings and therefore persons.

I.

A crucial objection to both the crude materialist and phenomenalist explanations of personality is that they fail to take account of what it is to be active as only a person can be, to be a rational agent, a maker of decisions, a deliberate initiator of change. In other words, they fail to distinguish, as they ought to do, between characteristically human actions which are purposeful and free, and behavior which is not purposeful and of which it cannot properly be asked whether it is free, in the sense in which only purposeful behavior can be so. This was the objection to "materialism" and "sensualism" that Rousseau, in his book EMILE, put into the mouth of his Savoyard Vicar, an objection made later by other writers, especially German writers, more coherently and clearly than by Rousseau.

The objection is not merely that the phenomenalist and the materialist assume that all human actions are fully determined just as all physical events are: that they too have their necessary and sufficient conditions, so that, given these conditions, the actions cannot but ensue. For, logically, materialists need not be determinists in this sense. A physical event that happens by chance is not inconceivable, even though, until quite recently, natural scientists have worked on the assumption that the phenomena they seek to explain are all fully determined. Also, not all philosophers who have insisted that deliberate human actions differ essentially from all other events, and do so in ways that materialists fail to notice, have held that they are uncaused, or not fully caused. If, as some of their critics say, they are logically committed to holding this, the fact remains that many of them have not recognized it. They have not conceded that the very idea of choice or decision entails that a choice or decision is not fully determined but could have been other than it actually was, though everything else had remained the same. They have not conceded that, if all human actions were determined in this sense, there could be no genuine choices or decisions, so that anyone who claimed to have made a choice or decision would necessarily be deceived about the true character of what he had done.

If a physical event happening by chance is conceivable, then so too is a mental event, and not the less so if there were no more to it than a Hobbes or a Hume takes into account. Though a decision, as Hume might describe it, could well be not fully determined, its not being so would bring it no nearer to being a decision, properly so called; the inadequacy of the description would remain. For Hume explains all mental operations in terms of sense-impressions and their faint copies in memory, and speaks of them as if they were merely ways in which these impressions and ideas combine or behave. Thus, on this view of it, a choice or decision is rather something that happens to a person than something he does, just as the movement of an inanimate body is something that happens to it. It is a coming together of ideas (or of impressions and ideas), a concatenation of events appropriately stimulated and having its appropriate effects. As Hobbes, or even Hume, explains purposeful action, it is not essentially different from responding or reacting to a stimulus. But, according to Rousseau or Kant or Hegel, to describe human actions in this way is to fail to take proper account of what is involved in being a rational agent, a person, a maker of decisions responsible for what he does.

(We must, of course, distinguish the way in which a

philosopher explains something and the way in which he speaks of it when he is not theorizing about it. As critics of Hobbes have often noticed, after explaining mental operations in his own peculiar ways, he goes on to speak of them as we all do, and therefore in ways which suggest that there is much more to them than his theory about them takes into account.)

There is not much point to calling the idea of the person of Rousseau or Kant or Hegel "non-materialist", for the word "materialism" has been used in too many different senses to make this negative term informative. There have been thinkers who called themselves "materialists", or who have been so called by others whose conception of personality was closer to Rousseau's or Hegel's than to Hobbes's or Hume's. For example, Marx.

If we say that acting purposefully is essentially different from responding to a stimulus, we are not committed to separating the mental sharply from the physical. For example, we are not committed to speaking of intentions as if they were mental acts temporally prior to the bodily behavior which is their effect. Sometimes, no doubt, we can be thinking, or even forming intentions (not to speak of desiring or feeling) without any physical behavior on our part which either we or anyone else detects. But, quite often, we express our thoughts or intentions (or desires and feelings) in the words we utter or gestures we make, without our being able to distinguish them as causes temporally prior to the spoken words or gestures that express or manifest them. And it is arguable that we can have thoughts and intentions properly so called (as distinct from mere dispositions to act or impulses) without giving outward expression to them, expression that others might observe and understand, only because in the past we have learnt to express them outwardly and to understand them when so expressed by others. So, too, it is arguable that we could not have feelings, desires and dispositions (even such as are common to us with other animals) which we do not express and yet are aware of having, unless we had in the past expressed them or understood others who expressed them.

As a matter of fact, even avowed "materialists" have spoken of mental operations, such as thinking, intending and deciding, as if they were distinct from the publicly observed, or observable, bodily movements associated with them. Hobbes, for example, did so. Mental operations may be motions in the brain or the heart, or in some other invisible and inaudible part of the body, or may be "appearances" of these internal and hidden motions but they are quite separate from the outward bodily movements they cause, and the "appearances" are private to the person inside whose body the internal motions occur.

According to Hegel, whose views about what it is to be a person greatly influenced Marx, only a self-conscious being aware of itself as the enduring subject of its actions and experiences can be rational and act purposefully. This does not mean that it could conceivably exist, as a bare subject, apart from its actions and experiences, or apart from the body through which it acts and has experiences; its acts and experiences have their necessarily physical aspects. To think of it as a collection or series of acts and experiences related to one another in ways in which they are not related to acts and experiences forming other collections or series; or to think of it as a collection or series identified by its close or immediate causal links with a particular body; or as a bare subject of actions or experiences; or as something purely mental associated with a body, controlling it and by means of it objects in the world outside it, and getting information of that world through it: these ways of conceiving what a self-conscious, rational and purposeful being is are all quite inadequate. Or, rather, these ways of explaining what a person is are inadequate: For a person is a self-conscious, rational and purposeful being, whose behavior attests that it does not in fact conceive of itself in these ways.

Hegelian metaphysics, or the Hegelian conception of the world, are unacceptable, and perhaps even in part unintelligible, to many people, including the writer of this paper. Nevertheless, Hegel makes about human beings, about persons as we know them, assertions, negative and positive, which make very good sense. He rejects the mind-body dualism of Descartes as much as the materialism of Hobbes and the phenomenalism of Hume. Self-knowledge in man is not prior to knowledge of other persons, but it is acquired in the course of the same practice and process of learning as knowledge of a world that is external to the knower and includes other persons besides himself. Man is essentially a social being in the sense that it is by learning to live and communicate with beings of his own kind, and by acting upon and reacting to what is external to him, that he develops his distinctively human capacities, and acquires his conceptions of himself, his environment and other persons.

This conception of man as an essentially social being, though not original with Hegel, was presented by him more elaborately and perceptively than by his precursors; it is an idea that he enriched, bringing out many of its cultural and psychological implications. But he expressed it in a vocabulary difficult to master, and as confusing as it is suggestive. Ideas similar to his about knowledge, self-knowledge, reason and will, about the distinctive attributes of a person, have been expressed

more clearly and carefully, and less pretentiously, than
he expressed them. But it is, I think, implicit in his
account of personality, that a person, a self-conscious
and purposeful being, does not first have direct knowledge of himself, and then later, observing other bodies
in the world behaving as his body does, infer that there
are other persons around like himself. He can, no doubt,
on this or that particular occasion, doubt whether there
is another person around behaving in typically human
ways, or whether some action of another person really has
the significance that he is inclined to attribute to it.
But, in order to have such doubts, he must already have
ideas, and among them the concept of a person, which he
acquires only in the process of engaging in inter-per-
sonal behavior, and coming to understand what such behavior is. He must find his bearings in a world in which
there are other persons besides himself in order to acquire both self-knowledge and knowledge of other persons,
and only then is he intellectually equipped to raise the
sort of doubts about other persons, the existence of the
significance of some of their actions, that can be settled by observation and inference, by the appraisal of
evidence. Which is not to say, of course, that there is
no sense, or no important sense, in which he has privileged access to his own experiences. It is to say only
that he cannot have such access unless he is already
aware of himself as a person among persons.

II.

A person is more than a rational agent, a maker of
decisions, who seeks to use and control what is external
to him the better to achieve his purposes. He is also a
social being whose ability to reason and to take decisions, whose ideas of himself and his environment--ideas
that largely determine his purposes--are products of social intercourse. He is a social being in a double sense:
he acquires his ability to form coherent purposes and to
pursue them deliberately in the process of learning to
live with others and to communicate with them, and his
purposes and methods of achieving them are themselves social in content. What he aims at achieving ordinarily
has no meaning outside a social context, or is connected
with an ambition which is essentially social, and the
same is true of his endeavors to achieve it. To get what
he wants he needs the assistance of others, and he gets it
by using social symbols and appealing to social standards.
He also uses them or refers to them in regulating his own
conduct. His methods of persuading himself when he delib-
erates _in foro interno_, his own principles and standards,
idiosyncratic though they may be, derive from, are variations upon, ideals and standards current in the communities and other social groups he belongs to, or has belong-

ed to. When he defines either his aims or himself, he uses ideas and standards he shares with others, and can achieve his aims and be as he thinks he is, only with their assistance and in society with them.

Writers on morals, especially the more utilitarian among them, when they try to explain what moral rules are and their social functions, sometimes use arguments which assume that human beings must already be rational agents who form coherent aims and act deliberately in pursuit of them, preferring some to others, before they can become moral agents. They speak as if social rules of all kinds, and therefore moral rules as well, arise among men to make it easier for them to achieve their aims; as if men acquired these rules (conceived of them and learned to use them as guides to conduct) the better to ensure that they should not thwart one another, but should rather assist each other in their endeavors to achieve their purposes. These rules, as these writers explain them, serve to limit interference and to regulate collaboration in such ways that, on the whole, rational agents are better placed to get what they want--to the extent that they know what they want and have clear preferences among their wants. These writers do not deny that social rules and standards, once they have arisen, affect men's aims, so that some of their aims could not even be defined apart from the rules and standards they accept or those accepted in the circles in which they move: as, for example, the ambition to be respectable or virtuous or to achieve a good reputation or any one of the many forms of excellence. But, though they do not deny this, the bias of their explanations of morality is to suggest, not just that rationality is logically prior to morality, but that human beings become rational before they become moral. Whereas the truth is that they ordinarily become both these things together.

To be sure, the prime concern of these Utilitarian writers is to explain, not the origins, but the functions of social rules and standards; and even then, not just what those functions actually are, but also--and perhaps above all--what ideally they should be. They are not much interested either in the social origins of morality, in how moral rules and standards arise or change in human communities, or in the process whereby a human child develops into a moral agent. The conclusions to which they argue--so it might be held--do not stand or fall with any assumptions they may make, tacitly or overtly, about the social or psychological origins of morality. If they were to go so far as to say that the aims whose achievement social rules and standards do, or ideally should, promote are not deeply affected by those rules and standards, they would be absurdly unrealistic. But this they ordinarily do not say, and are not logically committed to saying. Provided they confine themselves to asserting that, ideally, social rules should serve to promote the aims most widely shared and most highly valued by the persons required to observe them, then, though what they say may be questionable on other grounds, they do not take up a position that can be challenged on the mere ground that, as a matter of fact, people's aims are deeply influenced by social standards accepted before the aims were formed.

Nevertheless, when--as at present--we are considering what it is to be a person, we do well to take notice of some facts which may be obvious enough but which many writers on morals, and especially Utilitarians, have passed over in silence. A child as he grows up to become a maker of decisions, who pursues more or less coherent aims and regulates his own activities in pursuit of them, grows up at the same time to become a moral agent. Or, at least, this is ordinarily the case. He does not first become a rational agent and then afterwards a moral agent as well; he becomes both the one and the other <u>pari passu</u> and by stages. As he grows in self-awareness, as his beliefs about and attitudes to himself (and to others) become more discriminating, he develops gradually into a rational and moral being, a person, and learns to think of himself as such. He becomes, and thinks of himself as being, not a rational agent who recognized that he should also be moral because, living in a world which he shares with other rational agents, he sees that he cannot achieve his aims unless he conforms to rules which serve common interests, but a rational agent whose aims are permeated by his values, by the rules and standards he accepts, most of which he shares with others. Typically, his being moral is as much an essential part of his being a person as his being rational is--both in other people's eyes and his own.

Obviously, if he is to be reckoned a person, a man does not need to be virtuous or benevolent any more than he needs to be highly intelligent and a good reasoner. Just as, to be a person, he need only be capable of acting rationally and reasoning correctly, even though he often acts irrationally and reasons incorrectly, so too he need only understand what it is to have standards and to conform to them, to have obligations and to carry them out. To be sure, he will not understand what it is to have standards, unless he has some himself to which he occasionally conforms, nor understand what it is to have obligations, unless he recognizes that he himself has some and sometimes carries them out. But this is compatible with his often not conforming to his own standards, and his often not carrying out obligations that he recognizes, just as it is compatible with his having standards unacceptable to others.

I have not been arguing that a rational agent who is not also a moral agent is inconceivable, that rationality entails morality. My point is rather that, as we ordinarily use the term in referring to human beings, a person is both a rational and a moral agent. He is a social being who, in becoming rational, also becomes moral, and whose aims, when he is old enough to have any and to pursue them deliberately, are many of them aims that cannot be defined without reference to ideals and standards which have no meaning outside a social context. No doubt, human beings are "amenable to reason" as other animals are not, and this is largely what we have in mind when we call them <u>persons</u>. But their being "amenable to reason" does not consist merely in the fact that they can be persuaded by arguments which point to the relevant facts and their particular aims; it consists also in their responsiveness to moral appeals--appeals to standards they share, or are believed to share, with the persons who make the appeals. It consists in its being possible to move them by such assertions as that they have no right to do something, or have an obligation to do it, or that their doing it, though in their own interest, is against the common interest. To move them at least to the extent of their feeling the need to justify their actions or failures to act, even though only to themselves.

A rational agent who was not also moral could, no doubt, make useful rules to guide his own conduct, though they would be, not moral rules, but maxims of prudence. If he were in more than occasional contact with other rational agents, he might also find it convenient to make rules to govern his behavior towards them, and these rules too would be merely prudential. The other agents, with whom he came into contact, might also make rules to govern their behavior towards each other and towards him. The rules these agents made, each for himself, might be largely similar, and indeed would be so, if their needs and resources were much the same. And, in any case, whether they were similar or not these agents might get to know about each other's self-made rules and might take them into account when trying to predict or to influence one another's behavior. Occasionally, one such agent, seeking to persuade another, might point out to him that what he was doing went against one of his own rules. But if he did so, he would not be making an appeal to a moral principle, nor reminding the other of an obligation. He would merely be bringing a fact to his notice in the hope of changing his behavior by so doing, as he might do by telling him that, given his destination, he had taken the wrong road.

Men sometimes do - especially when they are enemies, though not only then - treat each other merely as rational agents and not also as moral agents. They do so when dealing with one another in situations in which there is no need to make moral appeals, or in which it would, in their opinion, be useless to make them. Presumably, in such situations, they do not cease to refer to one another as persons. So, too, if we were to come across such rational and non-moral beings as I have just imagined, we would probably not refrain from speaking of them as persons. Nevertheless, ordinarily, when we speak of one another or ourselves as persons, we do so in ways which imply that persons are moral as well as rational beings. Indeed, as I suggested when I discussed what is typically meant by saying that someone is amenable to reason, the word 'rational' is itself often used in the sense of 'rational and moral'; when such phrases as 'the dignity of the person' or 'respect for the person' are used, the dignity is held to pertain to, and the respect to be owing to, a moral and rational being.

III.

A person, in the sense of the word I am now discussing, is necessarily a rational agent, but a rational agent is not necessarily a person. This sense is, I suggest, the commonest of the senses in which the word is used, and I shall call it the "full sense". Nearly all rational beings we come across are persons in this full sense, and though there may be some who are not, we cannot be certain who they are, and so we do well to treat all rational beings as if they were persons. Nevertheless, the rational beings I imagined a moment ago would not be persons in this sense, for they would have no obligations to and no rights against one another. Their merely being rational would not be enough to ensure that they had obligations. Though they could take account of each other's intentions and maxims of prudence, they would still not understand what it is to be under an obligation. Human beings, we know, develop the ability to reason through social intercourse with one another, and through this same intercourse also acquire the ideas, feelings and dispositions that make moral beings of them. But it is at least conceivable that there could be rational agents who lacked these ideas, feelings and dispositions. To them such phrases as "the dignity of the person" or "respect for the person" would mean nothing, or at least something different from what they mean to us.

It is arguable that, if persons in the full sense were brought into contact with rational but non-moral beings whom they knew to be such, they would not have obligations <u>to</u> them but only <u>in respect of</u> them, as some writers on morals say is the case with men in relation to animals. According to this view of the matter, only persons can have obligations, and can have them only to other persons. If, then, the word 'person' is understood in the full sense, it follows that persons cannot have obligations to, but can have them only in respect of, rational beings that are not also moral. Just as, on this view, I do not owe it to my dog to treat him well - presumably because he cannot invoke this obligation against me, though he can respond to my kindness - so I would not owe it to a rational being to treat him well, if morality meant nothing to him. To be sure, I ought to treat my dog well, and can be justly blamed by other moral beings for not doing so, but this - so it has been argued - does not entail that I have obligations _to_ him. So, too, it might be argued, a moral being ought to treat a rational and non-moral being well, and can be justly blamed for not doing so, but does not therefore have obligations to him. If we argue in this way, we are not committed to denying that our obligations in respect of merely rational beings, if we came across any, would be much greater than our obligations in respect of animals. After all, whether we prefer to speak of obligations _to_ or _in respect of_ animals, we are ready enough to admit that we ought to do much more for some animals (for example, those we domesticate) than for others (for example, wild animals).

In any case, in the world as we know it, persons are well advised to treat all rational beings as if they too were persons, for rationality is more visible, easier to detect, than morality. We can never be sure that a rational being is not moral, or would not become so if it were treated as if it were moral. As it is, we treat children that are not yet rational or moral, or are only beginning to be so, as if they were so, or were more so than in fact they are, in the hope that our treatment will help them to become so more quickly than they otherwise would. In our experience, rationality and morality go so much together that we are inclined to treat anyone who is rational as if he were also moral, even when we have reason to believe that he is unusually vicious or indifferent to moral considerations. The appeal to reason, addressed even to a Hitler or a Stalin, is often also an appeal to justice or decency, an appeal to someone assumed to be a moral person to recognize and treat others as moral persons.

Though it is not inconceivable that there should be rational beings in more than ephemeral touch with one another who were not also moral, our experience inclines us so much to associate rationality with morality that a rational being who was not moral would seem to us a monster or a freak. We should have to assume that such beings, if they were not monsters or freaks, were born rational and not just born with the capacity to become rational, and that their needs were not social and could be satisfied with little or no assistance from others. If we assume that they are born merely with the capacity to become rational, it is difficult to imagine how that capacity could be developed in them, how they could learn to reason, unless it were in the process of learning to communicate with others, which they would hardly do if they did not have needs which required them to live and work together, many of these needs created by this cohabitation and collaboration. How could they live and work together, unless they acquired rules to guide their own and to control each other's conduct? And how could they use the rules effectively unless some of the rules were moral? Or, perhaps I should say, unless their ways of using them and their responses to breaches of them were characteristically moral? We know nothing of beings that are born rational and not dependent on one another, and whose needs are unaffected by their living together. The only rational beings known to us are through and through social, and the social conditions of rationality are conditions also of morality.

Psychiatrists tell us that there are freaks among us: men and women who do not acquire the capacity to "respond morally" to their own and other people's behavior, or who acquire it only to a very slight extent. Many of these people, perhaps most, are of sub-normal intelligence, but by no means all of them are so. Some are exceptionally intelligent, exceptionally good at reasoning, and are defective only morally. They come closest to resembling the imaginary beings I discussed earlier, who are rational and yet not moral; though with this important difference, that they live among moral beings, while the beings I imagined do not. They are "moral defectives" because they lack something that most people they consort with have.

These moral defectives, even when they are completely amoral, may not only learn to use social rules and take account of the fact that others use them; they may also acquire a sort of "external" understanding of attitudes they do not themselves share. For example, though, when they hear others passing moral judgments, they do not understand them as the makers of them do--for, _ex hypothesi_, they do not respond to the types of behavior, actual or contemplated, that the judgments relate to as the makers of them do, and their judgments express these "responses"-- they are nevertheless aware that judgments intended to influence action have been made and can take notice of the words or gestures that express them. For they are rational beings observing the behavior of other such beings.

They can also learn from experience that, when such judgments are made, certain consequences are likely to follow. They can learn, for example, that the makers of such judgments and those who agree with them are disposed to behave in a hostile manner to the persons about whom the judgments are made. Indeed, it is important to non-moral persons to take notice of these moral responses, even though they do not "understand" them as the makers of them do, for they have to learn to live with moral persons, who are the great majority. They have to learn what to expect of them, when they make certain utterances or gestures, even though the utterances or gestures do not mean to them all that they mean to those who make them. Also, it is important to them to learn how to evoke such responses, for this will often be useful to them in promoting their own aims. Rational but amoral beings can exploit the morality of others, can appeal to sentiments of which they are themselves incapable, and therefore do not understand as those who share them understand them, and it is greatly to their advantage to learn how to appeal to them. Just as it is greatly to the advantage of someone who is born blind, who cannot even imagine what others see, cannot know the world as they know it, to take notice of the behavior of people who would not behave as they do unless they could see. Though the blind man cannot see, his world is different from what it would be if he were living only among the blind. It is important to him to learn what to expect from those who are not blind, and to learn also how to get them to do what, by reason of his blindness, he cannot do.

There are, of course, important differences between the situations of the congenitally blind and the morally defective. A man born blind soon discovers that he is defective in a crucial respect, even though he cannot imagine what the world looks like to those who can see. He soon discovers that there are many things that other people can do that he cannot do, things that he would very much like to be able to do. He is aware that his blindness is a great disadvantage. But the non-moral man may not discover for a long time that there are feelings and attitudes that others have and he lacks. Indeed, he can learn to evoke these feelings and attitudes without knowing that he is doing so. As he sees it, he has learnt merely to influence other people's behavior by saying or doing certain things that he has noticed them saying or doing, and this he can do without discovering that he influences their behavior by evoking in them sentiments and attitudes that he does not share and cannot understand as they do. And when he does discover that there is something that others have and he lacks, he may think himself the better off for not being like others in this respect.

Also, the understanding that the blind man lacks is understanding in a different sense of that word from the understanding that the amoral man lacks. What the blind man does not see. and therefore cannot fit into his scheme of things as other men can. is part of the external world. His knowledge of the world, or his direct acquaintance with it, lacks a dimension that other people's knowledge does not lack. But what the amoral man lacks is not simply the ability to observe what in fact exists, outside himself or indeed inside. What he lacks is not the power to introspect, to notice in himself sentiments and attitudes which he really has, though he is unconscious of them. This is, indeed, a possible situation, for--as we all know--a man can be jealous or envious or selfish without recognizing that he is so; but it is not situations of this sort that we have in mind when we call a man amoral. We mean rather that he lacks certain kinds of feeling or the disposition to act from certain kinds of motives, or both the one and the other. Despite this lack, he understands many things that a man must understand if he is to be moral, but because of it he is nevertheless not moral.

He understands what a social rule is in the sense that he knows what it requires, what anyone to whom it applies must do or refrain from doing to conform to it, and he sees what the point is of having such a rule, what he and others gain by its being generally observed. Again, he understands that people use certain forms of speech or certain gestures to induce one another to observe such rules, and he too uses them for this purpose. When others use them in their dealings with him, he knows what they want him to do or refrain from doing; and he often does what they want because he thinks it in his interest to do so, or from some other "non-moral" motive. A man who knows or understands all this, who can use moral language and respond to its use by others, at least in his outward behavior, to this extent, might well be unaware that there was something lacking in him that others had, or might take little notice of it.

Indeed, some people might be inclined to say that, if a man knew and understood all this, he would be a person in the full sense, a moral as well as a rational being, on the ground that there is nothing more to being moral than just this. Some "behaviorists" have said this, but we need not be behaviorists to say it. Yet, surely, to say it is a mistake. There is more to being moral than the ability to recognize social rules (to learn that in situations of certain kinds one is required to do certain things or to refrain from doing them), or the ability to see that everyone stands to gain by the general observance of such rules, or to recognize and be able to use signs indicating that the persons to whom the signs are addressed are expected to observe the rules, on pain of incurring hostility if they do not do so. To be a moral person a man must, of course, be able to do all this, but he must also have attitudes to himself and to other people, when he or they do or fail to do what is required or expected, which go beyond recognizing these things. He must recognize that he and they have obligations, which involves more than taking notice of the general consequences of certain kinds of behavior and how people are likely to react (what they are likely to do) to anyone who behaves in these ways: more than seeing how one should behave in one's own interest, given these consequences and reactions, and more than knowing how to use social symbols to influence the behavior of rational beings. A man who recognizes obligations feels guilty when he does not carry them out, and because he treats others as moral persons who can recognize obligations, blames them for not doing what they ought to do. Recognizing obligations, feeling guilty or remorseful, blaming others: that the second and the third are not mere cognitive acts, though they involve knowledge (the ability to recognize the situations in which they are appropriate), is clear enough, but the same is true of the first. Recognizing obligations differs in kind from recognizing the consequences of actions and also from recognizing that there are social rules that one is required to observe. For a man can feel obliged to do what no social rule requires, or what he knows will expose him to hostility or punishment or even blame. So, too, he can feel guiltless, even though he has broken an important social rule, and has incurred the hostility or the blame of others by doing so.

IV.

The language of morals expresses such "feelings" and attitudes as these (the sense of obligation, guilt and blame) and appeals to them in others. It is therefore more than a language of "imperatives" and "warnings": a language used to indicate to people what they are to do and to refrain from doing, and how they are likely to be treated if they do not do what is required of them. To people--if there are any--incapable of these feelings and attitudes, it may be only a language of imperatives and warnings, but to most people it is much more.

It is said, sometimes, that a rational egoist, though he can be prudent, cannot also be moral, since he will not carry out an obligation when he believes that he has more to lose than to gain in the long run by doing so.

But I have not assumed that an amoral rational being is always an egoist, that he is never concerned to do good or harm to others except in the hope of some advantage to himself. It is not inconceivable that such a being should sometimes be genuinely benevolent, wanting to do good to someone else without hope of himself benefitting thereby. Yet there is, as writers on morals have often insisted, a close connection between morality and benevolence; or rather between morality and sympathy, or the ability to feel for others. And I have in mind here a connection closer than sympathy's being a condition of morality, in the sense that a creature incapable of sympathy would not acquire feelings, or attitudes that are characteristically moral.

We need not accept the account of the "moral sentiments" put forward by Hume or other writers on morals of his (and later) times to agree that only rational beings able to feel for others and to want for them what they want for themselves can acquire the attitudes, the feelings and dispositions that are typically moral. Affection for others, and perhaps also attachment to ways of life for what they are and not just as means to individuals' achieving their own aims, and even devotion to the communities one belongs to: these are sentiments which are not in themselves moral, and yet morality (the dispositions to act that constitute the sense of obligation, and the ability to feel guilt and the sensitiveness to blame that go along with this sense) is surely rooted in them. To say this is not to imply that a person is moral precisely to the degree that he has these non-moral sentiments from which morality springs but only that he will not be moral unless he has them. Nor is it to deny that there is a kind of self-absorption--often the most harmful of all--which is parasitic upon morality. This is the egoism of the man who is concerned above all for his own righteousness, and is lacking in compassion; whose strongest motive for doing good to others is that his own principles require this of him, and his self-esteem or hope of salvation depends on his abiding by them.

I have said that a person is a rational and a moral being. I might put the same idea differently, and more adequately, by saying that a person, as we ordinarily use that term, is more than just a rational being who recognizes that there are other such beings around, and who takes account of this in his dealings with them, treating them differently from the way he would treat them if they were not rational; he is a rational being who can "identify with" beings of his own kind, who can feel for them as they feel for themselves, and who therefore, because he can do all these things, can respond to certain kinds of appeal, the kinds that we call moral.

No doubt, a rational being who can feel for (or identify with) beings of his own kind can also feel for beings not of his own kind, for non-rational beings. And he can feel for them as they cannot feel for each other, simply because he is self-conscious and imaginative as they are not. A man can feel for a dog or a cat, as no dog or cat can do, and in doing so need not attribute to it wants and feelings that it does not have. Yet there is a difference between feeling for a dog and feeling for a fellow human being: a difference marked by our greater readiness to speak of sympathy in the second case than in the first. This, I suggest, is because the object of sympathy, as that word is ordinarily used, is assumed to be aware that he is an object of it, or at least to be capable of this awareness. To be sure, a dog can respond to the kindness of a man who feels for it, but this response falls short of knowing or believing that the man does feel for it. Where there is sympathy, in the sense that men have it for one another but not for dogs, both the sympathiser and his object identify with one another. There is compassion and a recognition of it by both parties for what it is; or else this joint recognition is possible. Sympathy in this sense is a recognition of one person by another, a recognition which would not be what it is unless both persons were (or could be) conscious of it.

A rational being may recognize that other beings are rational, and may treat them accordingly (that is to say, differently from the way he treats non-rational beings) because he finds it expedient to do so; and this he may do without sympathizing with them. Insofar as he does no more than this, he is not yet a moral being. To be one, he must be capable of sentiments or attitudes which presuppose, though they are not identical with, the capacity to "sympathize" with beings of one's own kind, to feel for them in a way that goes beyond feeling for a sentient but non-rational being. These two capacities, sympathy and morality, though they differ, are not connected merely in the sense that the first is a condition of the second, but also more closely than this. If there are, as there well may be, men and women who are "morally defective" without also being "rationally defective", this is presumably above all because they are incapable of "sympathizing" with their fellows. For, if they were capable of this sympathy, why should they be morally defective? It is perhaps not inconceivable that they should be, but it is difficult to credit.

We might, of course, imagine (as some Utopians have done) a situation in which rational beings were so completely in sympathy with one another, so fully aware of each other's wants and needs and so ready to help each other, that they could do without social rules to guide their conduct, without obligations and without a sense of guilt or a disposition to blame. In this state of blessedness they might be entirely good without also being moral. But if we suppose them so related to one another, psychologically and socially, that they cannot do without such rules, we can hardly suppose them capable of genuine sympathy without also supposing that they soon acquire characteristically moral attitudes to one another's behavior and their own.

Rational beings could, however, be rule-observing without also being moral. As I argued earlier, we may have good reasons for holding that the great majority of rational rule-observing beings known to us are in fact moral, though there may be a minority--and who knows just how considerable?--of "moral defectives" among them. Though the language of morals is not a language merely of imperatives and threats, but also a language of appeals to which moral defectives are largely or wholly unresponsive, they can still make use of it and be influenced by others who use it. They use it, as the others do, to influence behavior, and yet it does not mean to them what it means to the others, for they lack the feelings and dispositions (or attitudes) born of sympathy to which the language appeals. Their intention in using it is not to appeal to attitudes of which they are themselves incapable but merely to get people to behave as they wish them to, and yet their use of it would not have the effect it does have unless it were language that appealed to these attitudes. Thus, their use of it is parasitic upon the use of it by others who have these attitudes and who use the language with the intention of appealing to them. For what makes it specifically a moral language, and not just a language of advice, exhortation or warning, is precisely that people who have these attitudes use it to give expression to them and to appeal to them in others. The species, no doubt, first acquired the attitudes in the process of producing (or should I say evolving?) the language, and the individual acquires them in the process of learning to use the language. But the language can be used in more than one way to influence behavior, and not everyone who learns to use it effectively for that purpose acquires these attitudes.

May I repeat--to avoid possible misunderstanding--that in saying that a person is a moral being who would not have the attitudes that make him one unless he were able to sympathize with beings of his own kind in a way that goes beyond mere feeling for sentient creatures, I am not suggesting that it is a misuse of language to speak of a merely rational or a morally defective being as a person. I am only trying to explain what I believe to be the most usual sense of the word, from which many of the other senses derive. And I have called it the "full" sense because it connotes more than the others: because someone who is a person in this full sense is one also in most of the other senses, whereas the converse does not hold.

I do not even suggest that it is an abuse of language to call a human idiot, whose idiocy is irremediable, a person. We call such human beings persons because they belong biologically to a species most of whose members are persons in the full sense, though they themselves are not so. We can even hold, and not unreasonably, that we have obligations to these subnormal human beings far greater than any we have to normal specimens of non-human species, merely because they are biologically human, even though their defects are great and irremediable. The fact that such obligations arise in the first place among persons in the full sense does not entail that, once they have arisen, such persons have them only to others who are persons in this same sense. Part of what we mean when we insist that a human idiot is a person is that we ought to treat him, in some important respects at least, as if he were rational and moral, even though in fact he is neither.

V.

Persons are, typically, self-conscious beings, and they could not be self-conscious unless they were rational. And yet writers who, like Rousseau or Hegel, have insisted upon this self-consciousness have turned their minds to problems, psychological, moral and social, of which the rationalist philosophers of the eighteenth century took little notice. They are problems of identity, though not of the sort that I discussed in the beginning of this paper. They are not philosophical problems about the criteria of personal identity. Rather they are problems of self-identification, problems of the self about itself, or of a person about his personality. We must suppose someone to be a person before we can explain how he comes to have such problems about himself. A man faced with such problems might ask himself, What am I? but he would not be satisfied with the answer that he belonged to the species <u>homo</u> <u>sapiens</u>, and had such and such attributes.

In the DISCOURSE ON THE ORIGINS OF INEQUALITY Rousseau says of man, considered as a social being, that he is always "outside himself" and "knows how to live only in the opinion of others", and that it is "from their judgments of him that he draws the sense of what he is." Man as a rational and self-conscious being, as Rousseau conceives of him, is a product of society, of social intercourse. The ideas he uses in building up his image of himself, even though it should be an image he takes care not to disclose to others, are ideas he shares with his neighbors, ideas that belong to the language and the culture of the society he belongs to. So, too, the aims he pursues and the principles he uses to guide his conduct are determined largely by the standards and practices current in his society, and by the opportunities it affords to its members.

But this, unfortunately, does not ensure that a man is at peace with himself and at home in the communities and social groups he belongs to. It does not ensure that he has firmly held, consistent and realistic principles to guide his conduct, or coherent and attainable aims, or an idea of himself as he is or as he aspires to be that satisfies him. His being "outside himself" consists precisely in his lacking such principles, such aims, and such an idea of himself; in his not being, as David Riesman would put it, "self-directed". Since his principles and aims are inconsistent with one another and unrealistic and unstable, he is at odds with himself and at odds also with the social groups he belongs to, or not at his ease in them. A self-conscious being, if he is to be happy, must have ideas of what he is or wishes to be that are stable and satisfying, that are proof against the shifting opinions of him that others have; he must have what is sometimes called a "sense of his own identity". This, I take it, does not mean that he must be able to describe himself and his aspirations accurately, or that he must be inflexible, or must find no serious fault with himself. It means rather that he must either "accept himself" as he is (or as he appears to himself) or else, if he aspires to change himself and his condition, must know what it is that he wants to achieve and must believe it worth achieving. A man who has suffered disaster, not just materially but even morally or spiritually, and is intent on "remaking his life", has a "sense of his identity", has ceased to be "outside himself", provided he knows what he wants to do with himself and thinks the effort worth while, painful though it may be.

No doubt, however firm and realistic a man's principles, however coherent his aims, however stable and satisfying his ideas of or aspirations for himself, he does not produce them in the privacy of his own mind. He acquires them in the process of reflecting upon an experience that is essentially social. What is more, his ability to hold on to them, to use them effectively as guides to conduct, always depends greatly on how others treat him, on what they think of him. His cultural and moral dependence on others does not last only while he is acquiring the principles, aims and aspirations that shape his life and the dispositions that form his character; it persists throughout his life. Not only because he must rely on the assistance of others if he is to be able to live as he wants to live, or to be what he wants to be, but also because his faith in his own principles, ideals and aims depends considerably on his having friends or acquaintances who share them or who think them worth while. We all need spiritual allies. We all live, more or less, "in the opinion of others" and "draw our sense of what we are from the judgments they pass on us." We do so whether we are "self-directed" or not, whether we are on good terms with ourselves and society or whether we are, as Rousseau put it, "outside ourselves" or, as later writers have put it, "alienated". This is part of the human condition.

The need for a stable and satisfying sense of one's own identity, which Rousseau was the first to proclaim (though not precisely in these words), is a need of social man, and yet is a need that men often cannot satisfy in society as they find it. It is an essential need of persons who, in the world as we know it, become persons in the process of learning to live and communicate with one another, and yet the forms of life, the social activities they take part in, may be such that they are confused and bewildered or hostile and contemptuous in their attitudes both to themselves and to society. And this need, though men in all societies have it and no doubt often fail to satisfy it, is one of which they become acutely conscious only in sophisticated societies which are "progressive" at least in the sense that they are changing fast and are aware of doing so. Rousseau seems to have believed, not only that men are more conscious of this need in societies which, owing to the progress of the arts and sciences, are in process of rapid transformation, but also that in these societies the need goes unsatisfied much more than in simpler and more static societies. We may think him mistaken in believing this, and yet agree that people in these societies are much more conscious of the need and of the extent of the failure to satisfy it. It presents them with problems, psychological and social, that did not exist for people in earlier societies.

The society in which these problems come to the fore is also the society in which certain claims are made for the individual merely as a rational and moral being, as a person. These claims, when they were first made, were called natural rights, though nowadays they are more often referred to as human rights. They could just as well, and perhaps more accurately, be called rights of the person. For an idiot is a human being, and yet these claims are not made for idiots--or at least not all of them are. For example, freedom of speech and freedom of association are not claimed for idiots.

If these rights are no longer ordinarily referred to as natural rights, the reason, presumably, is that the writers who first claimed them for all men spoke of them as if they were rights, not merely of social men (men formed by society) but rights that men could have even if they were not social beings and lived in a state of nature. As a matter of fact, most of these writers did not think of the state of nature as a condition in which there were no enduring social ties between men, for they contrasted it rather with civil society than with society in general. Still, they were misunderstood, and it came to be thought less misleading to speak of human than of natural rights.

Human rights are rights that men and women are supposed to have merely by virtue of being persons, quite apart from any particular status, social role or occupation they may have. And yet in all societies, until quite recently, and in many even now, most persons have never heard of these rights, let alone demanded that they should be secured to them. And they have not on that account been any the less persons, any the less rational and moral beings. But this does not make it improper for us to speak of human rights, rights that human beings have merely by virtue of being persons. For the word "rights" is quite often, and quite properly, used to refer, not to claims that are actually made and recognized (acknowledged to be just), but to claims which the speaker thinks it desirable should be made and recognized. We may admit that quite often, social and cultural conditions being what they are, it would be unrealistic to the point of absurdity to expect people to make and to recognize these claims. But the admission does not logically preclude our holding that the making and recognition of them is desirable, and that the claims are to be justified on the ground that the beings for whom they are made are persons. Doctrines of natural or human rights, like the ideas about personality on which they are based, have their histories; they are sophisticated doctrines that flourish only where certain social and cultural conditions hold.

Still, these doctrines must be judged by their contents, by the claims they make and the arguments used in support of them, and not by the circumstances of their emergence and widespread acceptance. They do not refer to these circumstances, they do not make claims only for persons living in societies of the type that gave birth to them. They make claims for persons as persons. That, indeed, is what makes them radical and even revolutionary doctrines that men appeal to to justify extensive and often drastic social reforms. The attempt to realize them, to meet the claims they make, has led to great and rapid change even in the countries in which the doctrines first arose and gained wide currency. In other countries that are more "backward", as the industrial West understands progress, the attempt has led to much greater upheavals. But the profoundly disturbing influence of these doctrines is no condemnation of them.

Thus there have been in modern times two problems (or sets of problems) relating to persons--I mean other than philosophical problems about the criteria of personal identity. There are the problems to do with human rights, the claims made for persons merely as persons, regardless of social status, role or occupation; and there are the problems that turn on the need that a person has (or is alleged to have) to achieve a satisfying sense of identity, an idea of himself as he is and aspires to be that is acceptable to him, that gives him self-confidence and self-respect and the sense that life is worth living.

These two problems are closely connected, not only because much the same social and cultural conditions have given rise to them both, or because they make similar claims for the person. The champion of human rights argues that man as a rational being who forms purposes and pursues them deliberately in society with others, and also as a moral being responsible to other such beings for what he does (especially when his actions affect them), has a claim upon them that he should be allowed and assisted to seek a way of life that satisfies him, that he should not be a mere instrument of others but should be able to set up aims of his own and strive to achieve them. The doctrine of human rights is an elaboration of this claim. So, too, presumably, the need that Rousseau and others speak of (the need of a person to achieve a sense of identity acceptable to himself) is satisfied, if at all, in the process of setting up aims for himself and striving to achieve them. Given that a person has a claim on other persons, and on the community acting through its representatives, that they should allow and assist him to satisfy this need, does not this claim too in practice take the form of an assertion of human rights?

But, despite the convergence of these two problems-- how to secure the essential rights of the person, and how to satisfy the need for an acceptable self-identity--

there is also a divergence between them. Though both problems arise in the same type of society, there are reasons for believing that the first is easier to solve in that society than is the second, and even that attempts to solve the first can make the second more acute. Not only does the doctrine of human rights emerge and spread in a commercial and industrial society in which smaller and more closely knit groups and communities are breaking down or else weakening their hold on their members, but the security of these rights to the individual seems to require that they should, that society be more open and its members more competitive, that the individual should be freer to move from place to place, group to group, occupation to occupation. It requires that he should depend less for assistance on kinsfolk and neighbors and more on comparative stangers and on the state, on officials who deal with him, not as a friend or a neighbor, but as someone belonging to a category of persons whose claims and obligations are defined by law. Cultural and moral diversity, change, the proliferation of short-lived groups whose activities take up only a small part of the lives of their members, increased social mobility, dependence on hierarchies of remote officials, and "impersonality" in the sense of brief and formal dealings with a host of indifferent persons: these are features of the society in which the demand for human rights grows, and they are features that are accentuated rather than diminished by attempts to provide for all sections of the community the opportunities and services needed to secure these rights. But it is precisely these features which make it more difficult to satisfy the need of the individual for a self-identity acceptable to himself. On the one hand, we have the demand for a larger freedom, for wider opportunities for the person and, on the other, the demand for a kind of spiritual security, an acceptability--no longer to God--but to oneself and to others on whose good opinion one's self-respect depends. For some persons, no doubt, both of these demands are met, more or less adequately, but for others-- and who knows how many?--they are not.

HISTORICAL DEVELOPMENT OF THE PRINCIPLES
OF EQUALITY AND FREEDOM AND
THE CONCEPTION OF MAN

Maria Borucka-Arctowa

If we adopt for the starting-point of our reflections the question whether contemporary man avails himself of equality and freedom principles to a greater extent than his predecessors did, and whether these ensure more amply the full development of person, then we must take a historical perspective. We need to consider the historical interpretation and evolution of these principles from the time of their incorporation into the American and French Declarations, up to the Report about "Human Rights and Scientific and Technological Development," submitted by the Secretary General of the United Nations in 1973, the 25th anniversary of the Universal Declaration of Human Rights.

We shall also be concerned with the questions of whether the contents of these principles are interrelated with a certain conception of man, and whether the current store of knowledge about man can serve as a theoretical basis for activities aimed at fully implementing the principles of freedom and equality, activities which are a part of legal policy making and social engineering.

Attempts to provide a semantic explanation of the different meanings of the term "freedom" or "equality" may well avoid some misunderstandings or apparent problems, but they are not an adequate substitute for the axiological choice of a definite--this and no other--conception of freedom or equality, treated as goal-ideal within a given political program, or as model-ideal for comparison and theoretical valuation from the point of view of organization of social life.

W. Lippman in THE GOOD SOCIETY has written that "... the politics, law and morality of the Western world are an evolution from the religious conviction that 'all men are persons and the human person is inviolable'." The historical interpretation of the conception of freedom and equality indicates that this evolution has been effected within a very concretely conditioned social reality.

The Declaration of Independence, strongly influenced by the ideas of Locke, Paine and Rousseau, had spoken of man's inalienable right to life, liberty and the pursuit of happiness, and these ideas are amply reflected in the American Constitution. They are one of the most important and lasting elements of the natural law doctrine of that time, and express the continuous search for certain absolute ideals, which could account for the aspirations of the consecutive generations and their fight against social injustice.[1] At the same time these conceptions are a very concrete answer of the young bourgeoisie as an ascending class, corresponding to its need and the new economic, political and social situation. This was an answer by a certain milieu to historically determined dangers and threats in the language and stylistics of the ideology predominant in the given society. Those who formulated the content of the different laws, sought protection against those dangers which they considered to be most troublesome and formidable. It was but seldom that they were able to foresee the various complications which the application of these laws might involve, and to safeguard against.

In the American Declaration of Independence of 1776 and the French Declaration of the Rights of Man of 1789, the freedom of individual is an element of the struggle against the existing system. The authors and the intended readers of these Declarations, were affluent, educated people who did not need anything apart from freedom of action. Even the political rights to elect and control the government, were considered by them to be secondary to freedom of property and freedom of contract.[2] Typical of this period is the approach to law as a sphere of freedom, as "subjective right," and the conviction that rights are prior to the state, and the state and the law it enacts are subordinate to them. They have characteristics of natural, innate and inalienable rights.

This is, then, a conception of a freedom that is "defensive," "negative," conceived as assurance against interference on the part of the state, or of other units, into the wide sphere of human activity exempted from legal regulation. At the root of this view lies the conception of the "balance of interests," of self-regulation of economic and political relations." A counterpart of an ideal of liberty thus understood is the conception of a self-sufficient unit defending itself against the impact and interference of the environment. In the words of Adam Smith, "The natural effort of every individual to better his own condition, when suffered to exert itself with freedom and security, is so powerful a principle, that it is alone, and without any assistance, not only capable of carrying on the society to wealth and prosperity, but of surmounting a hundred impertinent constructions with which the folly of human laws too often incumbers its operation."[3]

The positivization of some determined rights and freedoms responds, however, rather to the conception of "liberties" than of "liberty." According to Friedrich Hayek, "Liberties appear only when liberty is lacking: they are the special privileges and exemptions that groups and individuals may acquire while the rest are more or less unfree. Historically, the path to liberty has led through the achievement of particular liberties. ...The difference between liberty and liberties is that which exists between a condition in which all is permitted that is not prohibited by general rules and one in which all is prohibited that is not explicitly permitted."[4] Nevertheless, it should be stressed that values protected by such liberties acquire a greater significance when the sphere of individual freedom is kept outside the scope of legal regulation, and any interference by the state into this sphere constitutes a violation of fundamental inalienable rights and freedoms. In this lies their ideological, persuasive role.[5]

An interesting line of development is revealed by the conception of the natural rights of the individual: From the rights alienable and transferable through social contract to the sovereign endowed with strong, centralized power (Grotius, Hobbes, Spinoza), through inalienable rights, limiting the range of the state power transferred to the sovereign only in a delegated form (Locke), to Rousseau's construction of transferring all rights to the community to safeguard fully the sovereignty of the people. In Rousseau's conception of the natural right to freedom, as a share in the shaping of the general will, are inherent the origins of a conception which was to be extensively developed by the Jacobins, which treated liberty as an actual possibility of active participation in the carrying of collective power, and as a necessity to subordinate fully to this power, a conception founded on the conviction that the discharge of power by the people gives sufficient guarantee of liberty.

An opposite conception is the liberal concept of freedom, as an innate and unlimited right to individual independence and the ensuing restriction of state power and the individual's right to oppose the authorities in power in case of challenge to his stipulated sphere of freedom. M. Duverger has seen in the two opposed concepts of liberty--"liberté participation" and "liberté resistance"--the starting point for the two later entirely divergent conceptions of democracy.⁶

In France, as a result of the furthest-reaching radicalization of the popular masses the line of division between the bourgeois doctrines of the right wing and the left democratic current was the most marked. The starting point of the basic divergencies (similar to the differences between the levellers and the diggers in the doctrines of the English revolution) is the different interpretation of the principle of equality which in the bourgeois views is reduced to the postulate of equalization of the bourgeois with the nobles, and in the doctrines of the left wing assumes the form of the postulate of general economic equalization and leads to an entirely different approach to the key problem of property. The disproportion between the postulates of the popular masses and the ofjective conditions for their effectuation at that time introduces Utopian elements into the ideology of the left democratic current.

In the seventeenth and eighteenth centuries natural law philosophy was the predominant world outlook, and at the same time the leading and creative political idea upon which the two Declarations, the American and the French, were based. Simultaneously the middle class was trying, in conformity to its professed outlook, to subordinate men to nature's general regularities and to find theoretical justification for the new natural order, equivalent to nature's order. The postulates expressed under the guise of "natural order" are presented as an absolute pattern of life and social system, juxtaposed to the "perennial feudal order" and to the arguments of theological authorities. The abolition of this old order needed justification in the form of a non-historical, universal doctrine, representing the new regime as determined by the traits of the "human species," and hence irreversible and necessary.⁷

To the thinkers of that period nothing was more certain than "human nature," which was also the wellspring of law. Some principles of justice are "natural," i.e., rational and unalterable, they can be known by the mind only. The basic premise was the "state of nature," which preceded the social state. It was this state of nature that made possible the discovery in man of persistent unvarying elements, which could serve as indices for behavior obligatory for all men. Such a code of the unchanging natural law had to be a central, controlling system. Positive law was to be judged from the viewpoint of its convergence or divergence with natural law.

In the natural law doctrines of this period we find the formulation of the basic principles of the new capitalist society, individualism, utilitarianism and constitutional postulates that will contribute to the development of liberalism (inalienable, individual rights, harmony of interests, limitation of state power). The double role played by the natural law doctrine, namely, the destruction of the old system and the creation of a new social structure corresponding to the needs of capitalism, leads to an inevitable contradiction. It is manifested on the one hand in the atomization of society and in radical individualism, accounted for by inalienable individual right which is to oppose and break the compact organization of feudal society, and consequently leads to political and economic liberalism. This individualism gives birth to utilitarianism interpreted at first as mere selfish interest of the individual.

On the other hand, however, the necessity of closely uniting and welding the new capitalist society gives birth to the conception of social utilitarianism preferring the common welfare to individual interest (Helvetius, Holbach). The place of natural harmony is taken over by the postulates of a conscious shaping of man's living conditions by legislation. This postulate constitutes a consistent development of the thesis on the influence of environment on man, a thesis undoubtedly incompatible with the conception of immutable human nature and "natural order."[8]

The breakup of the natural law school was a result of inner contradictions as well as limitations in solving newly arising problems. The main attack was to be undertaken by utilitarianism and historicism, two trends which had originated within the traditional natural law. The change in the approach to the human individual and to the traditional liberalistic interpretation of freedom, was also due to the development of social sciences. Sociology breaks away from a picture of man who is isolated, self-sufficient and defending himself from the impact of his environment; it shows him linked with a group, with the society. David Riesman calls this process "revaluated individualism whose essential characteristic is the recognition of group-conditionedness."[9] This approach is also necessary for understanding modern democracy and the mediating activities of the widest range of groups. This fact, writes Georges Burdeau, also has significance from the point of view of human rights.[10] "Group allegiance (including organizational allegiances) had to come vigorously to the fore just in the fields primarily affected by society's economic structure. Such are associations in the field of production, marketing, and labour."[11]

The stress laid by Marxist theory on economic relations made it necessary to analyze the effects of technical progress on human rights. The relativization of human rights was effected along with changes which took place in man's world outlook under the influence of evolution theory, of extensive anthropological research on the variety of culture. Comparative research led to rejection of the conception of our own path of civilization as the "natural" one.[12]

The concept of "natural law with changing content" was an attempt to solve the difficulties involved in establishing general and unchanging verities about man and society, and in repulsing the attack upon the absolute and dogmatic character of the natural law doctrine. In Stammler's analysis this concept is related to the Kantian categorical imperative, because it states the formal conditions of just behavior, even though, as a result of differences in time and space, the same behavior could be classified as just or unjust. Thus the variability of the concrete manifestation of demands is recognized. (We are dealing with a concept of objective, but not absolute justice.[13]) This conception has exerted a marked influence upon contemporary trends in natural law, and we will therefore return to it in the concluding parts of our paper.

The undermining of the immutability of human rights favored their gradual revalorization. As would soon be seen, the liberal conception of "defensive freedom" is favorable to maintaining a certain established <u>status quo</u> and the ensuing actual privileges of certain groups. It became evident, too, that freedom is relevant and of value only when the citizens have enough material means to make use of it, and that the individual may be threatened not by another's interference into his life and personal affairs, but just by the lack of this interference, by his own weakness as an economic subject.

The economic depressions, the unemployment during the period between the two World Wars, and these wars, all had contributed to the final collapse of the "defensive freedom" conception and the adoption of the program of social security, which is opposed to individualistic conceptions.

Entirely new elements of the concept of human rights and of the interpretation of freedom and equality were brought about by the October Revolution and the development of the USSR. The 1918 Declaration of the Working People contains new liberational rights--freedom from exploitation, the condemnation of racial and national discrimination, the right to self-determination by the peoples of the Soviet Union. The economic and social rights outlined in the Declaration were to be developed to a large extent in the 1936 Constitution of the USSR and in the later constitutions of different socialist states.

To a great extent the new rights had been taken into account by the Mexican Constitution of 1917, and by the Weimar Constitution of 1919. Following the example of the latter, some of the West European countries introduced economic and social rights to their respective constitutions. But the process did not run smoothly or without opposition. Objections were raised not so much against the content of the rights (which in view of the decline of liberal conceptions and the observation of changes which had taken place in Russia, and apprehension that the revolutionary claims might spread, were actually voted for in the respective Parliaments) as against raising these rights to the range of constitutional norms.

A period highly significant for the shaping of economic and social rights in the United States was that in which some substantial legislative changes were introduced by the New Deal. The result of the fact that these rights were not settled in a constitutional way, was a protracted struggle of the Supreme Court against attempts at introducing reforms in the regulation of working time and wages. Significant, too, was the resistance shown by Congress to the idea of establishing economic rights--formulated by Roosevelt and repeated by Truman--which were only partly enacted.

The inclusion of economic and social rights in the constitutions, or their preambles, in Italy, France, and German Federal Republic after the end of World War II was closely connected with the formation and swift evolution of the "Welfare State" model. It manifested itself not only in the incorporation of economic and social rights in the constitutions, but also in the emphasis laid on the "social character of the state" and in the introduction of changes to the concepts of the traditional rights of property and liberty.[14]

Of outstanding importance also was the incorporation of economic and social rights along with the traditional liberty rights in the Universal Declaration of Human Rights of 1948. These rights, for the first time expressed in the by now traditional form of Declaration, were included in the Charter of the United Nations, and not just in a constitution of a given state, as had been done before.

The character of the Declaration is universal (as shown

in its title), which is an expression of its being accepted by states of different ideology and culture, different political and economic system. Its decisions relate to the human individual regardless of his being a citizen of this or that state, or of any state at all. This is, then, the fundamental catalog of human rights. But this universal character involves also a number of difficulties connected with differences in culture and religion, in the tradition of the respective countries, as, e.g., is the case with the question of freedom of religion in the countries where religion is one of the elements by which the citizens' political status is determined, or the problem of equal rights for women, and its connotations with the culture and traditions of the given community.

These difficulties were revealed in the course of the work carried out by the Commission which prepared the text of the Declaration, and then in the work on the Covenant on Civil and Political Rights and the Covenant on Economic, Social and Cultural Rights. The fact that two different Covenants were passed, which permits ratification of only one of them, shows that the two categories of rights are treated differently. One can detect differences in opinion both in the interpretation of the binding force, and in the manner in which the economic and social rights comprised in the Covenant should be carried into effect, as well as in the system of domestic law of the respective countries. There is a marked tendency to treat these rights as program ones and to make the degree of their achievement dependent on the economic possibilities of the different countries, which creates a certain vagueness in the control criteria.[15] These rights are carried into effect in a different manner than is the case with the liberty rights, their achievement being conditioned by the issuing of a number of regulations making the state activity in this field more concrete. A vital but difficult problem is the judicial guarantee of these rights. This problem, however, calls for more detailed research. Apart from the abovementioned difficulties connected with the differences in the achievement of economic and social rights, one must not overlook certain reluctance in their application, as well as certain bias in their interpretation on the part of officials or judges, which in some countries are recruited from social groups of unequivocally conservative views.[16]

The economic and social rights bring a change to the previous conception of freedom and equality. The distinction between "freedom to" and "freedom from" becomes quite manifest. "Freedom to" without actual equality becomes the privilege of a certain group only. But what

does "actual equality" stand for? The point lies not in
abolishing every difference between men, but only those
which might give birth to differences in the situation
of men; not in a real levelling of conditions of life,
but in the equal opportunities offered by the given social system, in the actual possibility to realize the
rights comprised within the given legal system, in creating equal and possibly ample opportunities for self-
realization of persons, while maintaining the different
forms by which this self-realization can be manifested,
dependent on the individual person's sex, cultural milieu, skill, and predilection. The fact that every citizen is given equal chances to attain a certain educational standard, to take a learned degree, to make a career, to achieve a high social position is not tantamount
to every citizen being able or willing to attain these
things.

Many difficulties are still involved in the securing
of an equality principle (equality of chance, possibility). For example one might cite the problem, widely discussed and investigated in legal-sociological research,
of access to law and of manifold "barriers" hindering
this access, as well as measures taken to overcome these
barriers.[17]

The growth of state activity is linked up, however, not
only with the steps taken by the state in order to secure
the economic and social rights to its citizens. Along
with advances in technology new domains comprised within
state activity are expanding. The growing force of arms
is becoming one of the most essential outcomes of advance
in technology, since it renders any attempt at withstanding power (the classic "right to resistance") much more
difficult.

The constant expansion of state activity, taking over
more and more new agencies, and the growth of administration have other far-reaching consequences. The modern
state, to quote M. Duverger, "gouverne mieux mais il
gouverne plus."[18] It increases the number of those who
have to be obeyed. Instead of a single "tyrant," thousands of "petty tyrants" have come into being. In their
manifold activities they can be likened to the Lilliputians who tied Gulliver with a great many troublesome
bonds and deprived him of freedom. The question of bureaucracy, its deformations and the ensuing social consequences is absorbing every modern state. The individual's actual freedom frequently turns out to be just a
fiction. While on the one hand the individual is increasingly subjected to the decisions and patterns of behavior set by social organizations (THE ORGANIZATION
MAN),[19] on the other he becomes wholly dependent on the

public services performed by the state, services whose efficiency is necessary for the achievement of his basic needs.

The fact that the state has monopolized the mass media, cultural institutions, and education, brings into limelight the now much discussed question of "manipulating" the individual, a challenge to his independence of valuation, his free choice. According to M. Duverger the conflict between citizen and state in the "overdeveloped" countries ("pays surdeveloppes") becomes again a basic one. The symptoms of a certain crisis of liberties in the oldest Western democracies have been detected also by other authors who point out the intolerant attitude of those in power towards the opposition, the persecution of radical movements, and the spread of violence and use of force which, incidentally, is sometimes just a response to the similar fighting methods reverted to by various social groups.

Scientific and technological development has opened new prospects for economic, cultural and social progress. Certain aspects of this development may require constant attention.[20] Scientific and technological development is not always accompanied by an equally rapid process of man's adaptation to new circumstances. In considerations about the future of mankind, as justly pointed out by Rudolf Bystricki, one hears much about research, science and technology, but too little about man himself.[21]

The so-called genetic revolution, the achievements of modern chemistry, the new pharmacological drugs with unknown side effects, organ transplants, pollution of the biosphere; these are problems linked above all with man's biological existence, but they are also strictly tied with ethical and legal problems. Most of these problems can be settled only on an international scale. This is manifested in the constant search for new forms of international cooperation, in the growing interest taken by the UN in questions of the effect of technical progress on human rights.[22] This issue is bound to contribute towards making the man-in-the-street realize that he is part of the global society, not just of a local community, a region or a state.

The authors of the Universal Declaration of Human Rights were unable to foresee such development and such effects of the scientific and technological revolution. Therefore, when viewed from the standpoint of present-day needs, the Declaration displays a number of lacunae and calls for completion, just like the domestic legislation of member states. This refers equally to the problem, or rather to the whole set of problems, connected with the legal protection of the sphere of privacy of life against intrusions committed by a public servant or private persons. The right to privacy and the rights of personality (the latter expression being mostly in use on the Continent and especially in the German speaking countries), are closely linked with the conception of "defensive freedom."

The main legal problem and basic difficulty was to determine how to define and protect the private sphere of life. The isolation of the sphere of privacy, of its range along with the conviction about the necessity to defend it--and about the sound grounds upon which this defense should be based--is a view connected with a given culture and given economic-social conditions. It was unknown to the primitive community, founded on clans and tribes. Stig Strömholm[23] characterizes the changes which have taken place in the last centuries remarking that in small communities 150 years ago, the delimitation between "private" and "public" was of secondary interest. Anonymity and seclusion are the results and needs of an urban and industrial civilization. The middle class of the metropolitan civilization took over and developed the "My home is my castle" principle. The legal protection of that sphere became a creed which fitted well into the individualistic pattern of liberalism, considered as "natural rights of the personality," the protection of individual against state.

The decline of liberalism and the great rise of modern state activities, the ideal of the supremacy of the community over individual interests introduced various forms of intervention. The press and, more generally, "mass media," have to adapt their style of information to the taste of an increasingly broad public, disregarding the anonymity of the person concerned. The trend towards the deprivatization of human life, towards commercialization (the sphere of privacy just like any goods is to be had for money) evokes a contradictory trend which certain authors call "humanization of law," a trend of development characterized by Savatier as "L'avènement de la personne au centre du droit contemporain."[24]

These new tasks were solved by the use of a concept which was not new, but which earlier had a meaning not so precise, being used rather in the language of legal philosophy as "rights of personality." Independently of the birth and evolution of "privacy" the principles which protect individual freedom, were developed in different branches of law on the base of the Rule of Law ideology. "What was new," according to Strömholm, "was in the first place the synthesis of interests, the vision of a complete protection granted by essentially uniform rules of law to a set of non-pecuniary interests...and secondly the recognition of those interests in the field of private law."[25]

A comparative study permits us to find and formulate common denominators of the actions which fall within the notion of "invasion of privacy" and which also constitute a violation of "right of privacy:" (1.) Intrusion into an area which a person has an interest in keeping for himself; (2.) collecting material about a person felt to be unfair; (3.) using material about a person, whether lawfully or unlawfully obtained, for publication or specific purpose, e.g., as evidence against that person.[26]

All efforts at making a clear distinction between private and public sphere fail to bring a satisfactory solution and prove that a uniform principle for such a determination cannot be found. Conflicts arise between the rights to privacy and Article 19 of the Universal Declaration of Human Rights, the freedom to "seek, receive and impart information and ideas through any media and regardless of frontiers." The problem of the intrusion of the press was largely discussed by the American and European writers and found its recognition in positive law.

The question of the impact of recent significant and technological developments on human rights and the protection of the rights of privacy is the object of great interest and various activities of the United Nations (the Covenants adopted by the General Assembly in 1966, the Conference in Teheran 1968, the UNESCO Conference on Right to Privacy in 1970 in Paris, the General Secretary's report based upon international interdisciplinary studies carried out in all the member countries, 1973). The invasion of the sphere of privacy has been made possible to a considerably larger extent by the development of mass media, the rapid advances of recording and other newly introduced techniques which are divided in the U.N. study into three categories: (1.) Auditory and visual invasion; (2.) psychological and physical invasion; (3.) invasion by data surveillance.

The factor which is typical of all these modern devices is their clandestine use, whether they are a telephoto lens, wire tapping and telephone espionage, polarized glass and, perhaps before long, a piece of mini-equipment to trace and spy on the behavior of an individual at any time and without his knowledge.[27] These techniques were often primarily used as an important tool in the treatment of mentally disordered patients, or as efficient methods to fight with organized crime in the interest of national security. Despite their beneficial use such methods of surveillance can be used equally well in ways which threaten and endanger the privacy of an individual.

It should also be stressed that a specific "erosion of privacy" is the result of the growing use of personality tests and aptitude tests and the lie detectors evolved from the new techniques of psychology and psychiatry. The Director-General of the International Labor Office in his report entitled "Technology for Freedom,"[28] summarized the question of personality tests. Tests which make it possible to place a worker in a job corresponding to his skills and aptitudes are to be welcomed since they can make for improved productivity and for greater job satisfaction. They can be most useful tools of personal policy and ensure greater equity in selection; still, they involve two types of problems: (1.) There is evidence that testing may discriminate against culturally disadvantaged groups; (2.) tests and interview techniques which seek to penetrate personality and measure such imponderables as emotions, attitudes, mental equilibrium, may be designed to have the subject reveal his political views or his attitudes on intimate, religious, political, and sexual matters. The gruelling interview and tests may place the candidate under great mental stress and he may be under observation without being aware of it.

Another group of problems is connected with interference in the privacy of the individual through the use of computers, the central problem being that they will make it possible to bring together all the recorded information, often of private and personal nature about a particular person in a way that would never have been practicable before.[29] Some writers consider that there is a danger of serious interference with the personal development of the individual resulting from the way in which he is presented by the computer. He may even modify his real personality to suit that presentation. There is a further danger that the individual will become too conscious of his past rather than of his future, leading to a "pollution" of the human personality. The computer may become, as summarized by A. Miller, "the heart of a surveillance system that will turn society into a transparent world."[30]

The destruction of trust, as stated in the U.N. study, is one of the major dangers to a free society. The detailed questionnaire, psychological tests, lie detector; all these devices produce a pervasive insecurity which suppresses individuality, discourages responsibility, and encourages conformity. The many-sided set of problems relating to the right to privacy which had been presented here in a very summary way can be interpreted as an indicator of a revival of liberties. It should be stressed, however, that the object of protection of these liberties is a different set of values. For, in fact, the point is here not so much securing the freedom of economic or political action, of voicing certain opinions, against the interference of the state or other persons, but in creating security against a clandestine, often unnoticed, intrusion of other people, whether private or connected with some organization or with the state, into the sphere of personal private life. The protection of this sphere is also often linked with preventing unintentional effects of the application of certain means and devices, or of their disapproval depending on the purpose they are to serve. It is, therefore, a sphere of certain experience often of psychological character, feelings, values, recognized to be relevant by the law, for their loss might damage a man's personality, destroying trust. Hence the positivization of the right to privacy is being defined as a manifestation of the humanization of law.

The problem of doubts as to the existence of an inevitable alliance between scientific advances and human progress has been already presented by some authors in much stronger terms--as the new slavery of man. "Just as primitive man was helpless before natural forces, modern man is helpless before the social economic forces created by himself."[31] The problem of alienation comes to the fore in philosophical, sociological and psychological considerations. Alienation and the interrelated loss of belief in the possibility of solving certain questions, in finding contentment despite the appearances of prosperity and success. Man had achieved "freedom from" without yet having achieved "freedom to"--to be himself, to be productive, to be fully awake. Thus he tried to escape from freedom. His very achievement, the mastery over nature, opened up the avenue for his escape.[32]

Freedom as liberation from the previous ideals, from the cultural patterns, as a return to unpolluted nature, to the simplicity of life, to love among people, occurs in the postulates promoted by certain youth groups of anarchical orientation, not free from Utopian elements. They seem to be a manifestation of a certain crisis in the ideology of modern post-industrial society and of the consumers' society model. A most characteristic similarity is shown between the postulate set forth by Western authors (Fromm, Riesman), the postulate for struggle for the autonomy of man, "to choose oneself in work and in play" and his "productive orientation" in which creative work takes the main place, and a similar emphasis put by Marxist philosophy of man on the role of creative work, which is the form of the expression and self realization of man, revealing his preferences and talents and shaping his personality.

In the light of this complex set of problems how have we fared in our attempt at replying to the question asked in the initial paragraphs of the present paper: Whether contemporary man avails himself of freedom and equality principles to a greater extent than his predecessors did, and whether these ensure to a larger degree "the full development of person." It is very hard to give a straight answer. Every age brings new chances for the development of man, but also new forms of threat to his personality, and the new situations call for new solutions. The adjustment between the rights of individual and the community, between freedom and a certain interference to this freedom must be therefore a matter of changing needs and conditions.[33]

Our age will pass into the history of political thought as one which has incorporated economic and social rights in the catalogue of human rights and has put a special emphasis on them, just as the period of early capitalism and the young bourgeoisie has become memorable for its political rights and freedoms. One must not, however, disregard certain processes characteristic of modern societies which lead up to again setting forth "liberties" as the basic issue. Is this contemporary interpretation of the freedom and equality principles backed by some uniform world outlook, some uniform conception of man, as was the case during the birth of the American and French Declarations? To what extent is the interpretation of these liberties encumbered with the traditions of the natural law approach?

The term "nature" holds still its old fascination.[34] In modern legal philosophy one can clearly observe a revival of natural law concepts. However, the scope and impact of this trend is by far not the same as during the time of its triumphs in the Age of Enlightenment. Jacques Maritain, one of the philosophers most engaged in preliminary work on the Universal Declaration of Man, has written that people with opposed theoretical opinions can reach a purely practical agreement on the principles of the catalogue of human rights, while any attempt at a common substantiation of these practical injunctions for all concerned, would be hopeless. He thinks that such a common solution is made possible by the common practical knowledge reached by the common development and common experience of mankind.[35]

In the reviving natural law doctrines one can detect a distinct division into a relativist trend, connected with natural law with changing content, and an absolutist trend, defending the possibility of a knowledge of the permanent human characteristics and the rules governing the society. Both versions look for support to the contemporary social sciences.[36] The concept of natural law with changing content is nowadays more disseminated than the absolutist trend, and in most cases it is being linked with the modern version of human rights.[37] The basic difficulty lies in determining the method of establishing the content of the "just" law. Attempts to solve these problems follow two directions. One can be termed "institutional" and is a consequence of including natural law into constitutional texts which leads to empowering certain state agencies to determine those principles. This solution has some essential political consequences. The second seeks assistance in sociology whether by referring to certain theoretical statements or by making use of such research techniques as polls of general public opinion or of some chosen experts as well as analyzing of press material, shorthand records of discussions, etc. The concept of natural law with changing content has met with criticism, objections being made not only against incoherence, but, above all, the possible effects of relativism are anticipated and reprehended.

The absolutist version[38] speaks in defense of the "essence" of natural law, the possibilities of a knowledge of universals, of the permanent human characteristics achieving solutions increasingly closer to the ideal organization of social and political life. The realization of these tasks calls for an integrated knowledge of those disciplines which concentrate on studies of human behavior, attitudes, values, and social organization. Special challenges are linked with sociology. Integrated knowledge should be still enriched by axiology. Using his intelligence man can determine the aims he has to pursue. Because these aims correspond to the essential features of his nature, they are the same for everybody.

The question arises whether this contemporary version of natural law, relating to the present stage of the development of research, from the methodological point of view is more correct than the classical version. In both contemporary trends of natural law the starting-point, just as in the classical version, is human nature. By "human nature" all representatives of the doctrine understand the complex of certain characteristics, needs and dispositions common to all men. "Human nature" is presented either as a descriptive or as a normative model.[39] The strict distinction between these two models in the opinions of the particular thinkers is far from easy, for often nature interpreted in a descriptive way serves to build on its base certain aims which man should strive to achieve, and the whole system of behavior norms is justified by these aims.

The descriptive model gives rise to serious doubts because of the selection of characteristics included in the description of human nature. In spite of appearances of descriptivity such predicates as "natural," "human" have an evaluative character and constitute very vague criteria for including some characteristics and omitting others. Also a choice of some definite aims and principles and their recognition as natural are equivalent to a certain valuation, and all attempts to base it on objective criteria have failed so far to give satisfactory or generally accepted results. The attempts at justifying such choices often lead to the thesis that it is possible to establish the rule most disseminated and recurrent within different legal systems and to acceptance of the argument that such norms are of a higher value and should be generally applied. It should be realized, however, that we make here an axiological choice of a principle stating that such generally accepted norms should be taken into account and serve as signpost because there is no logical foundation to proceed from descriptive sentences to norms.

The normative model of human nature might, from the methodological point of view, constitute a source of behavior rules, but this would call for adopting some definite philosophical premises connected with the recognition of certain objective values, premises contrary to the empiric version of science. Without these premises this model will constitute but a construction which gives special meaning to certain preferred ethical and legal-political opinions, with a tendency to make them absolute.

The tendency to combine research and the results of different sciences concerned with man, is by no means isolated or proper only to the modern natural law doctrine. It has found expression in the general trend towards inter-disciplinary research in social sciences as well as in the concept of "behavioral science." The difference lies in the fact that the studies of natural law dissociate themselves from positivist orientation, which is characteristic of the contemporary social sciences, and make attempts to enrich the study of man by axiological elements. The development of the social sciences leads to extending the knowledge about man, and to the possibility of efficient action in a direction designated by the adopted aims, in which the full development of person may be included. (Moreover, this aim calls for a stricter definition).

The rejection of natural law premises, and of the possibility of founding behavior rules upon them, does not mean renouncing the valuating of law, nor is it tantamount to a conviction that the formation of legal norms is a completely arbitrary process. A jurist has not only to refer to, but also to contribute to, the shaping of value systems. It is essential, though, that he should realize and be aware when he is valuating, and when just stating facts, and be clear about it. The fact of discarding the possibility of providing a theoretical justification to the choice of supreme aims need not necessarily lead to skepticism as regards the pertinence of scientific approach, which helps to eliminate aims unrealizable under given circumstances, or which involve excessive "expenditure," thus making the choice more rational.[40]

The norm-giver must always rely on some sum of information concerning the biological and social existence of man (frequently termed "human nature") and all these facts which are linked with the object of legal regulation. This information is provided increasingly by the detailed disciplines both in social and in natural and technical sciences. This information can and should be used by the norm-giver in the making of norms: He must, however, choose and decide as to what aim he is pursuing and about the means by which this aim can be achieved. This choice depends on axiological premises related to the accepted world outlook, ethics, political system and founded upon "primary," "categoric" evaluation (evaluation which express the scale of particular values accepted by us, what we prefer, to what we attribute greater significance and not related to a system of norms or instrumental evaluations) when the basic aims are concerned and frequently as well as far as the means are concerned, for sometimes in choosing certain means we cannot base ourselves only upon teleological evaluation, also some ethical problem can be at stake, e.g., of the permissibility of some particular means or another.

"Neither nature nor any _vis major_ will be able to solve on our behalf the aims to pursue or to establish the rules of behavior. These aims and rules are formed by people in a way more or less conditioned, but also transforming nature, society and themselves."[41] Reference to nature as the supreme authority justifying the given order, the rules of behavior, is appropriate to the heteronomous conception of man. This is particularly manifest in the theistic trend of natural law. The heteronomous conception of human individual posits the existence of some superhuman forces of which the human individual is a product or emanation, not only in a physical sense but also, and perhaps primarily, in the sense of his attitude and behavior, based on a system of values built from outside.[42]

The autonomous conception of man rejects the existence

of any super human forces as responsible for the creation of the human individual and his behavior. There are two opposed varieties of this approach. One of them takes as its starting point the individual interpreted as "spiritual monad"--the individual is lonely, isolated, "doomed to freedom," or "doomed to choice," and has no help or assistance (e.g., Sartre's existentialism). The second approach bases human autonomy on society and social relations. It takes as its point of departure the individual man, who thinks and acts and always cooperates with others within some social framework, but is a distinct individual. His autonomy is only relative. He is a social individual because he is unable to live without society. He is shaped by society and is its product physically and spiritually, but at the same time he is able to set in motion those social forces which are capable of removing social barriers to individual development and self realization. In this approach, the emphasis is laid not only on his social aspects but also on the social way to the implementation of individual aims (Marxist philosophy of man).[43]

The sociological theory of personality, despite the differences in the formulations made by the respective authors, examines the question of individual in the light of his relation to society. The individual at his birth is an organism with biological equipment which can be made actual and developed only within the process of socialization. Man's personality is a product of society and its culture. "Man is not born human" but becomes so in the process of associating with others and of internalization of culture.[44] Social structures, according to Riesman, differ very much in the degree to which they evoke a social character that in the process of socialization fills up, crushes, or buries individuality.[45] The challenge to the full development of person and his individual evolution within the frame of certain definite structures is nowadays in danger, because of the development of mass society, mass culture and increasing conformism, as a predominant attitude.

Of very great importance for the philosophical concept of man are the data of contemporary psychology. The analysis of the driving mechanism and needs founded on psychological and neuro-physiological data brought about a modification of the traditional view on psychology. The structure of man's regulative mechanism, i.e., his drives and psychological needs, is determined by the character of the social environment he lives in. Therefore contradictions between needs of an individual and social requirements are, in their essence, contradictions of human society, contradictions in the culture produced by mankind.[46]

The characteristic sociological and psychological conceptions of man are opposed to the image of man full of inner contradictions and vainly struggling to solve unsolvable conflicts, an image outlined by the philosophers of existentialist orientation. They are also hardly reconciliable with the psychoanalytical conception of man, as a creature to whose natural cravings are opposed the absolute restrictions imposed by civilization, i.e., a picture of man which stresses the inner contradictions in man's mind tracing them back to the biological and psychological construction of man.

The idea that human personality, its needs and aims are the result of social activity of man, that, therefore, the contradictions of human nature are in fact contradictions of human society, is at the same time pessimistic and optimistic. Pessimistic because it shows man's frequent helplessness in face of himself. Optimistic because it permits us to hope for a harmonious shaping of human universe which will create harmonious personality and adequate conditions for the full development of person and its self realization.[47]

FOOTNOTES

1. W. Friedman, "Legal Theory," London 18, 1947.

2. M. Sobolewski, "Prawa i wolnosci obywatelskie w wysoko rozwinietych krajach burzuazyjnych" (Rights and Freedoms of Citizens in Economically Developed Capitalist States), in PRAWA I OBOWIAZKI OBYWATELSKIE W POLSCE I ŚWIECIE (Citizens' Rights and Duties in Poland and in Contemporary World 59, Warszawa 1974.

3. A. Smith, AN INQUIRY ABOUT THE NATURE AND CAUSES TO THE WEALTH OF NATIONS, 42, Book IV, Ch. II.

4. F. Hayek, THE CONSTITUTION OF LIBERTY 19, Chicago 1972.

5. S. Wronkowska, "O znaczeniu i socjotechnicznej roli terminu 'czyjeś prawo'" (On the Meaning and Sociotechnical Role of the Term 'Somebody's Right') 6 Państwo i Prawo 1079, 1969.

6. M. Duverger, DROIT CONSTITUIONNEL ET INSTITUTIONS POLITIQUES 204, Paris 1956.

7. K. Grzybowski, HISTORIA DOKTRYN POLITYCZNYCH I PRAWNYCH (HISTORY OF POLITICAL AND LEGAL DOCTRINES) 413, Warszawa 1967.

8. M. Borucka-Arctowa, PRAWO NATURY JAKO IDEOLOGIA ANTYFEUDALNA (THE LAW OF NATURE AS AN ANTI-FEUDAL IDEOLOGY) 292, Warszawa 1957.

9. D. Riesman, SELECTED ESSAYS FROM THE INDIVIDUALISM RECONSIDERED 12-27, New York 1954.

10. G. Burdeau, LES LIBERTÉS PUBLIQUES 177, Paris 1961.

11. K. Kulcsar, "Social Factors in the Evolution of Civic Rights," in SOCIALIST CONCEPT OF HUMAN RIGHTS 144, Budapest 1966.

12. Ph. Selznick, "Sociology and Natural Law," 6 Natural Law Forum 92, 1961.

13. J. Wróblewski, "Realtywistyczne teorie prawa" (Relativist Theories of Law), 1963 Państwo i Prawo 215.

14. S. Zawadzki, PAŃSTWO DOBROBYTU (Welfare State) 217, Warszawa 1964.

15. A. Michalska, "Pakty człowieka a katalog praw obywatelskich" (The Covenants of Man and the Catalogue of Civil Rights), 1973 Państwo i Prawo 47 ff.

16. E. Savona, "Civil Trial and Social Justice," Report for the Conference on Sociology of Judicial Proceedings, Bielefeld 1973.

17. P. Morris, R. White, Ph. Lewis, SOCIAL NEEDS AND LEGAL ADVICE, Bristol 1973.

18. M. Duverger, SOCIOLOGIE POLITIQUE 348, Paris 1966.

19. M. Whyte, THE ORGANIZATION MAN, New York 1956.

20. Materials of the Teheran Conference on Human Rights.

21. R. Bystricki, "Quelques remarques sur la révolution scientifique et technique et les droits de l'homme," dans: R. Cassin, AMICORUM DISCIPULORUMQUE LIBER: PROBLÈMES DE PROTECTION INTERNATIONALE DES DROITS DE L'HOMME, Paris 1969.

22. A Nartowski, "Ochrona Praw Człowieka a postep nauki i techniki" (The Protection of Human Rights and Scientific and Technological Progress), 1972 Zeszyty Naukowe Uniwersytetu Jagiellonskiego 88 ff..

23. S. Strömholm, RIGHT OF PRIVACY AND RIGHTS OF THE PERSONALITY 16, Stockholm 1967.

24. S. Stromholm, op. cit., pp. 16-19

25. Ibid., p. 32.

26. Ibid., p. 60.

27. P. Juvigny, "The Right to Privacy in the Modern World," paper presented at the International Meeting on the Right to Privacy, organized by UNESCO in 1970.

28. Geneva, 1972.

29. International Social Science Journal, 1972, No. 3 presents the results of the UNESCO comparative study of the laws relating to the legal protection of the right of privacy.

30. Courrier, 1973, No. 7, p. 17.

31. E. Fromm, "The Sane Society," p. 362.

32. Ibid., p. 355.

33. W. Friedman, op. cit., p. 432

34. Ibid., p. 62.

35. J. Maritain, L'HOMME ET L'ETAT 67, Paris 1965.

36. M. Borucka-Arctowa, "Sociology and the Contemporary Concepts of the Laws of Nature and the Rights of Man," 1968 Polish Round Table 57 ff.

37. ENTSTEHUNGSGESCHICHTE DER ARTIKEL DES GRUNDGESETZES, Jahrbuch des Offentlichen Rechtes der Gegenwart, Neue Folge, Bd 1, 1951, 45 ff., 52 ff.

38. Amongst the followers of this trend one meets representatives of different philosophical ideas as well as of different disciplines, as, e.g., the Neo-Thomists J. Maritain, J. Leclercq and the sociologist Ph. Selznick.

39. J. Wróblewski, "Natura a reguły postepowania" (Nature and Rules of Behavior), 1970 Etyka 51 ff.

40. M. Borucka-Arctowa, "The Problem of Evaluation in Legal Sciences," 1969 Archivum Luridicum Cracoviense 12 ff.

41. J. Wróblewski, op. cit., p. 76

42. A. Schaff, "Marxism and the Philosophy of Man," in E. Fromm, ed., SOCIAL HUMANISM 142 ff., New York 1966.

43. Ibid., p. 145.

44. J. Szczepanski, "Filozoficzna i socjologiczna koncepcja czlowieka (Philosophical and Sociological Conception of Man), 1969 Studia Filozoficzne 19.

45. D. Riesman, THE LONELY CROWD 275, New York 1953.

46. J. Reykowski, "Natura ludzka a potrzeby" (Human Nature and Needs), 1970 Etyka 32 ff.

47. Ibid., p. 47

A GLIMPSE OF PERSONS
IN THE MODERN LEGALISTIC WORLD

Mitsukuni Yasaki

I.

The network of law surrounding us now in Japan is often said to originate in the modern Western world. The legal thoughts and ideas underlying them and connected with them, freedom of the individual for example, have often been referred to as decisively influenced by the Western legal tradition. J. Shklar once used the expression "legalism" to indicate a rather broad area (including law and legal thought) in the Western world. In this short paper, the writer will borrow, in somewhat modified form, from her for the purpose of referring to connections between law and legal thought concerning the idea of the individual.

At first glance, a similar situation to the West's is found in Japan, since she accepted from an early period some features of the Western legal tradition. But Japan is also a country in which a long Asian tradition of legal culture has been maintained. Taking this into consideration, the question must be raised as to how and to what extent law, legal thought, and legal concept of the individual which originated in the West could be accepted in Japan. Even though society in Japan has been developed and modernized with the heavy structural changes since Meiji restoration (1868), it becomes an urgent issue to understand the possible connections between the Western and Japanese legal traditions which so many people previously attempted to resolve in various ways according to their different situations.

It is, accordingly, the problem of the individual which I wish to consider as the criterion by which to measure the mutual relationship of both traditions. I should like to suggest that legalism was not developed in Japan with so intimate a connection between the idea of the individual and the concept of freedom and equality as it was in the West. The purpose of this short paper is mainly to make a survey of their complicated relationships, at first in the West, then in Japan, and to add a few simple remarks. In so doing, the paper is limited merely to introducing some of the scholarly achievements connected with these issues, and to offer a simple reconstruction for further discussion.

II.

"What is legalism?" J. Shklar explains it as follows: "It is the ethical attitude that holds moral conduct to be a matter of rule following, and moral relationships to consist of duties and rights determined by rules... It is also a very common social ethos, though by no means the only one, in Western countries."[1] "At one end of the scale of legalistic values and institutions stand its most highly articulate and refined expressions, the courts of law and the rules they follow; at the other end is the personal morality of all those men and women who think of goodness as obedience to the rules that properly define their duties and rights. Within this scale there is a vast area of social beliefs and institutions, both more and less rigid and explicit, which in varying degrees depend upon the legalistic ethos."[2]

It is obvious that legalism is treated by her in a broad sense rather than in a sense of legal formality. Certainly she recognizes the fact that "the core of that Western tradition, for those who have discovered it, is essentially legalism, the rule of law."[3] But she gives her attention not only to such a formal aspect of the matter, but a material aspect so to speak, for example, the compatibility of legalism with democracy or liberalism.[4] Legalism, thus understood, implies much more than the idea of the rule of law. Referring to various ideological characteristics of such ideas, she goes on to point out the importance of social diversity and to raise the question of Max Weber's view of formal rationality, that is, the unique cultural traits of the West. "The question is whether it is valid to extract a quintessence of 'the West' by subtracting from its history all that it shares in various degrees with the rest of mankind."[5] This is the kind of question which is one of the most interesting topics in this paper and which shall be mentioned again.

There is another way to take legalism--as an attitude of legal fetishism, in other words, an attitude, confined to its formal aspect, that is, to identify laws with the Law. For the purpose of this paper, however, the writer will use "legalism" not in this sense, but in the broader sense above.[6]

A few words should be added about the material aspect behind this formal aspect of legalism from a historical perspective. It may be summarized in the idea of human beings in general entitled to be free, and equal. This idea, as is well known, had various expressions in modern history, typically in The Virginia Bill of Rights (1776), The Declaration of Independence (1776), Déclaration des droits de l'homme et du citoyen (1789). Attempting to trace back this idea to its origin, it will be found not only in medieval documents, but also in ancient ones. For instance, Corpus Juris Civilis, Digesta, I, 1, 4 says: "Cum jure naturali omnes liberi nascerentur..." Similar ideas were repeated in the medieval period not only in scholars' writings but in official documents, for example in the "preamble" to the order concerning the liberation of serfs issued by Louis X in 1315, and repeated by Philippe V in 1318,[7] or in the Preface of "Sachsenspiegel", though differing from the former in regard to the Roman Law style.

The proposition that according to natural law all human beings are free by nature sounds attractive. But, however attractive these expressions appear, these were either limited in scope and real function, or remained mere lip-service to conceal existing problems. It was in the modern period that the idea of the equal right of persons to be free was settled and advanced in legal thought and institutions. At this stage, two points should be kept in mind: (1.) The limited and abstract character of the idea of freedom and equality, and (2.) The liberal and absolutist views of the mechanical application of the law.

(1.) In the modern Western world, the idea of individual freedom and equality was increasingly asserted in a generalized form (mainly by modern natural law theories). Generalization was followed by abstraction. Here we may find relevant matter which continues to be much discussed even today. G. Radbruch vividly explained it in his book.[8] R. M. Unger wrote as follows: "The legal order which sees in man a legal personality abstracts from the total context of the individual; it regulates different aspects of his social situation under different rules. Because these 'rules' are necessarily both limited in scope and abstract in formulation they deal with specific transactions rather than with the total social and economic context in which men find themselves. The restatement of social relations as rules permits: (i) The resolution of the total social situation of an individual or group into isolated parts, the relation between which and therefore the total structure of which is thereby obscured; (ii) Specifically, the distinction between man as a legal person and as a social being. This distinction makes possible the coexistence of political equals as social unequals."[9] This is also the case with general and abstract ideas of freedom and equality.

(2.) The legal order corresponding to such an idea of individual freedom and equality, it has often been held, was realized by means of its mechanical application. It may be enough to refer to names of Montesquieu, Beccaria, Bentham, A. Feuerbach. Assertion of the mechanical application of legal rules here perhaps came from a liberal point of view, but also from authoritarian or absolutist point of view. E. Ehrlich correctly pointed out both sides of the question. "In the basis of German, French, and Austrian jurisprudence lies always the legal outlook of law as emanating from the State (staatliche Rechtsauffassung): Man insists that the judge should always decide according to the laws (Gesetz). The significant thought here is the enhanced thought of the outlook of law as emanating from the State: It should be either the legislator's intention in a sense of absolutist legal outlook, or the expressed will in the form of laws of members of society according to liberal outlook of law, which determines judge's lawfinding."[10]

Each of these two points are indispensable for understanding our problem, individuals in the modern legalistic world, especially in Japan. From each point, however, may result other serious issues which will be mentioned later.

III.

Problems in Japan

Traditional social relationships and ways of thinking still remained in Japan after the Meiji restoration, 1868. It became, however, essential for Japan to modernize herself to reach the level of Western countries for several reasons. One of the symbolic features of this trend is seen in her continual effort to have a modern legal system in order to suggest a modern national state unified and established. Codifications of various types of positive law, fundamentally constitutional law (Dainihon Teikoku Kenpo, cited below as the Meiji constitution) on the one hand, and legal provisions concerning rights of the individual to be free, legislated in various codes on the other, were both made and originally established during the middle period of the Meiji era (1888-1898), and are simple examples of this effort.

It may still be questioned, however, how far Japan was actually modernized in regard to legal and social traditions when we remember the fundamental pervasive strength of the older traditions. Perhaps we can illustrate this situation by saying that various ideas and institutions basically coming from the extended family system remained such a powerful barrier that newer ideas and institutions of popular freedom on the one hand and of the nation state on the other were more or less formulated and developed under their direct or indirect influences. Keeping this in mind, we shall briefly outline some trends in the fields of institution, legal thoughts and ideas.

The family system is common to all human beings as far as they live together within ties of community life as a social unit.[11] This is naturally also the case with Japan. But what is characteristic for the traditional family system in Japan is that it was really an extended family organized around criteria of lineage, etc., the ethics of which was impressively treated by Kamishima Jiro as corresponding to "ascetic" ethics of protestantism concerning calling in the Western modern world explained by M. Weber.[12] Scholarly achievements indicate several types of extended family systems and ideas existed here. For example, Kawashima Takeyoshi pointed out an interesting combination of patriarchal and feudalistic elements in the family by reference to parent-children relationships as supported not only by the absolute power of the father, but by his authority giving obligation in return for children's filial duties.[13] Such an extended family system based on criteria of lineage, etc. was retroactively connected with ideas of ancestor worship, and socially more and more extended, by analogy, to various social relationships and even to the State. It is for this reason that the idea of the extended family system was cited by early thinkers as a firm ground for their assertions.

Institutions Concerning the Individual

Some symptoms of the change will be seen in the fact that social relationships based on social status in the Tokugawa feudalistic period (1600-1868) was formally abolished and transformed since Meiji Restoration, so that people came to be treated as subjects on the one hand, and as persons able to possess property and so on in private law on the other. Starting from the confiscation of feudal land and land tax reform regulations, the change took place in regard to people by treating them as no more bound by different social statuses, but principally liberated or emancipated from them by legislation. This followed soon after the Meiji Restoration (1868-1872).

Institution Concerning Nation State

As to cabinet, local government, and the military system, too, changes had occurred under the decisive influences of German (Prussia at that time) institutions, which was disastrous for Japan's road to modernization. The promulgation of the Meiji constitution (1889) was externally a kind of monument to show how intensively Japan was modernized in the public field as well as in the private. It was also internally not only to show the establishment of fundamental law, but to serve for Japan as something like a guide for appeals to people's traditional moral sense, while law in the Western world including the constitution, generally speaking, had been regarded as separate from morality, which had been, it may not be overemphasized, left up to each person's own conscience, that is, each one's internal world.[14] Here is one of the contrasts between Japan and the West.

Meiji constitution guarantees in Chapter II entitled "Rights and Duties of Subjects" different rights and liberties: Liberty of abode and change thereof (Article XXII), freedom from illegal arrest, detention, trial or punishment (XXIII), right of trial by lawful judge (XXIV), the inviolableness of the individual's home (XXV), the right to the secrecy of the mails (XXVI), the inviolable right of property (XXVII), freedom of religious belief (XXVIII), the liberty of speech, writing, publication, public meetings and associations (XXIX), and the right of petition (XXX).

These rights and liberties, however, were privileged and qualified: (1.) Privileged for the reason that these "were not based on the natural rights philosophy found in the French declaration of human rights or on

that found in the American declaration of independence,
but were bestowed by a benevolent act of the Emperor."
Ito Hirobumi in his Commentaries on the Constitution
paraphrased it as follows. The "Emperors have made it
their care to show love and affection to the people,
treating them as the treasures of the country; and the
people have ever been loyal to the sovereign, and have
considered themselves as happy and blessed." (2.) Qualified for the reason that "while these rights might be
a guarantee against the executive power, they did not
serve as security against the legislative power...any
right or liberty set forth in the Constitution could
be violated by legislative act wherever the guarantee
of the liberty was qualified by such phrases as "within
the limit of law", "unless according to law", and
"except in the cases provided for in the law". (3.) "In
connection with the abridgment of individual liberties,
mention should be made of two other constitutional
provisions. Though never invoked, Article XXXI declared that the constitutional rights of the people
would not affect the exercise of the powers appertain-
ing to the Emperor in times of war or in cases of a
national emergency, and Article XXXII made it plain
that only those constitutional rights not in conflict
with the laws or rules and discipline of the Army and
Navy would apply to men in military service. These
two Articles stand in remarkable contrast to the post-
World War II Constitution in which it is asserted that
the fundamental rights guaranteed by the Constitution
are everlasting and inviolate, and that these rights
were gained over a long period of time by man's efforts
to acquire freedom (Article XI, LXXXVII)."[15]

That the Meiji constitution was given a role influen-
tial for the internal world of individuals is partly
due to these reasons, particularly to specific view of
people as "subjects", subjects as "treasures", which
perhaps was related to a symbolic view of Emperor as
father and subjects as children--symbolic in a sense
of the extended family system idea with patriarchalism.

Legal Thoughts and Concepts

Western legal thought and concepts of the right of
the individuals were increasingly translated into
Japanese and accepted by Japanese, who were scholars,
writers, politicians, bureaucrats. Such translated
thoughts and ideas covered an extremely wide range, as
those from Roesler, Laband, von Stein, Boissonade,
Rousseau, Austin, Bentham, Mill, etc., and appeared in
a variety of attitudes from conservative to liberal.
At one end of the scale of acceptance of the West stand

people, typically scholars like Hozumi Yatsuka (1860-1912) or Motoda Nagazane (1818-91), deeply interested in educational administration, whose ideas were mainly affected by both traditional patriarchal authoritarianism connected with Confucianism in Japan and German legal positivism or Austinian imperative theory of law. At the other end stand scholars or writers and often social reformers like Nishi Amane (1829-97), Fukuzawa Yukichi (1835-1901), Nakae Chomin (1847-1901), Ono Azusa (1852-86), Oi Kentaro (1843-1922) and so on, whose ideas, being mainly influenced by the Western liberal tradition, perhaps represented by Rousseau, Bentham, Mill, came to emphasize the liberation of individuals from social traditions and their identification as individuals as such. One anecdote tells us that a translator preceding Meiji Restoration felt it terribly difficult to translate the word "individual" into adequate Japanese since there was no adequate substance fitting to this word.[16]

The borderlines between these extremes, however, are not so rigid that ideas of both sides were not as a matter of fact implicitly supplemented by some elements from the other extreme. Among these men, Inoue Kowashi (1843-95) may well be considered, because, though he was not a real scholar, but rather a high officer in the bureaucracy, his legal thought itself indicates some complex features of our subject. Indeed he actually participated in the formation of much of the groundwork for the Meiji social political system. In accordance with the interpretation given by Ohashi Tomonosuke, we shall outline Inoue's thought in regard to two issues, keeping in mind his relation to other contemporaries.

(1.) <u>Conformity</u> <u>of</u> <u>legal</u> <u>control</u> or <u>generality of law</u> (<u>application</u>) were emphasized by Inoue from the earliest period of his career. His stay in Europe (mainly in France and Prussia) from 1872 to 1873 to learn the judicial institutions there was perhaps enough to impress upon him the importance of such elements and "the rule of law" underlying them. Perhaps because of this, he continued to take the same position in principle, though he as a bureaucrat often expressed occasionalistic opinion. For example, according to the decree (Fukoku) issued by Great Council of state (Dajokan), 1874, a person even though originally coming from family of nobles (Kazoku), or gentry (Shizoku), is to be treated as commoner (Heimin), if he sets up a branch family. The decree, however, tended to be disregarded in real politics.

Inoue in his opinion expressed in 1879 questioned: Is it right to make an exception regarding the "distinguished" noble in the case of the application of the decree? "If so, validity of the decree would be lost and its letter would be merely empty."[17] He pointed out here the necessity of the general application of the law (decree at that time). The same is true in the case of order (Meirei) provided with penalty. Under the Meiji constitution naturally there are laws (Horitsu) concerned with legal penalties such as imprisonment, etc. "If such matters are not controlled by laws passed through approval of parliament, it would inevitably result that parliament has no more real function and constitutionalism has no more substantial meaning."[18]

By pointing this out in 1890 Inoue gave special attention to the significance of the constitution against some dangerous tendencies of constitutional interpretation. His approval of the constitution may partly be due to the fact that he was really one of the makers of the draft for this constitution. It may also be interesting to compare this to a similar controversy soon after the Japanese constitution was issued (1946, enforced 1947), that is, the problem concerning whether it is legally permitted to issue orders with a wide range of punishment by means of legislative delegation. In order to grasp his constitutionalistic or legalistic attitude, it is enough to mention the defense of judicial independence he made in regard to the famous case "Otsu Jiken", which may afford a key for understanding his distance from other contemporaries', for instance, Hozumi Yatsuka's attitude (conservative).

(2.) Inoue paid little attention to the notion of human rights and civil liberties, in spite of his constitutionalistic attitude. Presumably he attained knowledge in this field from his experience in Europe, foreign books, foreigners who were employed by Japanese government as advisers or by colleges as teachers. Nevertheless, he referred seldom to individual rights, and he rather seemed to take people (commoners) as subjects controlled by politics than as subjects participating in politics, still less in freedom and civil rights--for example those expressed by Rousseau as universally valid principles. In a different sense from that above, it makes a sharp contrast to his contemporaries' attitude. Ono and Oi should be remembered here. Western liberal tradition of legislation, it has been often asserted, was based on principles of liberty, equality, and inviolability or sanctity of property underlying them, as typically stated in Déclaration des droits de l'homme et du citoyen. Among these two Ono, in referring to civil law, found the ends of legislation to be subsistence, abundance and equality as did Bentham.[19] Indeed he learned much from Bentham and was under the influence of the English school of thought of this sort. Partly owing to this fact he could assess with keen insight and ask for freedom and civil rights as being indispensable for the modern capitalistic society.

Similarly, Oi stressed the meaning of liberty and equality as well, but he differed from Ono not only in the way of thinking, particularly influenced by the French school, but in the argument of nationalization of landed property, which he held was unsuitable for private ownership.[20] It will be clear that the argument implies a kind of socialistic idea more than mere recognition of the bourgeois capitalistic society. Both of them were practically oriented social thinkers as well as writers. Differences inherent in their ideas and arguments, however, led them in different directions in the highly changed and rapidly changing atmosphere of the early Meiji period.

Why, then did Inoue think about human rights and civil liberties in such a different way, as mentioned above? One of the main reasons perhaps may be due to the fact that he still held the older view stressing the importance of the gentry (Shizoku). Being born in a gentry family and brought up on the teaching of Confucianism would seem to determine the internal background of his moral outlook, even after he attained his knowledge of Western culture. Thus in 1875 he pointed out the importance of the gentry. In 1876 he stressed hierarchical family order in the case of (laws of) succession,[21] and then even in 1887 expressed his opinion with some bias towards the older hierarchical outlook of society based on agricultural production.

As pointed out by Maruyama Masao, two distinct principles characterized the early Meiji period, that is, the principle of political integration and the principle of political diffusion toward the base.[22] The former was connected with ideas oriented to state power and the latter with ideas oriented to civil rights. These opposite principles and ideas were arranged by Inoue in the direction of absorption of civil right into state power. Again here lies the hierarchical outlook of society, which came from the older cultural tradition in Japan, to undermine those ideas. Thus we may summarize the points of issue in this way: Opposite principles and ideas on the one hand, and opposite ways of thinking (Western or Japanese) on the other, were so confused with each other in the actual development of Meiji history that most of the thinkers of the time were faced with a complicated labyrinth. At this point it is very impressive to find most of the enlightened were more or less involved with the traditional Japanese way of thinking, and this is so even among those who were not oriented to state power, but to freedom and civil rights. Hence it is often remarked that writers or scholars who at first glance looked liberal, e.g. Ono, or Minobe Tatsukichi, were still influenced by traditional views, and the opposite was true of those who at first glance looked conservative or ultranationalistic, e.g. Hozumi Yatsuka or Uesugi Shinkichi.[23] This may be enough perhaps to indicate the tangled connection between ideas, institutions, and social structure.

IV.

Transition and Prospect

R. M. Unger marked the social structure of societies undergoing liberal capitalist development by reference to three trends: "(a.) Decline of the extended family as a fundamental unit of social organization... (b.) the decline of traditional intermediary groups organized around criteria of lineage, rank of locality and the use of new intermediary organizations..., oriented to more or less specific political and economic function; (c.) A questioning of ascriptively determined roles, statuses and other inequalities of social and economic position, accompanied by a transformation of group relationships which may or may not result in an increase of social mobility."[24]

Regarding the situation in Japan, he continues to maintain that it is distinct from modernization in the West. "A further complication is introduced by the fact that 'traditional' social groups may adapt themselves to industrialization and the economic or technological changes which are associated with or precede it. In some modernizing societies (e.g., Japan), the perpetuation of traditional groups is part of a more or less conscious strategy of development and is closely related to forms of social consciousness, political authority and normative order different from those which characterize liberal capitalist development."[25] He calls the development or modernization of this type "traditionalistic".[26]

"Traditionalistic" modernization in Japan has been much discussed from various points of view and still remains a basic and controversial issue, especially for writers or scholars interested in that period. Unger's remarks may serve as an illustration of the situation we saw, in which a case of state power (rights) v. civil rights was going to be settled in the direction of predominance of the former over the latter by means of the complicated maintenance of the traditional elements beneath the surface of Western imported ideas. The situation is the same with the later period of the Meiji era, and, it may not be an overstatement to say, even with the Taisho (1912-1926), and Showa (1926-   ) eras.

In examining the modernizing process in detail, we surely may not overlook some trends with really modern characteristics which result directly or indirectly in emphasizing freedom and civil rights, that is, so called citizen-oriented modern legal theory (Shiminhogaku) and so on, which are increasingly established. Theories of Minobe Tatsukichi in the field of public law, and of Suehiro Izutaro in the field of private law can be counted as examples of these trends.[27] On the other hand, it must also be kept in mind that social structure in Japan has undergone gradual changes mainly resulting from capitalistic development, so that it has been faced with several contradictions, e.g., concerning landlord-peasant relationships or labor problems raised by employer-employee relationships in the industrial field, etc. Even these scholars, however, as remarked above, were not free from traditional conflict in their way of thinking.

The situation after the second world war considerably influenced Japan's development. The new constitution was notable for two distinct features, antimilitarism and antifeudalism. It includes fundamental rights of "people", not "subjects", most of which concern both rights of persons to be "free from" and of "participating in" as well as other modern Western constitutional concepts. It also includes "the right to maintain the minimum standards of wholesome and cultured living", etc., as a right to live in contrast to the right to be free cited above. The rule of law, presupposing such a constitution as the supreme law within a system of laws, appeared to be so firmly established in this occasion, that it made legalism in a formal and material sense possible.

The situation since then, however, indicates some crucial difficulties, new and old, common and peculiar.

These difficulties, on the one hand, result from the older traditional elements peculiar to Japan, mainly based on the family system, though the extended family system to some extent has changed into the nuclear, and, on the other hand, from capitalistic development, not only peculiar to Japan, but common to the modern Western world. The latter is related to various problems raised by characteristic changes of contract system in capitalistic society, increased interference of political power in the private law field, labor-relationships involving unemployment, etc., and current problems raised by housing, social circumstances, public nuisance, etc.,[28] all of which may lead to an atmosphere of "self-alienation" of human beings in society. It is, needless to say, a problem carefully to be examined. The former difficulties, however, concern the special problem of the relation of Japan to the West.

Japan is not only geographically located in East Asia, but it maintained its own cultural tradition as distinct from the West. The author has attempted merely to expose it as a somewhat older tradition and something to be modernized for the purpose of clarification of sharp contrasts. But this may be only one aspect of the matter as seen in the light of modernization. As pointed out by Shklar, the direction of Western-styled modernization in regard to law and legal thought was naturally realized in the West and it has had adequate reasons. Is such a direction, however, strictly applicable to the non-Western world including Japan? We must consider whether we should keep our own different standards, even in connection with law and legal thought, as suggested by scholars such as D. M. Trubek, R. Unger and others.[29] Then, what other aspects should Japan maintain? Is there something else indispensable for the development of law and legal thought which is proper to the Japanese character and way of thinking? If so, what influences does it exert upon the subject: the individual in the modern legalistic world? Here is a great problem to be discussed.[30]

FOOTNOTES

1. Shklar, LEGALISM, 1, 1964.

2. Ibid., p. 3.

3. Ibid., p. 21.

4. Ibid., pp. 14-16, 20.

5. _Ibid._, p. 22. cf. M. Weber, GESAMMELTE AUFSÄTZE ZUR RELIGIONSSOZIOLOGIE, 1 ff, Bd. 1, 1920.

6. Ohashi Tomonosuke's article cited below, p. 11; Yasaki Mitsukuni, LEGAL PHILOSOPHY AND SOCIOLOGY OF LAW (HOTETSUGAKU TO HOSHAKAIGAKU) 258 ff, 1973.

7. Takahashi Kohachiro, HISTORY OF FORMATION OF MODERN SOCIETY (KINDAISHAKAI SEIRITSUSHIRON) 29 ff.

8. G. Radbruch, DER MENSCH IM RECHT, 1961.

9. R.M. Unger, THE PLACE OF LAW IN "MODERN" SOCIETY, 57 f, (undated, unpublished study, on file in Harvard Law Library).

10. E. Ehrlich, DIE JURISTISCHE LOGIK 181-182, 2 Aufl., 1925.

11. Cf. Nakane Chie, HUMAN RELATIONSHIPS IN SOCIETY CHARACTERIZED BY VERTICAL LINE (TATE SHAKAI NO NINGEN KANKEI), 1972.

12. Kamishima Jiro, INTELLECTUAL STRUCTURE OF MODERN JAPAN (KINDAI NIHON NO SEISHIN KOZO) 265.

13. Kawashima Takeyoshi, SOCIETY IN JAPAN BASED ON FAMILY CONSTRUCTION (NIHONSHAKAI NO KAZOKUTEKI KOSEI) 104 ff, 1950.

14. Isomura Tetsu, CITIZEN ORIENTED MODERN LEGAL THEORY 19, (_Shiminhogaku_), in series of history of development of modern law in Japan (Nihon Kindai Ho Hattatsushi Koza), cited by Yasaki, _op. cit._, pp. 81-83.

15. JAPANESE LEGISLATION IN THE MEIJI ERA, edited by Ishii Ryosuke, translated and adapted by William J. Chambliss, 1958, (JAPANESE CULTURE IN THE MEIJI ERA, vol. IX).

16. Kamishima, _op. cit._, p. 259. As to similar problems, cf. Inoue Shigeru, STRUCTURE OF LEGAL ORDER, (HOCHITSUJO NO KOZO) 411 ff, 1973.

17. I HISTORICAL DOCUMENTS OF INOUE KOWASHI 189 (ed. by Kokugakuin Daigaku Inoue Kowashi Denki Hensaniinkai); Ohashi Tomonosuke, "Inoue Kowashi's Outlook on Law," ("Inoue Kowashi no Horitsukan") 13, Annual of Legal Philosophy, ed. by the Japan Association of Legal Philosophy, 1970.

18. II HISTORICAL DOCUMENTS OF INOUE KOWASHI 255; Ohashi, _op. cit._, p. 15.

19. Ono Azusa, OUTLINE OF CONSTITUTION (KOKKEN HANRON) 48, cited by Nakamura Kichisaburo, LEGAL HISTORY OF MEIJI ERA, (MEIJI HOSEISHI) 64, 1955.

20. Oi Kentaro, OUTLINES OF LIBERTY, (JIYU RYAKURON) 453, cited by Nakamura, op. cit., p. 86.

21. I HISTORICAL DOCUMENTS OF INOUE KOWASHI 84, cited in Chashi, op. cit., pp. 21-22.

22. Maruyama Masao, "Intellectual Aspects of Meiji State" ("Meiji Kokka no Shiso"), in: HISTORICAL STUDIES OF JAPANESE SOCIETY 183 ff, (NIHONSHAKAI NO SHITEKI KYUMEI, ed. by Association of study of history).

23. Nagao Ryuichi, "Legal Philosophy of Minobe Tatsukichi," in 82 The Journal of the Association of Political and Social Sciences 157 f, (Kokka gakkai Zasshi).

24. R.M. Unger, op. cit., p. 45.

25. Ibid., p. 48.

26. Ibid., p. 136.

27. Cf. Isomura, op. cit., vol. 7, 9, 10.

28. Shimoyama Eiji, "History and Prospect of Human Right," ("Jinken no Rekishi to Tenbo") 7, 1972.

29. D.M. Trubek, "Toward a Social Theory of Law: An Essay on the Study of Law and Development," 82 Yale L.J. 2 ff, 1972; Unger, op. cit.; Nakane, op. cit., pp. 18 ff.

30. Cf. R.H. Minear, JAPANESE TRADITION AND WESTERN LAW 9, 1972. Also, see Nagao Ryuichi's review, Kokka gakkai Zasshi, vol. 84, no. 1, p. 112. I am grateful to Prof. V. Held for her criticism of my paper on two points; first, concerning women's problem in the historical development of Japan, and second, the role of legalism. I plan to consider both in the future.

LES DEUX FIGURES DE LA LIBERTE AU XVIIIème SIECLE

Simone Goyard-Fabre

En décembre 1784, Kant, répondant à la question <u>Was ist Aufklärung?</u>, déclarait que les Lumières débarrassaient enfin les hommes des "grelots attachés à leurs pieds"[1]. Entendons qu'en déracinant de l'esprit les superstitions et les oppressions de toutes sortes, les Lumières arrachaient l'homme à son état de minorité. Et pour cela, disait le philosophe, "il n'est rien d'autre requis que la liberté."

En effet, dans la sensibilité toute neuve du XVIIIème siècle, les idées de bonheur et de liberté sont devenues la "dimension d'espérance" sans laquelle la vie ne vaudrait pas d'être vécue. La liberté, voilà, comme diront Kant et Fichte, "la destination de l'homme". C'est aussi la même foi militante, le même engagement, le même combat. De cette anthropodicée chaleureuse, la fête révolutionnaire sera la manifestation populaire: Liberté, Egalité, Fraternité. L'homme refuse les chaînes et revendique ses droits.

Mais célébrer la liberté en des chants d'allégresse, ou combattre, et même mourir pour elle, n'indiquent que l'éminente valeur qu'on lui reconnaît. Connaître ce qu'est la liberté en son essence, comprendre ce qu'elle implique en son existence exigent autre chose que des poèmes et des sacrifices. Savoir ce qu'est la liberté est aussi difficile que la conquérir.

"L'invention de la liberté"[2] est inséparable du mouvement général des idées des Lumières dans lequel, d'une part, l'idée de NATURE prend un nouvel essor tandis que, d'autre part, s'affirment la force du DROIT et l'autorité de la LOI. Aussi bien les philosophes découvrent-ils l'amphibologie de la liberté: la liberté a deux figures; elle est, d'un côté, la liberté sauvage et sans loi de l'individu à l'état de nature et, d'un autre côté, sous la loi civile et par elle, la liberté du citoyen. Et ces deux figures de la liberté définissent deux styles de vie entre lesquels l'homme du XVIIIème siècle, parvenu à un tournant de l'histoire d'Occident, doit choisir.

I.

Que l'idée de liberté soit dichotomique n'a rien d'étonnant dans la pensée du XVIIIème siècle attachée aussi passionnément aux puissances et à la beauté de la nature qu'à la rationalité et à la vertu de la loi. Il convient cependant d'écarter ici une erreur toujours possible car, si la philosophie du XVIIIème siècle dessine deux figures de la liberté, elle ne répète pas pour autant l'antinomie classique du monde cassé pensé par Platon ou Descartes pour qui Nature et Liberté s'opposent en une altérité irréductible. Les philosophes du XVIIIème siècle rectifient le schéma dualistique de la métaphysique traditionnelle au point que Kant, en qui l'on peut voir l'aboutissement et la synthèse du siècle, enracinera la liberté dans la nature même[3] et n'en concevra l'accomplissement qu'à travers l'effort historique de la Kultur par lequel l'homme rejoindra enfin la Natur.[4] Ce disant, le génie kantien établit le bilan d'une réflexion en laquelle la problématique de la liberté se pose, non pas en opposition à l'idée de nature, mais dans sa liaison à l'idée de nature. C'est pourquoi, plutôt que de parler, comme Hobbes ou Spinoza, de deux libertés, nous préférerons dire, comme Rousseau et Kant, que la liberté a deux visages.

Dans ce contexte où la liberté ne désigne plus une Idée du ciel intelligible, mais une réalité effective, la première figure que l'on découvre à la liberté est "naturelle".

1. Certes, la liberté de nature n'est pas une découverte propre au XVIIIème siècle puisque déjà Hobbes assimilait la liberté à un droit individuel de nature. Cependant, de Pufendorf à Rousseau, s'éclaire progressivement cette notion de la liberté de l'individu à l'état de nature.

En effet, il apparaît d'abord que décrire cette liberté n'est possible qu'en vertu d'un principe méthodologique dont la portée, au XVIIIème siècle, est générale: c'est, à savoir, que toute question sur l'homme exige une remontée à ses origines, une démarche régressive et archéologique. L'anthropodicée du XVIIIème siècle procède donc, ou bien de façon empirique, à une recherche des moeurs primordiales, ou bien, de façon réflexive, à une quête des fondements de la société; d'une façon ou d'une autre, la question de l'homme ramène toute philosophie à ce que l'on appelle, depuis Grotius, l'état de nature, qu'on le conçoive comme réalité historique ou comme fiction méthodologique. Lorsque, ce principe admis, l'on

s'interroge sur ce qu'est l'état de nature, on accepte
la définition qu'en donnait Hobbes: c'est la condition
des hommes quand elle ne comporte ni art, ni navigation,
ni commerce, ni institution, ni pouvoir civil, "comme
s'ils ne faisaient maintenant que de naître, à la façon
des potirons."[5] Que l'on s'appuie sur les récits des
explorateurs ayant observé les "sauvages" d'Amérique et
les "primitifs" d'Océanie, ou que l'on se fie à la logique du raisonnement conjectural, on estime que la vie
de l'homme en l'état de nature n'enveloppe que simplici-
té et pureté. Cela, toutefois, ne signifie pas que la
négativité caractérise l'état de nature. Assurément,
l'homme, en cet état, est dépourvu de tous les acquêts
que l'on dira bientôt de "culture" ou de "civilisation";
mais il n'est pas voué au chaos car la loi de nature, divine et sacrée, le gouverne. La liberté de nature en
l'homme ne peut donc point être assimilée à la négativité de l'anomie ou de l'anarchie. Loin de se dessiner en
creux comme privation ou manque essentiels, elle se pré-
sente comme une dimension de la vie originaire, donc,
comme une modalité positive de l'existence et, à ce titre,
elle possède un sens anthropologique d'un grand intérêt.

2. Dans la vague de primitivisme qui envahit la littéra-
rature et la philosophie, l'important n'est pas que l'on
célèbre la jeunesse du monde ou que l'on cisèle le mythe
du bon sauvage; ce n'est même pas que l'on se demande si
l'état de nature entraîne le malheur de la guerre et de
la violence ou, au contraire, un bonheur paradisiaque.
La séduction et la chaleur poétiques laissent place en
effet à la gravité de l'interrogation philosophique. Et
l'important, c'est que l'on découvre la modalité existentielle du Huron de Voltaire, des Patagons de Bougainville
ou du sauvage de Rousseau: tous vivent dans l'indépendance la plus radicale. C'est dire que si la liberté est
bien reconnue comme la différence spécifique de l'homme,
elle est, en tant que liberté de nature, essentiellement
la liberté de l'individu. Elle est donc l'antithèse de
la sociabilité naturelle décrite par la philosophie d'o-
bédience aristotélicienne. La liberté de l'homme à l'é-
tat de nature est l'index de la solitude. Hobbes, déjà,
l'avait compris, dont l'individualisme fait de l'existence naturelle des hommes un rapport de forces: et c'est
la guerre. Toutefois, la guerre n'est qu'un effet et,
comme telle, elle est seconde, et secondaire. L'essen-
tiel, Spinoza, ou Rousseau, ou Kant, le retiendront.
Pour Spinoza, la liberté naturelle coïncide avec la puissance du désir en tout un chacun; selon Rousseau, elle
fait de l'homme primitif et même, dans une certaine mesure, du bon sauvage, un solitaire; et selon Kant, la liberté de nature, qui va de pair avec "l'insociable sociabilité", est privée de la dimension de synthèse, donc, de
cette puissance d'union tendue vers l'universel, qui con-

fère à l'homme sa vérité.

Ainsi, l'homme libre naturellement existe par soi seul et pour soi seul. Son indépendance de solitaire confère à sa vie une unité immédiate aussi simple que naïve. Le sauvage, dit Rousseau, "vit en lui-même", tandis que l'homme socialisé et civilisé, toujours hors de lui-même, "ne vit que dans l'opinion des autres".[6] Mais cette unité existentielle, dans l'immédiateté du donné naturel, est l'indépendance de Robinson dans son île: la liberté d'une monade sans portes ni fenêtres. Ce n'est qu'une simple adhésion à soi: la liberté individuelle de nature conduit à l'amour de soi et à l'égocentrisme; elle est fermeture sur soi. Certes, dans cette suffisance à soi, l'homme peut être parfaitement heureux: telle est du moins la thèse de Rousseau dans le DISCOURS SUR L'ORIGINE DE L'INEGALITE. Mais ce bonheur originel n'a de sens que rapporté à sa liberté individuelle, c'est-à-dire à son indépendance qui est sa solitude. Ce bonheur, tout de spontanéité, d'immédiateté, est un bonheur sauvage, neuf et naïf, qui n'a ni règle, ni frein, ni ordre, ni discipline. La liberté individuelle de nature a la subjectivité pour principe et pour fin.

Aussi bien, paradis originaire ou guerre primordiale sont-ils, paradoxalement, des effets de la même "liberté sauvage et sans loi", comme dit Kant. Ce n'est pas une contradiction, mais un signe: le signe que l'homme du XVIIIème siècle, qui se réfléchit aux lumières de la raison et avance vers la conscience de soi, décèle dans la suffisance immédiate de la liberté naturelle une insuffisance pour la conscience réfléchie. En effet, en prenant conscience de soi, l'homme, loin de s'enfermer dans une subjectivité monadique, découvre qu'autrui est toujours à l'horizon de sa vie. Dès lors, il lui faut tenir compte de la rencontre des libertés. Assurément, cette rencontre peut être la guerre de tous contre tous, comme l'ont pensé Hobbes et Pufendorf; mais la coexistence peut être aussi--et elle doit l'être--, en même temps que l'ouverture du sujet, la mutation de la liberté.

II.

Cette mutation fait de la liberté civile la seconde figure de la liberté. Mais, dans l'espace mental du XVIIIème siècle, les deux visages de la liberté ne correspondent ni au dualisme métaphysique traditionnel, ni à une antinomie logique. Ils symbolisent deux modalités de l'existence; et l'avènement d'une liberté civile qui permette de restructurer l'existence humaine obéit à un principe de nécessité qui enveloppe une double dimension anthropologique: à savoir l'éclatement du solipsisme et une conscience de responsabilité, qui sont une double attestation de majorité.

1. De la "découverte métaphysique de l'homme" par Descartes à l'anthropodicée du XVIIIème siècle, la question de l'homme s'est enrichie et nuancée. En effet, l'homme de Descartes est l'egocogito, sujet ontologique en qui triomphe la raison. Dès lors que se pose--si tant est vrai qu'il se pose effectivement pour Descartes--le problème de l'expérience ou de l'existence à plusieurs, le Je et le Tu sont "nivelés",[7] l'Alter est induit à partir de l'Ego. Il n'y a pas vraiment d'intermonde ou d'intersubjectivité pour Descartes: la liberté demeure une liberté intérieure, spirituelle, une liberté métaphysique. C'est pourquoi le problème social et politique n'a guère de place dans la philosophie cartésienne. Hobbes modifie sensiblement le registre cartésien: il n'ignore pas le problème d'autrui et, pour cette raison, il est hanté par la question sociale avant que de l'être par la question politique qui, très vite, d'ailleurs, ne feront qu'un en sa pensée. Seulement, pour l'auteur du LEVIATHAN, l'existence de l'autre est un enfer pour moi tant que subsiste l'état de nature; elle signifie l'imminence de la guerre et l'horrible timor mortis. Autant dire, contre Aristote, que la vie sociale n'est pas possible dans l'état de nature. L'anthropologie du XVIIIème siècle, même lorsque Voltaire, Diderot ou Rousseau critiquent âprement Hobbes, en recueille pourtant l'héritage en ce sens qu'elle consomme l'éclatement du solipsisme. Elle voit l'existence comme une existence plurale, elle la comprend comme co-existence. C'est pourquoi elle porte la question politique au-devant de la scène philosophique. Dès le début du siècle, il y a, comme dit G. Gusdorf, "une fascination newtonienne"; le paradigme de l'attraction est transporté du monde physique au monde humain, non seulement à l'univers psychologique avec Hume qui veut être, par l'association des idées, le Newton de l'espace mental, mais aussi à l'univers social avec Montesquieu, Morelly, Helvetius, Rousseau, Kant... qui, tous, utilisent le schème épistémologique de l'attraction newtonienne. Qu'est-ce à dire, sinon que, seul, l'homme ne l'est jamais? La liberté naturelle du solitaire devient un mythe. Elle va de pair avec l'utopie d'un âge d'or passé, et dépassé. Dans l'existence réelle et actuelle, la liberté de nature s'avère a-topique et a-chronique; elle n'existe en aucun lieu, en aucun temps. Donc, elle n'a de sens que selon des coordonnées mythiques; elle est une fiction.

La conséquence s'impose: sous peine de cataclysme, il faut que la liberté soit démythifiée: elle doit être transmutée dans l'existence concrète, spatio-temporelle, qui est existence à plusieurs. Dès lors, le principe de nécessité qui sous-tend la seconde figure de la liberté s'éclaire. Tandis que Hobbes cherchait, dans sa mécanique politique, à substituer la paix civile à la guerre naturelle, les penseurs du XVIIIème siècle, moins pragmatiques mais plus profonds, ont avant tout souci de la liberté des hommes. Ainsi, pour Rousseau, le "problème fondamental" est "de trouver une forme d'association ... par laquelle chacun, s'unissant à tous, reste aussi libre qu'auparavant"[8]; et, selon Kant, conquérir la liberté est un devoir: "Tu dois, en raison de ce rapport de co-existence inévitable avec tous les autres hommes, sortir de l'état de nature pour entrer dans un état juridique."[9] Renoncer à la liberté individuelle de nature selon laquelle "chacun n'en fait qu'à sa tête"[10] est un commandement de la raison pure pratique, la quatrième formule de l'impératif catégorique.

En même temps qu'éclate la fausseté du solipsisme, s'affirme une exigence de responsabilité, qui est une attestation de majorité. Rousseau, par exemple, montre combien l'homme, au sortir des mains de la Nature, est enclin à l'amour de soi et comment cet amour de soi, dès qu'il va de pair avec l'attraction et la répulsion vis-à-vis des autres, dégénère en une attitude comparative qui est l'amour-propre: de l'égocentrisme, l'homme, en se socialisant, passe à l'égoïsme. Or, cet égoïsme fait obstacle à l'épanouissement: il est introversion, repli, donc, rétrécissement de la sphère d'existence; comme tel, il est une erreur existentielle. L'homme qui n'adhère qu'à soi escamote sa propre vérité qui est de répondre de soi devant ces inévitables témoins que sont les autres. Ainsi se découvre l'exigence de responsabilité dont Beccaria, et Voltaire qui le commente, ont fait, avant Kant, l'un des requisits essentiels de la modalité éthique de la vie. En se reconnaissant capable de répondre de soi devant autrui comme devant sa propre conscience, l'homme témoigne de sa maturation existentielle: l'approfondissement de soi-même qu'il réalise ainsi marque son progrès vers la conscience de soi qui enveloppe la conscience de l'altérité. L'existence insulaire et l'irresponsabilité naturelle doivent être surmontées.

2. Comme l'enseigne Rousseau, le problème primordial de l'existence plurale est que l'homme "reste aussi libre qu'auparavant".[11] Mais, à l'heure où les préoccupations philologiques commandent au philosophe de savoir ce que parler veut dire, il est fort utile de soumettre le mot de liberté aux exigences de la lexicographie. Il convient en effet de lui ôter toute équivocité car, dit Montesquieu, "il n'y a point de mot qui ait reçu plus de différentes significations et qui ait frappé les esprits de tant de manières."[12] Dans sa polysémie vertigineuse, le même mot ne désigne-t-il pas le pouvoir d'exercer la violence et de déposer le tyran, le droit de porter longue

barbe, la possibilité de suivre ses inclinations, de faire ce que l'on veut, d'élire un représentant, de promouvoir telle forme de gouvernement, d'en condamner telle autre...? Le débat ne consiste plus, on le voit aisément, en une joute scolastique où l'on discute de la liberté d'indifférence, du libre-arbitre, du pouvoir de choix, de la liberté spirituelle ... dans un registre abstrait et impersonnel, imprégné de mystère ontologique. Montesquieu le dit de façon expresse: n'est point libre celui qui fait ce qu'il veut. "Il faut se mettre dans l'esprit ce que c'est que l'indépendance et ce que c'est que la liberté. La liberté est le droit de faire tout ce que les lois permettent."[13]

Nous sommes au rouet. Et c'est l'épreuve de vérité pour l'homme. En effet, si le mot de liberté ne désigne pas la pauvre indépendance naturelle de l'individu, il se rapporte à un droit de l'homme, et ce droit, loin d'être inhérent ou consubstantiel à la nature naturelle de l'homme, est défini par la loi. La liberté est une oeuvre de la loi, plus précisément, de la loi constitutionnelle, qui la constitue et la protège. L'homme n'est libre que par la loi et sous la loi. C'est pourquoi, au regard de Montesquieu, la seconde figure de la liberté--la vraie--ne se trouve que dans certains types de gouvernement, selon les termes de leur Constitution. La Constitution d'Angleterre, issue de la Glorious Revolution, porte témoignage de la dimension juridique de la liberté. La liberté a un caractère institutionnel; ou alors, elle n'est qu'illusion de liberté. Elle est impensable en dehors de la détermination légale de l'existence.

Ainsi, l'équivoque est levée: la liberté sauvage et sans loi de l'individu naturel n'est qu'une pseudo-liberté, sa fausse figure. Ce n'est que sous la loi, dans l'état civil, que la liberté prend son vrai visage.

Dès lors, Rousseau peut montrer que la liberté civile a pour condition de possibilité, selon le mot de Diderot, "la dé-naturation de l'homme".[14] N'entendons point par là, cependant, qu'en l'homme dénaturé, l'être et la liberté de nature sont niés; ils sont sublimés. Certes, explique Rousseau, en fait, du point de vue de l'histoire empirique, la socialisation apparaît mauvaise et corruptrice au point que "l'homme, né libre, est partout dans les fers": pour qu'il échappe à la faiblesse et à la vulnérabilité de son état primitif, l'homme n'a besoin de rien de moins que la contrainte et le poids des règles. Mais, en droit, du point de vue de l'idée, la socialisation est bonne car elle permet la promotion de l'individu en citoyen et en personne morale. La dénaturation rend possible l'élévation à la sphère politique et éthique. C'est une métamorphose. L'homme n'a pas renoncé à sa nature; ce reniement serait tout simplement la négation de l'humanité en lui.[15] L'homme a changé sa nature, ce qui est proprement l'engagement et l'acte politiques par lesquels il affirme sa vérité d'homme. Ce "changement très remarquable" le prive, concède Rousseau, "de plusieurs avantages qu'il tient de la nature"; mais, "d'animal stupide et borné" qu'il était, l'homme devient, grâce à lui, "un être intelligent et un homme".[16] Ainsi, la dénaturation de l'homme en est la promotion. Son être est transformé; son existence, sublimée; sa liberté devient vraie: de simple différence spécifique, elle devient témoignage de perfectibilité et gage de progrès. Avec elle, peut s'accomplir une restructuration existentielle et se dessiner le mouvement de l'histoire.

Le contrat social, en effet, est non seulement instituteur de la société civile, mais, tout ensemble, la cheville ouvrière par laquelle les rapports intersubjectifs sont radicalement transposés. Au lieu de l'agrégation ou de la juxtaposition d'une pluralité d'individus insulaires, une "communauté" se trouve instaurée, faite de l'union de tous en un corps commun et un: le corps public comme totalité une et unie est né. En son sein, chaque citoyen, au lieu d'être à lui seul un tout indépendant et autarcique, est devenu membre de la Res publica, c'est-à-dire "partie indivisible du tout".[17] Il s'agit bien d'une restructuration de l'existence au niveau de l'intersubjectivité; et cette nouvelle structure, qui est l'ordre politique même, est l'index de l'essence du politique: au lieu d'une multitude de volontés particulières, s'exprime désormais la volonté générale--qui n'est pas la volonté de tous, non pas une somme, mais une intégrale. Par voie de conséquence, le citoyen, qui n'est tel que parce qu'il participe à la volonté générale, au lieu de céder à la spontanéité, donc, à l'esclavage des appétits, obéit à la loi qu'il s'est prescrite: il est "contraint d'être libre". Rousseau rejoint Spinoza disant que l'homme est plus libre dans la Cité où il obéit à la loi que dans la solitude où il n'obéit qu'à lui-même; mais il radicalise et affine le rationalisme politique de Spinoza car, selon lui, l'insertion dans le tout public, où chacun est lié à tous dans la légalité et par elle, fait bien plus que compenser la perte de l'indépendance naturelle: en gagnant ainsi la liberté civile,[18] l'homme introduit dans son existence un ordre rationnel qui est déjà vertu. Si l'homme, déclare Rousseau, s'entêtait à refuser le moment originel de la communauté politique, il ne serait jamais ni homme, ni citoyen: il ne serait en somme rien.[19]

3. En ce point de l'analyse, théorie et pratique ont une commune naissance. D'un côté, la volonté générale signifiant que l'homme est l'homme de l'homme, il faut,

selon le mot de Chamfort, "recommencer la société humaine":[20] Danton, Robespierre, les Jacobins voulurent faire vivre ce projet et la Révolution fut un tournant de l'histoire. D'un autre côté, montrant qu'il ne saurait exister de césure entre théorie et pratique et condamnant la Révolution moins en raison de ses abus extrémistes que de ses carences logiques, Kant, dans la voie de la réforme, cherche à fonder la liberté civile dans l'exigence pure de la raison pratique.

L'idée d'une liberté civile fondée en raison n'est pas inédite et, déjà, se devine chez Thomas More. Mais Kant quitte la sphère de l'utopie pour élaborer une logique transcendantale du droit politique qui dévoile les fondements de la liberté civile. Dès 1784, Kant condamnait la liberté de nature en ce qu'elle correspond aux tendances égocentriques et empiriques de l'individu; ce naturalisme psychologique est des plus faux car il est établi, depuis la CRITIQUE DE LA RAISON PURE, que rien n'est moins subjectif que le sujet. Si, dans la théorie comme dans la pratique, l'homme peut dire Je--Je pense, Je veux--, c'est que l'universel est en lui. Aussi bien Kant peut-il, au miroir de la République platonicienne, repenser Rousseau, le corriger en l'approfondissant. Car Rousseau a raison contre le pragmatisme eudémoniste de Hobbes: moins que le bonheur et la paix, c'est la liberté que veut l'homme, ce pourquoi il s'insère dans un état de droit. L'homme est libre parce que "sa dépendance procède de sa volonté législatrice propre."[21] Cependant, Rousseau n'a pas su voir dans le pactum unionis civilis l'expression de l'impératif catégorique du politique; car, selon Kant, le contrat social n'est rien d'autre qu'une Idée a priori de la raison[22] qui s'impose de façon absolue. Dès lors, la citoyenneté et, partant, "le droit des hommes soumis à des lois de contrainte publique"-- c'est-à-dire la liberté civile--ont le caractère inconditionnel d'un devoir fondé en raison.

L'idée de la liberté comme devoir a une double signification. Si, d'une part, Kant pense, comme Rousseau, qu'Etat et liberté, loin de s'opposer, se fondent réciproquement en une exigence rationnelle, le formalisme kantien aboutit à d'autres conclusions que la philosophie de Rousseau: en effet, la liberté selon Kant implique le respect inconditionnel de l'ordre, ce qui entraîne, en même temps que la subsomption du droit privé sous le droit public, la condamnation, au nom de la logique, du droit de sédition et de révolution. Ainsi s'exprime, dès la fondation rationnelle de la liberté, l'idéalisme universaliste de Kant,[23] radicalement étranger au conservatisme ou au féodalisme qu'on lui a volontiers prêté. C'est pourquoi, d'autre part, l'homme libre sera selon Kant, dans un avenir encore à construire il est vrai, citoyen du monde. La liberté veut le fait international. Sans doute est-ce là une Idée, et Kant le savait. Ce qui est sûr, c'est que, pour lui, la liberté, en sa dimension d'universalité, n'a rien d'une Schwärmerei car les principes qui servent à la penser définissent des devoirs: les tâches mêmes par lesquelles l'homme se fait homme et accomplit, par sa culture, la destination que lui a assignée la nature. L'homme libre est celui qui assumera, dans la fédération prochaine des peuples, en une tâche infinie, l'universel devoir de liberté.

Ainsi voit-on l'immense optimisme juridique du XVIIIème siècle faire éclater le naturalisme individualiste et promouvoir la liberté civile au rang de valeur existentielle. Assurément, une valeur peut toujours apparaître, selon les coordonnées d'une autre pensée, comme une non-valeur. Quoi qu'il en soit, la liberté civile symbolise, en ce siècle d'espérance, l'idée que l'on se fait de l'homme vrai. La pensée libérale puise son inspiration dans le thème de la perfectibilité de l'homme, auquel fait écho l'idée du progrès universel et illimité. Ce progrès, que résume alors le terme de "civilisation", marque, à travers les idées de respect, de tolérance, de justice, de responsabilité, de "droits de l'homme" ..., la conquête de la conscience de soi de l'humanité. En cette "révolution copernicienne", l'homme a acquis la certitude que la liberté n'est pas l'indépendance. Comme le dira Fichte au lendemain de la Révolution de France, puisque l'homme, désormais, est à lui-même sa propre fin, "l'histoire ne fait que commencer".

FOOTNOTES

1. Kant, LA PHILOSOPHIE DE L'HISTOIRE 84, traduction Piobetta, Aubier, 1947.

2. J. Starobinski, L'INVENTION DE LA LIBERTE, Skira, 1964.

3. "Conjectures sur les débuts de l'histoire humaine", in PHILOSOPHIE DE L'HISTOIRE 162.

4. Ibid., p. 164.

5. LEVIATHAN, Chapitre XIII; DE CIVE, chapitre VIII, § 1.

6. DISCOURS SUR L'ORIGINE DE L'INEGALITE, Bibliothèque de la Pléiade, tome III, p. 193.

7. Merleau-Ponty, PHENOMENOLOGIE DE LA PERCEPTION 408, Gallimard.

8. LE CONTRAT SOCIAL, Livre, chapitre VI. Remarquons que cela implique que l'on reconnaisse la "grande différence" entre une multitude ou agrégation et une société ou association.

9. DOCTRINE DU DROIT, § 42, p. 188.

10. <u>Ibid.</u>, § 44, p. 194.

11. LE CONTRAT SOCIAL, Livre I, chapitre VI.

12. L'ESPRIT DES LOIS, Livre XI, chapitre III.

13. <u>Ibid.</u>, Livre XI, chapitre IV.

14. Article "Droit naturel" de l'Encyclopédie, in OEUVRES POLITIQUES 35, édition Vernière, Garnier.

15. LE CONTRAT SOCIAL, Livre I, chapitre IV.

16. <u>Ibid.</u>, Livre I, chapitre VIII.

17. <u>Ibid.</u>, Livre I, chapitre VI.

18. <u>Ibid.</u>, Livre I, chapitre VIII.

19. EMILE, Pléiade, Livre I, p. 250.

20. MAXIMES, en OEUVRES COMPLETES, édition Auguis, tome I, p. 445.

21. DOCTRINE DU DROIT, § 47.

22. THEORIE ET PRATIQUE, section II, p. 39.

23. DOCTRINE DU DROIT, REMARQUES EXPLICATIVES 255.

It is almost a truism to say that the thinkers of the XVIIIth century were fond of liberty. Indeed, at that time, not only a wind of liberty blew over the occidental world, but the idea of liberty occupied the center of philosophical works.

However, it is still interesting to remark that the philosophers of this century think there are two very different forms of liberty. The first one is the liberty of nature, wild and without law, which is an individual independence. The other is the citizen's liberty, under the civil law, and by it. And, from natural independence to civil liberty, the social contract realizes a mutation. This metamorphosis, by means of which man leaves his savage life and prefers a civil existence under the power of civil law, has a metaphysical meaning; it is a sign of his existential truth.

THE LEGAL STATUS OF THE INDIVIDUAL: PRESENT AND FUTURE

A. F. Shebanov

I.

The legal status of the individual and the relationship between man, society and the state are among the important problems in the world today. The individual is a man who acts as a bearer (subject) of certain economic and other social relations. Of course, in examining an individual, science can not abstract itself from the natural, biological features of man. Yet it is the social factors which play a decisive role in the development of man as a member of society, insofar as any society is a definite totality of social relations based on a particular type of production relations (e.g., socialist relations of production). Any individual in a state-organized society belongs to a particular social class or section and in this capacity acts as a participant in social relations.

In a state-organized society the individual/society relationships are primarily expressed in relations between the individual and the state as an official representative of that society. The individual/state relationships are in their turn largely expressed in the content and application of legal norms which lay down and regulate the individual's status under a particular legal system. So a measure of the freedoms and possibilities of the individual in a state-organized society is established by the state in the norms of law.

This approach to the relationships between the individual and the state (resp. law) leads us to the conclusion that the nature and content of these relationships depend on: (1) The socio-economic basis of society; (2) historical and social type of state and law; (3) the social status of the individual himself, i.e., on what social group he belongs to.

II.

The state is a most powerful instrument which influences the individual in modern society. To enforce this influence, it uses law. Some legal norms point to the citizens' particular mode of behavior as lawful and commendable by the state, and cause in the people's consciousness decisive stimuli in favor of such a conduct. Other legal norms forbid certain actions and call upon the people to refrain from commiting them. The state uses the whole system of various bodies, institutions and organizations to see to it that citizens observe legal prescriptions.

Needless to say, numerous factors (both material and spiritual) determine the nature of the influence exerted by the state and law on the individual. Of decisive significance are, in the final analysis, the socio-economic factors, the material possibilities of society. Yet the result of this influence depends to a certain extent on the individual who is not passive to it. He may see this influence as just or unjust, useful or futile. This recognition determines the formation of the motives of the individual's actions: Conviction in the necessity or irrelevance of a particular mode of behavior; Indifference; Stereotypic patterns; Fear of punishment, etc.

Thus the economic and social conditions are in themselves insufficient to ensure the freedom of the individual and the possibility for its development. What is required are definite legal norms and a definite mechanism to enforce the realization of these legal forms. The legal status of the individual is his social status expressed in law. It is the individual's basic legal condition in society predetermined by the existing economic and political system and the operative system of law.

International pacts on the economic, social and cultural, as well as civil and political rights of man unanimously approved by the Twentieth Session of the United Nations General Assembly, play an important role in defining the content of the individual's legal status in contemporary society. These pacts were the first of their kind to lay down the range of human rights which, as a minimum, should be accorded to all citizens and states. The Soviet Union took an active part in the elaboration of these pacts. Guided by the idea that the freedom of the individual is the goal and indispensable condition of social progress, the Soviet Union was one of the first to ratify international pacts on human rights in 1973.

Therefore, in considering the legal status of the individual in the present and the future it would be advisable to consider these pacts, and the states' domestic legislation which determines the legal status of the individual and his legal guarantees in various spheres of public affairs.

III.

According to the International Pact on the economic, social and cultural rights of man, the legal status of the individual must include a wide range of socio-economic rights and freedoms. The right to labor is of special significance. To implement it the United Nations member states pledged themselves to adopt and carry out programmes of vocational training, for achieving steady economic, social and cultural development and full industrial employment, and other measures.

The Constitution of the USSR establishes the citizens' right to labor as the right to get guaranteed work and wages according to its quantity and quality. This constitutional provision is developed in the Code of labor laws as the right to freely choose the place and type of work, freely to conclude a labor contract at a particular enterprise, the right to wages in proportion to the quality and quantity of labor expended, the right to healthy and safe conditions of labor, to free vocational training and raising qualifications, the right to take part in production management, etc.

Stated by law in the USSR are also principles of the social security of all citizens (including social insurance), free medical service and free education, including university training. These principles are also contained in the International Pact. In the USSR they are realized in laws on state pensions, public health and public education. The provision of the International Pact on the guaranteed right of trade union organizations to exercise their functions is realized in the USSR through the labor codes and trade union rules. These laws enable the trade unions to function unhindered, without any restrictions and in conformity with their aims.

It follows from the International Pact on human political and civil rights that an important place in the legal status of the individual must be held by his right to take part in administering public affairs, the right to life, to personal freedom and other political and civil rights and freedoms. In the USSR these rights of the citizens are established in the Constitution of the USSR, in laws on local Soviets, on the Deputies' status, and also in civil and criminal codes, etc. Art. 122 of the Soviet Constitution lays down that "any direct or indirect restriction of the rights of, or, conversely, the establishment of direct or indirect privileges for citizens on grounds of race or nationality, and likewise any advocacy of racial or national exclusiveness or hatred and contempt, shall be punishable by law." This legal provision on the equality of citizens is further specified in family, civil and procedural laws. For instance, both the Fundamentals of Civil Procedure and the Fundamentals of Criminal Procedure state that justice in the USSR, in both criminal and civil cases, "shall be administered on the principle of equality of all citizens before the law and the court, regardless of their social, property and official status, nationality, race or creed."

Special significance attaches to the implementation of the principle of the equality of women with men. The legal status of women in the USSR provides for granting them all rights and freedoms of men and also several additional rights arising from their functions as mothers and educators of children.

The International Pact prohibits a deliberate and unlawful interference in personal and family life, arbitrary or unlawful encroachment on the inviolability of dwelling, secrecy of correspondence, and honor and reputation. The Soviet Communications Charter establishes that the content of all types of postal and telegraph correspondence is a secret protected by law. According to Soviet legislation evidence obtained through unlawful searchings or overhearing telephone talks can not be recognized as such by the court. According to Art. 135 of the Criminal Code of the RSFSR, the violation of the secrecy of the citizens' correspondence shall be punishable by corrective labor of up to six months. The law makes it incumbent on the prosecutor to see to it that the details of the citizen's private life irrelevant to the case in which he is brought to criminal responsibility shall not be made public.

Of special significance for the legal status of the individual are the guarantees of the rights and freedoms included in that status. In its laws the state must establish a procedure for the work of its organs with the view to preventing and cutting short encroachments on the rights of citizens, restoring the violated rights and punishing the guilty persons.

The citizen's right to lodge a complaint or statement concerning the actions of any official that violates legality, rights and interests of citizens protected by law, is an important legal guarantee of the freedom of the individual in the USSR. The lodging of a complaint makes it incumbent on an appropriate state organ to promptly and properly consider and solve the question. Special legal guarantees are also contained in all branch codes. For example, the citizen's right to labor is guaranteed by a provision of the Labor Code according to which a worker can be discharged by the administration only with the consent of the trade union committee.

Describing the legal status of the individual, Soviet legal science assumes that the active obligations of the state correspond to the socio-economic, socio-cultural and political rights as elements of the legal status of the individual in the USSR. While it establishes the legal status of the individual (in the Constitution and other laws), the Soviet state thereby pledges itself to satisfy these rights and create conditions for their realization. For basic rights included in the legal status of the individual serve as the lawful precondition for the citizens' possibility to apply to appropriate state organs with the request that they should perform specific actions to realize these rights (for instance, persons of the age and with the labor record required by law have the right to apply to the state organ for the old-age pension).

IV.

The future development of the legal status of the individual in all countries depends on both internal and international conditions. The basic human rights and freedoms can be implemented most effectively only in conditions of peace and cooperation between countries, on the basis of the respect for sovereignty and customs of each country. That is why the Soviet Union, which played a decisive role in the defeat of German nazism during the Second World War and suffered the largest human and material losses, is pursuing a consistent policy of preventing new wars, the peaceful coexistence of states with differing social and economic systems, and creating favorable international conditions for the implementation of the basic rights and freedoms of man. The recent years have been marked by a continuing detente greatly contributed to by the Soviet-American summit meetings. The further development of the relations of peace and cooperation between different states will undoubtedly promote the dissemination of democratic ideas and principles on the international and home scene and this, in its turn, will be a good basis for the expansion and consolidation of the legal status of the individual.

On the home scene, the further development of the legal

status of the individual is connected with the democratization of public affairs insofar as the individual's position in society, his rights and duties are directly dependent on the essence and the degree of development of democratic institutions and the type of democracy. The higher the level of democracy, i.e., the actual participation of the broad mass of people in administering the affairs of society and the state, and the better the social and economic conditions of the population, the more opportunities emerge to expand the scope of rights and freedoms, comprising the legal status of the individual and the more feasible it is to implement them. Social equality, granting all members of society equal opportunities to work, get education, have spare time and use the benefits of culture and civilization, are indispensable for ensuring the genuine legal status of the individual.

The examination of the further development of the legal status of the individual also gives rise to the question about the possibility of restricting human rights and freedoms by the state. It follows from the international pacts on human rights that the rights and freedoms comprising the legal status of the individual must be used in conformity with the interests of the overwhelming majority of the population. These rights and freedoms can be limited in the interest of the respect for the rights and reputation of other persons, state security, public law and order, and health or morality. This means that public law and order which ensures the interests of the overwhelming majority of the population can not be violated at the whim of a particular person or group of persons who hold a different view. On the other hand, public law and order which denounces human rights and freedoms as such (e.g., the fascist regime in nazi Germany or that of the military junta in Chile) is essentially anti-democratic and turns the legal status of the individual into a fiction. In these cases the legal status of the individual can only develop if a genuinely democratic political regime is established.

PERSONA Y ESTADO

Agustín Basave Fernández del Valle

Decir persona es decir auto-posesión, auto-dominio. La persona comanda su naturaleza y la naturaleza. Llámase persona--para expresarlo de la manera más sencilla-- al individuo dotado de inteligencia. En lugar de ser llevada--como los otros individuos lo son--, la persona tiene la facultad de determinarse a sí misma. El individuo espiritual se polariza hacia instancias ajenas: la verdad, la belleza, la justicia....A diferencia del animal, la persona--como lo ha visto agudamente Max Scheler --objetiviza su contorno, lo convierte en "mundo", le otorga significación propia e independiente. La persona es propósito y designio de unidad, voluntad de coherencia, libre afirmación del valor.

Afirmar que en el hombre coinciden el individuo y la persona, no es identificar en otros casos estos dos conceptos. Un tigre es un individuo de su especie, pero a nadie se le ocurriría decir que es una persona.

Aunque no sea animal ni cosa, la persona es también substancia. ¡Voluntad libre--capacidad plena de elección y decisión--, comunicación con el mundo de los valores, religión, belleza, bien; he ahí, la esfera propia de la persona!

Siendo el hombre por naturaleza un ser social, evidentemente estará ordenado--parcialmente--hacia el Estado. Pero el Estado es un medio natural del que el ser humano puede y debe servirse para obtener su fin, por ser el Estado para la persona, y no a la inversa. La personalidad y las prerrogativas fundamentales no pueden perderse en aras del Estado. El individuo tiene sus deberes como ciudadano, pero sin menoscabo de sus privilegios y de sus modalidades de persona. La sociedad civil es la gran perfeccionadora de la vida de la persona. ¿Cómo pretender, de otra manera, realizar el bien común, sin desarrollar la personalidad humana? A la sociedad va el hombre en busca de protección y de realización, nunca de opresión y de frustración. De ahí que si la sociedad no protege los derechos naturales y el legítimo ejercicio de los mismos está falseando su estructura y su fin. Es preciso afirmar que el Estado no absorbe todas las finalidades del individuo. Hay órdenes en los cuales no cabe la competencia estatal. Familia, Iglesia y sociedades menores cumplen finalidades, perfectibilidades humanas que el Estado no podría cumplir.

Entre persona y Estado no hay--ni debe haber--oposición. Ni la persona puede repudiar o aminorar los derechos del Estado, ni el Estado puede desconocer las prerrogativas inherentes a la persona humana: "derecho a la vida, a la integridad del cuerpo, a los medios necesarios para la existencia; derecho de tender a su último fin por el camino trazado por Dios; derecho de asociación, de propiedad y del uso de la propiedad" (DIVINI REDEMPTORIS, No. 27. E. Pesch). Puede el hombre, en ocasiones, llegar hasta sacrificar su vida por la sociedad política, pero nunca su alma. Porque la vida, al fin y al cabo, no es el máximo valor. La vitalidad en sí misma--como existencia vegetativa--no tiene polaridad moral, no es buena ni mala. Su valor todo, depende del fin que la oriente. El valor de la vida es, pues, subalterno, instrumental. Contra la proclamación de la vida-fin (de sí misma), proclamamos la vida-medio. Sólo al servicio de un valor que la incite y la guíe, cobra la vida contenido y plenitud. La vida es misión, es ofrenda a algo meta-vital. Y precisamente el sacrificio de la vida generosamente consentido-- en caso de extremo peligro de la patria--es realizar plenamente el destino y la vocación del hombre.

Mientras que la persona es para el Estado relativamente, el Estado es para la persona absolutamente. Si los individuos se reúnen en el Estado para realizar su finalidad individual, la sociedad civil será, en último término, un medio para auxiliar a sus miembros. El bien común, que se traduce en bien común distribuido, es un fin intermedio--finis quo--por medio del cual cada miembro del cuerpo político obtiene su bien propio. El hombre se ordena, parcialmente, al Estado. Algo hay en él que no es parte de la agrupación política, algo que trasciende todo lo político y social. Menester es recordar siempre que el Estado no existe ni "por" sí ni "para" sí, sino "por" las personas y "para" las personas. No puede el Estado pensarse, ni pensar el mundo exterior, ni querer, ni tener una vocación eterna. El tipo de su realidad es accidental, temporal. En el dominio de lo privado --privatístico de la persona--el Estado no puede tener ninguna ingerencia. Todos los valores que definen el ser humano se sustraen al imperio de lo político. Los derechos esenciales de la persona son--deben ser, por lo menos--inviolables e inalienables. La facultad de procrear o la libre elección de las vocaciones--ejemplifica Jean Dabin--no están ni a la disposición ni a las órdenes del Estado. Y es que los derechos individuales--anteriores y superiores a toda concesión estatal--son naturales en el sentido de que hunden sus raíces en la propia naturaleza humana. En principio, estos derechos deben ser universales, aunque en determinados casos de repercusión social se reserven, en su plenitud, a los ciudadanos. Corresponde al Estado dar concreción y garantía a los derechos naturales de la persona humana. Las exigencias del bien público piden poner coto a las libertades puramente negativas, a los libertinajes y a los actos hostiles para el bienestar de la comunidad. Pero de aquí, no cabe concluir en una libertad dirigida que, en última instancia, es negación y pérdida de la libertad.

Yo no creo que exista el homo juridicus, pero estoy convencido de que hay una dimensión jurídica del hombre. Por eso el Derecho responde a una profunda necesidad humana enraizada en los estratos ónticos del ser humano. Porque el hombre es un animal insecurum busca la seguridad en el derecho. En este sentido, el Derecho está al servicio--aunque no exclusivo--de la seguridad de la existencia humana. No puede haber vida social sin orden. Sabemos que hay conflictos, aspiraciones que se entrecruzan, pasiones que se desbordan; pero queremos, no obstante, tranquilidad en el orden, firmeza en nuestras posiciones, previsibilidad del comportamiento--y de sus defectos--, seguridad para saber a qué atenernos. Cuando el poder del amor disminuye y no vincula una comunidad, el orden jurídico evita la lucha caótica del homo homini lupus. Gracias a la dimensión jurídica del hombre las relaciones humanas se clarifican y se tranquilizan. No es que el Derecho agote la cultura pero es que la cultura no podría existir sin el Derecho. Y aunque hasta ahora no haya podido eliminar, del todo, la violencia, la arbitrariedad, el odio destructor, por lo menos les ha puesto sitio desde la fortaleza de su juicio. Acaso nunca lleguemos a establecer, en la tierra, un continuo y verdadero orden de paz. Pero seguiremos intentando regular las relaciones humanas en el marco de la familia, del Estado y de la comunidad internacional.

Aunque alguna vez haya estado ligado a valores tribales y étnicos, el derecho emerge y cobra importancia desde la personalidad del individuo. La autoconciencia de la dignidad personal en la vida social es el genuino hontanar del Derecho. La dimensión social de las comunidades--que nunca llega a ser del todo impersonal porque lleva la huella de la persona--hace crecer al Derecho. Adviértese que en la autoconciencia de la dignidad personal en la vida social se da una veta jurídica junto a vetas morales y religiosas. Y cuando la dimensión jurídica del hombre llega a su cabal desarrollo nos encontramos, en su núcleo esencial, valores morales operantes: justicia, seguridad, bien común, respeto al prójimo, libertad, lealtad, veracidad, dignidad personal. Esta dimensión jurídica se enfrenta con la voluntad de poder--individual y grupal--, con la opresión en todas sus formas, con la injusticia socio-política. Porque el Derecho no se limita a mandar, sino que enseña la vida justa, indica el comportamiento debido, cualifica la acción. Podemos imaginar una ley, privada de sanción, que siga siendo ley: "Pacta sunt servanda". Al fin de cuentas, el Derecho es primordialmente dirección y secundariamente coerción. El acento se desplaza del (ius quia iustum) al Derecho como rectitud jurídica. Y es que el Derecho no se reduce a mandato ni radica, primordialmente, en la voluntad; sino que es acto de inteligencia: regla de vida social, medida de comportamientos. Partiendo de su normatividad axiológica calificamos acciones particulares, situaciones y hechos concretos. El Derecho es práctico y es lógico, manda y cualifica. Claro está que no todo mandato es una ley. De ahí la primacía de la vis directiva--elemento de justicia incorporada a la ley--sobre la vis coactiva.

De la dimensión jurídica del hombre surge el Derecho que llega hasta nuestros días, con todas sus complicaciones técnicas, con la prolijidad de categorías y figuras jurídicas dominadas--en esencial conexión--por unas cuantas y altas ideas éticas. Hágase el intento de suprimir estas ideas éticas o valores y se habrá acabado con la esencia del fenómeno jurídico. Si la vida del hombre tiene una textura ética, el Derecho no puede estar desvinculado del reino moral. Por imperativos morales nos sentimos impulsados a establecer un orden social libre y justo. Ciertamente el Derecho no agota la eticidad. Los valores jurídicos ocupan una modesta porción de la ética. Hay tareas morales de mayor envergadura. Pero estas mismas tareas morales requieren para su desarrollo libre, canales jurídicos. En el mundo de lo social, el Derecho se presenta como uno de los fundamentos de la moralidad. Las exigencias éticas de justicia, libertad y humanidad justifican la estructura jurídica. Mientras repudiemos el atropello, la violencia y la lucha caótica el Derecho tendrá mucho que decir. Nos obliga porque está ubicado dentro de la eticidad. En la medida y regla que impera en el campo social rastreamos, desde lejos y con nostalgia, el significado del absoluto. Al Derecho no le corresponde desentrañar la conexión significativa del todo.

La dimensión jurídica del hombre no puede desconocer ni la estructura permanente y general del ser humano--elemento nuclear--, ni el autoproyecto cambiante en situación histórica. Las leyes ontológicas del ser del hombre no son--no podrían ser--irrelevantes para el orden jurídico. La esfera cultural antropológica con sus cambiantes proyectos se refleja en las instituciones sociales.

La contemplación jurídica debe tomar como base una imagen ideo-existencial del hombre. El deber-ser--comportarse de una manera y no de otra--descansa sobre el ser del hombre--cuerpo, psique, espíritu--. Una antropología integral está en la base de una antropología jurídica. La estructura estratificada del hombre--estrato biológico, estrato psíquico, estrato espiritual--con su legalidad propia no puede ser desconocida por el Derecho. Hay un sector jurídico que regula el "ser natural" del hombre y hay otro sector jurídico que versa sobre el ser espiritual. Las normas jurídicas no pueden disponer comportamientos contra las leyes biológicas del hombre como ser vivo. Más aún, debe favorecer los legítimos requerimientos del bios. Los componentes psíquicos (base endotímica y estrato del yo) tienen particular interés para la estructura psicológica del comportamiento eficaz. Si el hombre es un ser abierto, no conformado por la naturaleza hasta el final, tiene que autodeterminarse en base al espíritu y sobre un orden jurídico. Responsable de sus hechos, culpable de sus transgresiones al orden jurídico, digno en cuanto persona, el hombre posee "a nativitate" el derecho a la libertad existencial, el derecho de autoconformación y los derechos esenciales a la persona. El hombre en estado de proyecto social da origen a la norma jurídica. Si el jurista no sabe leer en la óntica integral del hombre, no va a ver el Derecho, sino su sombra en la letra de los Códigos. Además de ser un ser axiotrópico, el hombre es un programa existencial valioso, un proyecto de poder y deber, una posibilidad de poder hacer y de poder exigir en el mundo, una libertad justamente delimitada por las otras libertades. Toda esta realidad de Derecho emergente, toda esta dimensión jurídica del hombre acaece antes de que las normas cristalicen. Hay un poder hacer y un poder exigir intencionalmente referidos a la justicia--no a la arbitrariedad--que estructuran el Derecho.

De los caracteres fundamentales de la naturaleza humana: individualidad física, libertad moral y responsabilidad, racionalidad espiritual, socialidad y religiosidad derivan una serie de derechos subjetivos públicos de la persona frente al Estado. He aquí algunas consecuencias que se desprenden de las características esenciales del ser humano: 1.- el cuerpo humano exige medios necesarios para mantenerse y cumplirse físicamente: a).- Derecho a la propiedad; b).- Derecho al trabajo; c).- Derecho al matrimonio y a la consecución de su objeto; 2.- La vida moral exige derecho a la seguridad jurídica y derecho a la participación en la vida pública; 3.- La vida intelectual exige prestaciones positivas por parte del Estado: a).- escuela; b).- libre enseñanza; c).- libre investigación; d).- libre educación; e).- libre formación religiosa. 4.- La sociabilidad humana trae como resultante el derecho a constituir todas las agrupaciones sociales que

son consecuencia lógica del desenvolvimiento personal;
5.- El derecho a rendir a Dios el culto debido--público y privado--es la cúspide de la vida moral y social.

Para que de un Estado se pueda decir que en verdad protege y garantiza los derechos individuales, no bastan las garantías generales (principio de legalidad, separaciones de funciones, sufragio) si no están complementadas por garantías especiales destinadas a funcionar, como mecanismos adecuados, frente a determinados ataques. Válganos como excelente ejemplo, el juicio de amparo: orgullo y honra de México.

Los derechos fundamentales de la persona humana emanan de su esencia y de su existencia. Trátase de primeras determinaciones de la justicia. Todo sistema político sano supone principios universales y categoriales de lo jurídico. Todo hombre tiene derecho a existir, subsistir, realizarse vocacionalmente, tender a la plenitud y tener las garantías públicas para el ejercicio de sus derechos.

Todos los derechos primordiales del hombre se basan en la exigente ética de respeto a la dignidad humana. Y esta exigencia está contenida en la idea misma del Derecho. No puede darse una comunidad jurídica ordenada sin el derecho general al respeto de la personalidad. Cada persona desea estar protegida en sus finalidades y en sus bienes. Todos los hombres son capaces de obrar porque son capaces de dirigir. Y si son capaces de dirigir y de obrar, son jurídicamente responsables.

El derecho de la personalidad se manifiesta en la cabal naturaleza anímico-corporal, en los instintos, aspiraciones y tendencias del hombre. En primer término existe el Derecho subjetivo de publicar y la existencia física. Cuerpo, vida y salud del hombre deben ser protegidos contra todo ataque de particulares y de órganos estatales. El derecho de tener bienes económicos para el propio uso y la propia disposición proviene de la naturaleza racional y previsora del hombre. La tendencia a la libertad se manifiesta, primigeniamente, en la "libertad de movimiento". Pero la personalidad humana, en sus más altos estratos, es libertad para autoconfigurarse, para elegir profesión, residencia, estado civil. Libertad de opinión, libertad de conciencia y libertad religiosa, son libertades espirituales en el sentido estricto. No podrían entenderse estas libertades sin el mandamiento ético de la veracidad. Sin veracidad se destruirá la convivencia humana.

El derecho a la igualdad se basa en nuestra común y esencial identidad de naturaleza, de origen y de destino sin mengua de desigualdades accidentales. Derecho a la educación, derecho al trabajo, derecho a la libertad de la miseria son nuevos derechos fundamentales que ofrecen los modernos catálogos de los derechos humanos. Y dentro de estos Derechos Fundamentales cabría hablar de una jerarquía espiritual: derechos al honor, derechos económicos.

Hasta aquí la descripción y explicación suscinta de los derechos fundamentales del hombre. Los mismos Derechos del Estado no podrían comprenderse sin comprender antes los Derechos del hombre. Porque el Estado, no hay que olvidarlo nunca, tiene un caracter instrumental. No se justifica, sino en la medida en que sirve de medio para garantizar los derechos del hombre como ser biológico, substancial y espiritual.

To state the term "person" is the same as declaring self-possession, self-responsibility and self-determination. The spiritual individual finds himself directed towards truth, goodness, beauty, justice, because he is the purpose and the design of unity, the will of coherence, the free affirmation of value.

Because a person is a social being, a person is responsible, at least in part, to the State. The individual has duties as a citizen, but the carrying out of the same should not be detrimental to the privileges and modalities inherent in a person. A being's personality and fundamental prerogatives cannot be lost as a result of providing service to the State. The civil society is the great improving force in the life of a person. Between a person and the State there is--there should be--no opposition. The person cannot repudiate or weaken the rights of the State. The State cannot disavow the prerogatives inherent in a human person: the right to life, to bodily integrity, to the necessary means for existence; the right to seek attainment of his ultimate goal; the right to choose ones associates; the right to own and use property. While a person partially "gives of oneself" to the State, the State gives its "all" to the person. The State cannot think, nor can the exterior world think, nor love, nor can it have an eternal vocation. The type of its reality is accidental, temporal. All of the values that define the human being are separate from the area of politics. The essential rights of the person--generating from his body and his spirit, from his individual and social nature --are, or at least should be, inviolate and inalienable.

The inherent rights of man are based on the ethical demand of respect for human dignity, which is contained in the very ideas of Law. Besides the general guarantees (the initiation of legality, the separation of functions, suffrage), special guarantees are required (the Mexican right to claim or enjoy legal protection, the Constitutional Guarantees of West Germany), which are designed to function as adequate mechanisms against determined attacks.

THE HUMAN PERSON AND THE LEGAL PERSON

Iredell Jenkins

This paper rests upon two general theses. First, that positive law inevitably transforms the goals of social justice in seeking to secure them. Secondly, that this transformation can become dangerous, if not fatal, to social order if law, either through its own arrogance or by the default of other institutions, assumes too dominant a role in the enterprise of justice. I mean to elaborate and test these hypotheses by examining the changes that occur when the idea of the human person is translated into the concept of the legal person.

Law is obviously an instrument of justice. Toward that end it seeks to assure social order, establish and control authority, protect person and property, guarantee equitable dealings, and in general promote human well-being. But just as obviously, law is not alone in this effort. It is the common task of an array of social institutions -- custom, morality, and tradition; the family, the school, and the church; professional, vocational, and economic associations of many sorts. The doctrine of the sovereignty of law has for long made the legal apparatus the dominant member of this partnership, exercising a broad supervision over its partners but also relying heavily upon their contributions. So law has been a supplemental instrument of justice. The question that concerns me is what happens if law too much supplants instead of supplementing these other institutions, imposing its standards and procedures upon them and intruding into their conduct of their proper affairs.

The most important element in the ideal of justice is the particular concept of the human person -- the image of man -- that animates and directs its pursuit. Justice (at least substantive or material justice as distinct from procedural or formal justice) is, after all, just a name for those conditions that will most effectively promote human well-being. So the terms in which the human person is conceived control the meaning of justice as an ideal end. And the transformation of the human person into the legal person controls the efforts of law toward this end.

I cannot here pretend to deal with all of the aspects of man's rich nature. So I shall focus attention on one of man's most significant characteristics, and one that is central to the pursuit of social justice. I refer to plasticity, which is so distinctive of man and marks him off from all other creatures. Plasticity has two important consequences. It means, in the first place, that every person is a nest of many potentialities, capable of diverse developments, but that none of these can be realized without upbringing, effort, training, and discipline. It means, in the second place, that individuals exercise self-determination, having a voice in their own futures and making a difference to the course of events. Each person must choose his role and prepare for it, he must anticipate the consequences of his decisions, and he must take pains that his actions do not harm others or defeat his own purposes.

The fact of plasticity thus imposes two demands if man, the biological creature, is to become a mature human person: Individuals must undergo cultivation and they must exercise responsibility. I now want to examine the meaning of each of these demands, the tasks they set, and, most particularly, the role of law in these undertakings.

The process that we call cultivation exhibits a duality that is deeply rooted in the human situation, for every human being is at once a social member and a unique individual. Cultivation is necessarily directed toward both of these dimensions of the person. It must form the character and conduct of people to accord with established standards and to fit into the social order; but it must also respect and foster the particular potentialities of each individual.

A society can remain sound and stable only by maintaining certain basic similarities among its members. The more important of these include a common language, tradition, and culture; generally acknowledged moral standards and practices; well established customs and habits; a general belief in the fairness of the social system; a recognition that different social roles require different skills, impose different obligations, and offer different rewards. So a society must take pains to initiate each new generation into its way of life and to prepare its now members to serve in useful roles.

But society must be equally solicitous of the differ-entiations that are so characteristic of human nature. People feel their individuality strongly. They value their own tastes and talents, seek to develop them, and insist on the opportunity to express and assert themselves. Furthermore, society is dependent upon these individual skills and energies for its own well-being. It must seek out and train human capabilities for the more and more varied tasks that are essential to its smooth functioning.

Cultivation thus has a double purpose. It seeks to make the human person at once an acceptable social member and a highly developed individual. The work of cultivation has traditionally been largely carried on by the various social institutions mentioned above, with law playing a relatively minor role. In this regard, three critical questions arise. What is the proper role of law in this enterprise? Into what other terms does law necessarily translate the twin goals of cultivation -- similarity and differentiation -- in seeking to promote them? What dangers ensue if law assumes too commanding a role in this enterprise?

There is not a great deal that law can do directly to assure the similarities or promote the differentiations that cultivation envisages. Law is simply not an effective instrument for the formation of human character or the development of human potentialities. There is, however, a great deal that law can do indirectly to promote these goals. Stated generally, it can keep a watchful eye on the entire institutional structure of society to see that its several elements properly discharge their responsibilities. This can be specified somewhat for each of the goals.

The legal apparatus can be particularly effective in assuring that the essential services and benefits of society are made available to all people in an equitable manner. There are needs that must be satisfied if peoples' lives are to be even minimally rewarding and if public order is to be secured: Adequate diet, housing, and health care; A general education and some form of special training; Employment under decent conditions; Access to the political process; Leisure and the means to put it to use. Law is ill-equipped to supply these services or make good these wants, or even to determine in any detail by what means and in what measure they are to be furnished to people. But it is admirably equipped to intervene and correct any form of exclusion, discrimination, or exploitation whenever it suspects that they are being practiced. Furthermore, law can set minimum standards and define broad guidelines to assure that other institutions do in fact provide the services for which they acknowledge responsibility and claim credit. In short, though law cannot secure the essential similarities that are necessary to a sound society, it can eliminate gross dissimilarities among individuals and groups.

Law is in a similar situation with regard to the human differentiations that seek expression in such diverse ways. Here again its proper function is protective rather than constructive. The legal apparatus is employed to guarantee certain basic freedoms and thus to secure the lives and affairs of men against undue intrusion or harassment by other institutions, or even by the law itself. Law can do very little to assure men of success in their undertakings or satisfaction in their lives. But it can do a good deal to guarantee them the opportunity, within broad limits, to pursue these as they see fit. Law makes men free to order their lives as they choose, so long as their choices do not obviously disrupt the social order itself.

We can now epitomize the changes that law effects when it translates the goals of cultivation into its own lexicon. Cultivation envisages an outcome in which essential human similarities are preserved and individual differentiations are shaped in accord with established norms and models. The machinery that law commands cannot directly further such an outcome. Instead, the legal apparatus transforms these goals into conditions that it can effectively promote. It does this by translating 'similarities' as '<u>equality</u>' and 'differentiations' as '<u>freedom</u>'. The shift that this brings about in interpretation and intention is quite radical. The similarities and differentiations that cultivation aims at are conceived in terms of human character and conduct. The equalities and freedoms that law aims at are couched in terms of social conditions and treatment. Law envisages an outcome in which equals will be treated equally and personal freedom will be protected.

The original role of law in the task of cultivation -- its relationship to other social institutions -- is that of a supplemental instrument. Law supervenes upon a social order in which similarities are being preserved, and differentiations are being fostered, by an array of established institutions. But this order exhibits inequities of two sorts: The services and benefits of society are being distributed unequally, with a good deal of discrimination and exploitation, and only certain approved differentiations are permitted, with many forms of individual expression being suppressed.

Equipped with its constitutional doctrines of equal protection, personal rights, and due process, the legal apparatus takes steps to remove these inequities and restrictions. Law requires that all persons be treated in the same manner, and have the same access to the services and benefits of society, unless good reasons can be advanced to justify differential treatment. And it requires that people be allowed to speak, act, and live as they choose, except when there is persuasive evidence that certain forms of expression and behavior are opposed to the public interest. The achievements of law in both of these respects have been very great.

We come now to the critical question. What is apt to be the outcome if, for any reason, the legal concepts of equality and freedom come to supplant instead of merely supplementing the social goals of similarity and differentiation? Where class discrimination and individual repression are prominent features of a society, the concepts of equality and freedom, backed by the doctrines of equal protection, personal rights, and due process, work very effectively as correctives. Further, these concepts and doctrines then work hand in glove -- arbitrary inequities are removed, throwing open the doors of opportunity to all alike, and the pressure to conform is relaxed, leaving more options to exploit. But as the concepts of equality and freedom cease to be weapons directed against an established order, and come themselves to define the goals of a new order, the outcome is very different. Equality is then taken to require that all be put on the same footing, and freedom that each be allowed to express himself and order his life as he chooses. Equality becomes a duty imposed on society -- the state -- to supply to all alike the services and benefits that they claim as necessary, while freedom becomes a right inherent in individuals to make what use they will of these. The logical outcome of this is that society becomes solely the bearer of duties, with no right to impose conditions on or to make demands of its members; and individuals become solely the holders of rights, with no duty to conform or contribute. Under such a dispensation, society would no longer be able to secure essential similarities or to discriminate among differentiations, since every one could equally claim the support of society in his freely chosen courses. The results to which this tends is chaos, not order. But this is only logic, and there is no necessity, only the possibility and the danger, that fact will follow in its footsteps.

The meaning of responsibility and the role of law in promoting it can be discussed more briefly. As man's nature becomes more plastic and he exercises a greater degree of self-determination, a sense of responsibility becomes indispensable if men are to live and work together. Individuals now have a larger voice in deciding what their actions are to be, and what they decide has repercussions, for others as well as for themselves. Men mutually rely upon one another's commitments in making their plans, and each counts upon the others not to harm or interfere with him. This being the case, men must adhere to their declared intentions, show respect for their fellows, and be concerned about the difference that they make. Man becomes a responsible agent to the degree that he takes pains that his actions should issue in good, not evil. To be responsible is to care and to take care.

Responsibility is a moral virtue, a state of character, a habit of attitude and action. As such, there is very little that law can do directly to inculcate it. This is primarily the task of other forces and institutions, notably the family, the church, the schools, the neighborhood, and common morality. But though law cannot do much to lead men to care about the differences they make, it can very effectively hold them accountable for these differences. It can bring them to task for the consequences of their actions. This is done chiefly through the law of torts and the criminal law, which create duties and proscribe certain forms of conduct, and establish penalties for any breach of these. Law enforces its edicts with fines, imprisonment, damages, reparation, and other forms of punishment. The legal apparatus cannot instill in men a spirit of benevolence and concern. But it can induce them to act with forethought and caution.

Law thus transforms responsibility into liability. It decrees an artificial order of antecedents and consequences with which to minimize and repair the harm men do when they violate the natural order of mutual respect and care. This is a necessary but second-best solution to the problem -- it does not make men good, but only inhibits them from being bad. If it is to achieve even this much, other social institutions must be working effectively to inculcate a sense of responsibility, so that the great majority of men will abide by and support the social order and it is only the few delinquent or vicious who will need to be brought to book for civil or criminal liability.

Suppose now that two things happen. For one thing, those institutions that have hitherto done the most to inculcate a sense of responsibility, such as the family, the school, and the church, begin to lose their effectiveness. Their hold on the minds and hearts of men weakens, and their own sense of mission is lost. They find it difficult to adapt to change social circumstances and human attitudes. As the current phrase puts it, they cease to be relevant. At the same time, the law itself, in its commitment to individual freedom, restricts the field of action of these institutions and frustrates their efforts to shape men's characters. As a result, men no longer grow up in an environment that instills responsibility as a moral virtue and demands responsible behavior. So the sense of concern and the habit of care are seriously eroded, with a rapid increase in actions that fall under <u>dolus</u> and <u>culpa</u>.

Secondly, and concurrently, law creates such mechanisms of protection against civil liability that there is no reason for men to exercise responsibility or even forethought and caution. Insurance, the doctrine of society as the ultimate bearer of risks, the concept of the Welfare State, and other developments, all combine to insulate men against the consequences of failure, whether this result from misfortune or their own fault. These same devices spare men from having to make recompense for the harm they do to others. Then, since one will not himself suffer from his carelessnes, negligence, incompetence, and wastefulness, there is no need for him to care or take care for the difference he makes or the future he prepares. In short, the sense of responsibility becomes redundant.

Again, this is but logic, and the facts will fit it only if we let them.

From their inception, equality and freedom have been revolutionary ideas, weapons in a struggle against intrenched forces of discrimination and repression. But like so many revolutionaries, when they have won the field, they are unable to exploit it; since they have no positive content, they cannot afford guidance or direction to the social effort. They simply assert that all men have an equal claim on the support of society in making of themselves, and doing with themselves, whatever they freely choose. Liability is a doctrine that is intended to protect the innocent against the depradations of the wayward or vicious, and its essential feature has always been the accountability of the party who was guilty of tort or crime. But now accountability is lifted from his shoulders and transferred to society. The logic of this movement has the odd result of leading from strict liability to no liability at all.

Up until the immediate present, the social scene has been largely peopled by human persons, whose character and conduct have been formed by other forces and institutions than the law. Law has merely intervened when these other agencies have treated the human person in an inhumane fashion. But now the scene is being more and more peopled by legal persons, whose character and conduct are defined only by the rights conferred on them and the duties imposed on society.

This is a radical change. I am not predicting that the notions of equality, freedom, and liability will sweep the board and issue in the logical outcomes depicted above. But if the antecedents should become real, the consequents will surely follow.

LAW, PERSON AND PRIVACY

Lincoln Reis

All societies are law bound, perhaps in varying degrees and in diverse forms. For, in associating law peculiarly with democracy, we tend to forget that democracy, after all, is a kind of power, one among the several kinds of power, but in any case power. And power in order to be effective must in one way or another communicate and define. Law is one form of communication. In addition, every society in order to exist and to continue in existence--to preserve its shape and direction--must exert control, and one most necessary instrument of this control always appears in the form of laws. For in order to exercise control, one must assert and indicate. This is done by the making of statements. These statements, which issue from the formal center of power, both define and direct what as a member of that society or community one is required, forbidden or even permitted to do or not do. Of course, these statements are not simply recommendations. Power ignored ceases to be power. And so the statement of any law carries another statement, implied or otherwise, that what the law says is to be heeded. And if not, the consequences will be painful. In any territory, then, where there is any kind of society, no individual, whether a member of that society or not, can ever hope to ever be free of its reach. The law may be present subtly or brutally, but it will be present.

The reality of human societies is such, however, that nothing in fact requires that any given society is necessarily moved from within, government and power may be external to the group over whom it is exercised. Or, even if government is internal, it may be distinct in interest and personnel from the major body of the group over which it exerts its power. But, so far as the source and making of laws is out of the hands of the members of the community, these members cease from decisions and actions and become passive subjects and therefore potential victims. More strongly, they are emasculated in their citizenship. In order not to be so emasculated, the people must have a real part in the formation and the making of the laws; (and also, in defining and ultimately controlling their enforcement). For a law badly enforced alters the character of society.

The making of laws is the making of statements, and for statements to be comprehended and accepted or rejected there must be discussion, argument, debate. Traditionally it is only in the liberal democracies that all have been able to discuss or debate, though in fact not all have either been able, or have wished, to so participate. In any case, however, no person can be regarded as wholly free in his society unless he has the opportunity both to speak and to be listened to, if not necessarily heeded. It is this opportunity which defines the wholly free political community.

In short, the individual whoever and such as he is can take part in weaving the net of law in which he seems to be forever enmeshed. By doing so he may to a degree define or construct or modify whatever conditions of his life involve or depend upon the coexistence of other members of his political community. Such conditions include the social, economic and political elements of his life, and even such internal and personal elements as may depend upon their institutional arrangements.

Public debate open to all which precedes the passage of law and of which the conclusions transform themselves into law is in fact the heart of a freely functioning democracy. But debate is a matter of proof and persuasion, of propositions to be proved or disproved. And even appeals to the emotions, such as fear, are at bottom only so many, even if illegitimate, forms of argument and so, possibly to be countered and even successfully opposed and dissipated.

Moreover, while threats and force through fear may prevent or secure action they fail to formulate, define or clarify. Laws are made for the sake of purposes, of actions to be taken for some good. But purposes to be accepted must first be proposed and sometimes explained. As to the actions taken to secure them, they must first be discovered and considered, their alternatives canvassed.

The substance of public debate is just this discovery of purposes and means in the shape of the formulation, discussion and final acceptance or rejection of whatever proposition it is that is intended to become law. It is clear that this same substance of public debate is part of the substance of law; the other part is execution and enforcement.

Free debate then entails reasoning; it makes use of observation, experience and ideas, obviously, as they are brought into the social pool. But observation, experience and ideas are always individual, personal and private performances. They cannot be otherwise, because they cannot be otherwise located. This public debate means the existence and participation of free individuals; since an individual is only free and only debates freely when his observation, experience and understanding are his own--whether as originating in him or as accepted by him.

With respect to the making of laws, then, the individual must be independent; nor strictly speaking, is this independence itself a matter of laws, since laws emanate from the individual as free. Nevertheless, the freedom of the individual is a socially relevant thing. It cannot be otherwise, for the individual, once within the ambit of any group (and in point of fact no individual exists really outside a group) exists as the term of a relation, and like the term of any relation cannot be defined nor understood without reference to those other terms. In point of fact, the group or the members constituting it will allow nothing else. It is this relation or others similar to it or based upon it, to whatever degree of complexity, which constitutes, is society itself--and which, too, defines the instruments of society.

These relations, however habitual they may appear, are in their origins never unconscious; they exist as communications between individuals. These communications as they eventuate in so many definite movements and directions one immediately recognizes as law, or, to be exact, the laws.

In short, factual human existence--whatever its form individual or social, private or public, is inseparable from law. More, it is identical with law, as anything is identical with whatever defines it, and without which it cannot totally exist.

Human existence then is integral with law--whether in the form of dominating and dominated or of equal with equal. Moreover, among societies whose members are responsible for those public decisions which eventuate in law, law is never alien nor imposed.

There is not, and cannot ever be any question of freedom and privacy as entities which exist outside of and independently of law. For the citizens in forming their discussions, and working for whatever is thought good, both define and implement their freedom. The free individual does not exist in opposition either to society, or community or the laws. The opposition between the individual and the community, the private, and the public, is either wholly imaginary or else not what it is ordinarily thought to be. The problem of individual and community, for example, is not as is so often imagined the reconciliation of irreconciliables. If there is a problem it is rather in the discovery of and agreement upon goals, it is in the knowledge acceptance and implementation of means. In short, it is in the formulation, determination and enforcement of whatever laws are thought to advance the good insofar as this is a social and communal matter of the ongoing individual life.

In the case of conqueror and despot, things seem clear, rule and determination are external impositions, the individual at first glance appears altogether passive, though as time goes on this is seen to be not quite so. Give and take, some form of mutual accommodation and understanding, begin to appear. The conquered whatever the appearance, discovers a voice, and this voice though not always admittedly so, is somewhat heard. Like Idi Amin, dictators have an ear to the ground they bloody. In a word, wherever and whoever men are in relation over a period of time certain mutual accommodations which have all the force of democratically arrived at law begin to modify structures and shape events.

With regard to democracies, the making and enforcement of laws as rules set and enforced by government and binding upon all, drive actively from the very same persons upon whom as subjects they are enforced. Nor does it alter principle to say that since in reality any enacted law may be the instrument and the effect of the workings of an interest or pressure group and is only in form and nominally the voice of society, then the dictates of government may not after all reflect the operations of the self, but may be just so many alien impositions not to be heeded in clean conscience.

For in a democratic society, whatever the particular interests behind them, whether as the effects of persuasions, trading or even threats laws do still require for their enactment, the public concurrence of bodies which constitute or represent the voting members of the group in question. As enacted, they stand forth as the public and appropriate hence official, instrumentalities which in fact they are.

In a democracy, laws made are laws accepted, or else become soon dead, and soon rejected. If they are seen to be and are agreed to be unacceptable, than that unacceptability may itself become the occasion of some newer model of acceptability. Democratic laws are seen to be what they must be, just so many instruments of social acceptability. Inevitably, acceptable should here be defined as absolutely making for good, or as better than any alternative in making for good.

The prime source and parent of the law under freedom then is the uninhibited flow of discussion preceding and leading up to the enactment and enforcement of laws. The objectives of such laws are the goods of free people or whatever free men can be persuaded to think of as either good are leading up to good--insofar as these goods are attainable in a communal way or through the community, i.e., fighting wars or building roads as against writing poems or painting pictures. Whenever, in any case, men discuss actions and proposals what they can only have in mind is some good; laws legitimate these proposals and make them common.

So, in a word, the enmeshment of the individual in the laws is only apparent. Upon closer inspection it turns out to be a representation, or more accurately expression, of the ways in which the individual working with or dependent upon others, or rather interdependent with them moves to the attainment of such chosen objectives as he identifies with his own good, and which he cannot reach alone.

As to the community or society it is not a person who thinks or feels; it has not the ability and so it cannot stand in opposition to the individual in the way of thinking or feeling or entertaining opinions and beliefs. As an existence it is simply a mode of the individual's own existence, it is the individual achieving with others what he cannot achieve alone--but these achievements are in essence the characteristic effects and performances of individual persons, as the work of an institute of mathematics is the work of its mathematicians.

Any community or society, viewed as a whole, or in the form of government appearing as an external, even alien force, may appear to the individual at any given moment as opposed to him wholly or to some particular wish. But again, in the case of a free society in the sense which we have assigned to it, any such opposition is a fiction. For how can one be totally opposed as a complete person to what differs from one in kind.

In a society of free men, then, the law, whatever its strength and force is not an alien imposition. The laws of such a community like all rules and regulations are the proposals and conclusions of free men, of arguments and reasonings open to all.

In a word, in a free community every statement which is to become law begins in the activity of the individual person. Agreement with this statement by members of the appropriate voting body, whether the whole community as in a referendum, or a specific body of legislatures, precedes the enactment through which the statement becomes and is called law, a statement which now, as directive, is to be heeded. Public agreement and acceptance of this sort though phrased in the first person plural, are in reality the agreements and acceptances, hence, the consents of individuals. In a democracy, the we is not as with a dictator different from the I; it is the I simply become plural.

Where the individual through his own thought and conviction participates in the making of the laws--and where the law making process is nothing without this participation, we have at once a democratic community and a government democratically based. In this sense government and community are not materially different. As we said laws not derived from the people participating in their making are laws imposed upon them. This is to lessen dignity--for dignity expresses persons in full and acknowledged possession of their powers and status.

To the degree that political decisions are made without the participation, representation or consent of those affected, one loses both one's freedom and that dignity which makes the agent. One is lessened as a person. Person is the operative term. To be precise, however, in this place the meaning is person-as-citizen. In a democracy as this paper speaks of it, person-as-citizen is at the center.

For while the term person has many senses, ranging from what it denotes the effective individual, to kinds of character, to role playing, and finally to the individual as source of action and making of divisions,

the political sense is always central. For since in
fact people as members of society must live involved
with others, this cannot be otherwise. Through the
law one's decisions with regard to the ongoing course
of society become effective; through the law the individual plays a part in the exercise of social control;
through the law the individual may at once determine
what he is and is to become; through the law he is
able to protect this determination. But this is the
definition of the free individual in a free society;
it is the person as citizen.

A man's other activities and roles depend upon the
security of this. But in order not only to act freely
but to act freely well one must be fully human. One
is body and mind and feeling; a state whose citizens
are emotional and mental cripples will never work well;
it is both unwise and impractical to envisage a guiding
hand from above setting all straight.

To be a citizen is to function as a human. It is to
have and to be in charge of all one's abilities, functions, actions and relations. Whatever interferes with
their growth or prevents their exercise not only deprives a man of his full personality so that he becomes
a kind of nothing like a man in a coma, but it also
weakens the state. It is interesting that Anna Freud
and Professor Goldstein, in their recent BEYOND THE
BEST INTERESTS OF THE CHILD, say that to allow a child
to become emotionally crippled is to injure society
itself and to deprive it of the use of its full
resources.

Thus to develop and to safeguard the person of the
citizen is to make it possible to work well toward the
good society and the state. There is an important
corollary, to do the former is at the same time to
develop and to enhance all those other activities,
roles and aspects of the individual which we identify
with his person in whatever sense. His person, whether
as son, father, lover, husband, poet, carpenter, in perspective are seen as so many dependencies of his primary role as free person and citizen. He helps to shape
the laws accordingly, and so he now has a prudential
guide; as free he will shape the law as appropriate to
his person, his being and his wishes; so too with his
institutions.

Law now becomes both the expression and the instrument of achievement. Man is no longer lost in the mesh
of law. Law is now seen to be what it can and must be
if we are to achieve a good and just society. It is not
a net in which the individual is caught. It is a web

in which the filaments are strands of the individual's own making, serving his purposes.

Person then denotes the individual as he is at the center and source of whatever action, and action itself is seen as emanating from him. To cripple him or to otherwise prevent him from acting is not only to cripple and to destroy him, but it is to cripple and to destroy society, or at the least, to make part of its membership not fully human. Such contemporary problems as omission of informed consent, human experimentation, exploitation and the rest are now to be seen correctly as contributing subsidiary and derivative problems. Common to all and in whatever form, is the transformation of the person politically and otherwise into a thing so ultimately and totally passive that he ceases not only to be a citizen, he ceases to be a man as well. And, in any case, this is not only impossible it is dangerous for unless one is physically dead, he will not be really passive. What appears as passivity will be a hidden and reckless and misdirected activity with destruction as its end.

Laws with respect to the person look to the development of man, and of protecting him as agent, as decision maker or as the voluntary participant in action, as in the case of pledges and consent. Privacy is a special and subordinate case of the concept of the person. It relates to the person in the form of defining whatever externally interferes with a man's appropriate functioning, or with whatever might alter him in his behavior or in the other conditions of his life without his participation or consent. This applies both to government and to other bodies or persons. This defining is the substance and the principle by which people may formulate and enact laws that relate to the protection of privacy.

While not every violation or invasion of privacy is a matter for law and government--one member of a family may intrude upon another without a legal case being made of it--it is clear that bugging, total memory banks, widely dispersible credit ratings may tend to or in fact constitute external control or other unwillingly received modifications of integrity, life and conduct.

Because of these and like dangers, especially as they emanate from government and such other powerful bodies which affect the individual in the direction and conduct of his life, it is necessary if the integrity of the individual's life is to be kept safe to make such suitable laws and regulations as may define and control all matters of this kind.

A free society entails privacy, privacy like the protection of the person is not only a legal as well as a moral right; it is an indispensible condition of the free society.

PERSONS, MALE AND FEMALE

Virginia Held

By now, few would deny that women are persons, although the U.S. Supreme Court in 1894 explicitly did hold that a state provision excluding women from the meaning of the term 'persons', and thus forbidding women from practicing law, was a reasonable provision.[1] But although women have for some time been thought to be persons in some sense, until very recently the radical disparity between the treatment of male persons and of female persons by the law and by society was seldom noticed and almost never taken seriously as a problem.

This spring, at a conference on political philosophy in New York City, when I briefly dealt with the problem of the unequal treatment of women, in a paper essentially on another subject, a groan of impatience went through a substantial portion of the male audience. I had not intended to deal with the subject at this conference, but the reception of my remarks this spring has led me to think I have an obligation to do so. For if those few token women at conferences such as this do not deal with this issue, there is the danger that it will remain as neglected as usual. For how often does one encounter at academic conferences discussions of how freedom and equality and their aspects and implications affect the female half of humanity?

Professor Yasaki, for instance,[2] makes no mention, beyond a vague and fleeting allusion to patriarchy, of the way the legal conceptions discussed have been twisted to allow for the extraordinary oppression of women that has characterized and still characterizes Japanese society. Hardly any of the books and articles of the distinguished scholars gathered here have dealt in any way at all with the injustices routinely suffered by women, or have raised questions about the conceptual and theoretical revisions that merely noticing the problem would require. A few exceptions can be mentioned. Stanley Benn did refer to unjust discrimination against women in his widely read book written in 1958 at a time when few others did so,[3] and Mihailo Markovic, in a recent article, examines with considerable understanding the sad fact that the oppression of women continues to be almost as serious in communist society as under capitalism.[4] But these few departures hardly lighten the picture of neglect, and the terrible rarity of women invited to give papers at conferences such as this make those with the most to say the least likely to be heard.

I would like to suggest in this brief paper a few basic requirements which _any_ discussion of equality and freedom must meet if _it_ is not to ignore and thus hold in contempt that half of humanity composed of female persons.

In the first place, there must be a recognition that the oppression of women is not the same as the oppression of other groups. The problem cannot be dealt with by merely examining the rights of the individual in relation to other individuals and to the state, and, after centuries of discussion and enactment of law giving expression to the claim that all _men_ are equal and all _men_ are entitled to freedom, belatedly extending the _term_ 'men' to include women, who will now be accorded various basic legal rights which they did not have before. Nor can the problem be dealt with by supposing that a transformation of the class structure will automatically take care of the problems of women, as if the problems of women were no more than the class problems of their worker husbands and fathers, and as if those women with bourgeois husbands and fathers shared no essential deprivations with working class women.

The denial of equality and freedom to women which still characterizes all social systems and which affects all women of all classes is a denial of equality and freedom to them _as women_. It is not merely a question of how women are treated as citizens or as workers, although these issues should of course not be neglected as part of the pervasive oppression of women. We have to become aware of the specific forms the denials of equality and freedom to women take, and of the ways they can be overcome. I shall discuss a few of these issues with reference to the United States. In many parts of the world the situation is incomparably worse. Rarely is it better, although the relative weights of the different forms of oppression may be quite different.

Consider a custom that may seem to many people somewhat trivial in import, the custom that a woman who marries takes her husband's name. In reality this custom concerns a person's identity as a person and is not trivial. A young woman, like a young man, struggles through childhood and adolescence to establish an identity. But when she marries, the person she had become vanishes, and she becomes merely Mrs. Man, an anonymous, replaceable occupant of a position whose identity depends solely upon her husband. This is not a free choice to adopt a new name, which choice should be a right of anyone, but a price of loss of identity which women are routinely expected to pay upon marriage.

This is the way a young woman describes the experience: "I was Terri Tepper for 22 years. I had gone to college, graduated, and been an elementary school teacher, and all of a sudden I was out of the phone book and off the mailbox. All of a sudden, I felt my identity was buried."[5] Occasionally, men find themselves in situations where their wives are momentarily better known or the object of more respectful attention than they, and in which they become, in effect though never in name, Mr. Woman. Men in such situations often describe feelings of discomfort and loss of self-esteem; they feel they are "mere appendages." Alas, women feel these same feelings, but they feel them throughout their entire adult lives and as a matter of course, and this systematic damage to women's sense of self-worth is so ordinary a custom it is hardly noticed, even by its victims. And if a person marries more than once, a male person retains his identity throughout these changes; a female person loses hers not just once, but repeatedly. Sometimes she is not even permitted by law to resume her maiden name upon divorce,[6] to prefer, that is, the name of her father to that of her husband.

Next consider the radically different expectations made upon persons who are mothers and persons who are fathers. Wives and mothers are expected by society and by the law to "render services in the home," as the law often puts it, no matter the price that these persons pay in independence and self-development. No requirement to share in these tasks is made by society or by the law upon persons who are husbands and fathers. Motherhood is over and over again regarded as an occupation, though an unpaid one, such that a person who is a mother must choose between parenthood and doing other work, while a person who is a father can normally and routinely enjoy both parenthood and a career.

The list of inequalities to which women are subjected could be extended a very long way.[7] A requirement forbidding discrimination in employment on grounds of sex was enacted in the U.S. in 1964 as a kind of joke;[8] enforcement of it has been weak and ineffective, and voluntary compliance almost nil, so that the earnings of women who work year-round and full-time still average only somewhat more than half those of men. But at least, if women can occasionally manage to acquire jobs in which they indisputedly do "equal work" they are entitled by law to equal pay. In England such a provision is only <u>this year</u> going into effect.

At present, some of the most blatant forms of discrimination against women are becoming formally illegal. The "equal protection" clause of the U.S. constitution has been used to overturn some grossly discriminatory practices, and the courts have ruled against financial credit policies that automatically discount the earnings of women and social security provisions that regularly deny women an equal right to earn benefits for their spouses. If the Equal Rights Amendment to the U.S. constitution is enacted, it will lead to further improvement in the legal status of women, though the recent campaign against it makes somewhat doubtful the enactment of even this ludicrously overdue minimum of legal equality.

But perhaps we suffer from a kind of legalism not emphasized by Professor Yasaki: The mistake of thinking that every social problem can and should be given a primarily legal solution. I have dealt with this issue in a number of articles.[9] In the case of women, even if women were the legal equals of men, and mothers the legal equals of fathers, we could be nowhere near genuine equality and freedom for women. While I do not share the Marxist view that law is merely part of a superstructure, I do believe that especially in the U.S., far too many hopes are pinned on mere legal reform. Some problems are only incidentally legal problems and attacking them through the law may not be the best way to confront them, or may be at best a small part of the problem. Mere legal reform for women may leave their social and economic and political and psychological mistreatment virtually unaffected yet may help to mask the degree to which true equality and freedom are denied. Of course, to be denied equality even in the law is profoundly degrading and damaging, but to enjoy equality before the law is a very long way from enjoying equality.

Consider the matter of the loss of identity suffered by women when they marry. In most U.S. states, women have had for some time a legal right to retain their own names, but the weight of custom and social pressure, of the expectations of economic and social institutions, of the psychological training of women from infancy to take a subordinate position, of the fact that keeping their own names will usually displease their husbands, combine to make it highly unusual for women to maintain that minimal possession, a name.

Or consider the role women occupy in housework and child care. To a large extent, the law does not intervene into the private decisions between husband and wife concerning how they will arrange their lives; often the law concerns itself with these matters only in the case of a breakdown of the marriage. Of course, knowing what the law will hold in case of a breakdown often functions as an effective threat against women's attempt at equality and respect within marriage. Still, much would be possible within existing law, but again, custom and expectation and the superior social, economic, political, and psychological power of men bring it about that women are burdened vastly more by family obligations than are men. In capitalist societies they are expected, if they are members of the middle class, to become happy serfs, dependent economically on their husbands and supportive of his strivings for success. In socialist societies, and if they are poor elsewhere, they are expected to perform two jobs--one in the factory or in menial service and another at home, while men are burdened with only one. The remedies for this exploitation are only minimally legal ones, although the law frequently reinforces still the exploitation of wives and mothers.

In many countries, women do not even yet have the right to vote, a fact which should provoke outrage and boycotts and demonstrations but which the overwhelmingly male leadership of all nations seems quite willing to pass over in silence. That a country such as Switzerland could accord women the right to vote only four years ago is an indication of the callousness towards women of which civilized men are capable. In those countries where women formally have a right to political equality, their power is eroded by the power of male-dominated institutions that influence political life--business corporations, state bureaucracies, the mass media. In the U.S., there is not a single woman in the Senate, and only 4% of the members of the House of Representatives are women. There have been only 3 women in the Cabinet in the whole history of the U.S.

and there has never been a woman Justice of the Supreme Court. In Yugoslavia, women compose only 6% of commune assemblies, less than 8% of the Federal Assembly, only 5% of workers' councils, and 3% of the management boards.

The conceptions of equality and freedom, if they are to encompass equality and freedom for women, will have to include revisions in the conceptions of these terms developed both by liberalism and by socialism. In many cases the revisions will benefit the poor and powerless, whether male or female. In other cases they will benefit everyone through the development of more human and more satisfying relations between persons, but only after an initial loss of privilege on the part of all men, who, no matter how deprived in the society at large, have always had the advantage of dominance within the family. That women have nowhere near an equal amount of leisure time throughout their lives is a worldwide phenomenon.[10] That men everywhere have educational advantages and more power to decide where the family will live and how their lives will be led and how its members will behave is too well known to need argument. Equality must mean equality for women within the family and equality with men in personal relations, however silent on these matters both liberalism and socialism have been, to the great convenience of patriarchy.

And freedom must include far more than freedom from state interference on the Lockean assumption that a man left alone would be self-sufficient and able to provide a living for himself and his family, an assumption still influencing such thinkers as Robert Nozick,[11] despite its absurdity when applied to industrialized society. It must include, as well, freedom from being jobless and helpless, unable to find work or the basic necessities of life for oneself and one's children. Socialism has traditionally recognized this economic aspect of freedom, but in many countries the price in lost political freedom and in free self-expression has been far too high. The concentrated tyranny of communist statism is not an acceptable alternative to the concentrated tyranny of corporate capitalism, though both represent, very clearly, a tyranny of men over other men, and especially over women.

Freedom will never be possible for women as long as the right to a job is not recognized and as long as employers, often with union acquiescence, keep women in the position of being last hired and first fired. And freedom will never be possible for women as long as caring for children and a home are not recognized as "work" and are not work that is shared in by men on a basis of equality. Freedom for persons requires that their rights to birth control and abortion be recognized, so that they may become parents by choice not by mistake, and it requires that society's responsibility to provide day care facilities and family support be realized.

The women's movement may at present be the most important of the social forces seeking political and economic freedom and equality. It has outlived and outdistanced all the other radical movements of the 1960's in America, and in the present climate it appears that the women's movement may bring about in the U.S. more of the aims of democratic socialism incidentally than the socialist movement will bring about intentionally. Freedom must never be equated with the freedom of the rich and powerful to hang on to their holdings with police protection no matter how those without holdings are prevented from acquiring the means to live. The best antidote to the acceptance of such popular arguments favoring the interests of the overprivileged and overwealthy may be arguments in favor of equality and freedom for women. For along with those at the bottom of and beneath the working class--the unemployed, the poor, those on welfare, those unable to work--women form the most substantial group of those who have never been allowed to acquire holdings to hang on to or to transfer, those who have lived their lives in dependence and nonfreedom.

An examination of the legal, political, economic, social, and psychological wrongs done to women presents a dismal picture of their lack of equality and freedom and of the degree to which they are still, systematically and everywhere, denied the moral rights of personhood. But I would not wish to end this brief discussion with a lament. I have argued elsewhere that a transformation in the relation between man and woman is a most promising model for a transformation in social relations generally.[12] In the process of gaining equality and freedom, women will not merely compete as the equals of men in the contest for self-satisfaction and the means to overpower. They will, I believe, begin to establish relations of mutual concern and respect with men, relations in which human persons treat other persons <u>as persons</u>. And we may thus develop human relations in which persons treat others--normally and routinely, and not only as at present in a very few rare and isolated moments--as persons of equal worth, whose freedom ought to be respected, and between whom trust and moral concern rather than power and self-interest will prevail.

FOOTNOTES

1. In re Lockwood, 154 U.S. 116 at 117, 1894.

2. Mitsukuni Yasaki, "A Glimpse of Persons in the Modern Legalistic World," Problem Paper for this Conference.

3. S. I. Benn and R. S. Peters, THE PRINCIPLES OF POLITICAL THOUGHT, The Free Press, N.Y., 1965. Originally published as SOCIAL PRINCIPLES AND THE DEMOCRATIC STATE, Allen & Unwin, London, 1959.

4. Mihailo Marković, "Women's Liberation and Human Emancipation," V The Philosophical Forum, Nos. 1-2, Fall-Winter, 1973-74.

5. New York Times, April 10, 1975, p. 48.

6. Harriet F. Pilpel and Theodora Zavin, YOUR MARRIAGE AND THE LAW 304, Collier, N.Y., 1964.

7. See especially Robin Morgan, ed., SISTERHOOD IS POWERFUL, Vintage, N.Y., 1970; Vivian Gornick and Barbara K. Moran, eds., WOMAN IN SEXIST SOCIETY, Basic Books, N.Y., 1971; Juliet Mitchell, WOMAN'S ESTATE, Vintage, N.Y., 1973; and Karen DeCrow, SEXIST JUSTICE, N.Y., 1974.

8. See Caroline Bird, BORN FEMALE. THE HIGH COST OF KEEPING WOMEN DOWN, Pocket Books, N.Y., 1969.

9. See Virginia Held, "Civil Disobedience and Public Policy," in REVOLUTION AND THE RULE OF LAW, ed. K. Kent, Prentice Hall, Englewood Cliffs, N.J., 1971; "On Understanding Political Strikes," in PHILOSOPHY AND POLITICAL ACTION, eds. V. Held, K. Nielsen, and C. Parsons, Oxford Univ. Press, N.Y., 1972; and V. Held, "Justification: Legal and Political," in Ethics, Vol. 86, No. 1, Oct., 1975.

10. See "Equality of Opportunity and Treatment for Women Workers," International Labor Organization study, 1975.

11. Robert Nozick, ANARCHY, STATE AND UTOPIA, Basic Books, N.Y., 1974.

12. See V. Held, "Marx, Sex, and the Transformation of Society," in V The Philosophical Forum, Nos. 1-2, Fall-Winter, 1973-74, reprinted in WOMEN AND PHILOSOPHY, eds. C. Gould and M. Wartofsky, Putnam, N.Y., 1976.

LIBERTAD, IGUALDAD Y DIGNIDAD DE LA PERSONA

Marcelino Rodríguez M.

I.

Hoy día parece sobradamente evidente que entre la noción de libertad y la noción de igualdad existe una relación muy estrecha. Hasta tal punto que en algunos de los principales Derechos actualmente vigentes la noción de igualdad se considera como una simple extensión teórica y práctica de la nocion de libertad.

Por otra parte, tanto respecto a la libertad como respecto a la igualdad, es preciso distinguir un plano abstracto y un plano concreto. En el plano abstracto entramos en ambos casos en el terreno puramente filosófico. De ahí que sea el punto en que se observan las mayores discrepancias al pretender esclarecer las nociones de libertad e igualdad, y más en particular su fundamento o raíz última y su contenido conceptual. En este plano existe sin embargo una notable diferencia entre las nociones de libertad y de igualdad. Y es que la libertad puede ser considerada en abstracto pensando sólo en una sola persona o individuo, mientras que en el caso de la igualdad, aunque sea contemplada en abstracto, se ha de pensar siempre en la existencia de varias personas, pues de lo contrario no podría haber igualdad. De todos modos, las discusiones sobre el fundamento de la igualdad o acerca del origen de la desigualdad entre los hombres, comprueban que se inserta plenamente en el campo filosófico.

En la esfera social y política, y por supuesto en la esfera jurídica, interesa mucho más el plano concreto, sin que esto implique restar importancia al plano de los principios abstractos. Consecuentemente, en estos campos adquieren mayor relieve las diversas concreciones históricas de aquellos principios, es decir, las llamadas libertades <u>concretas</u> e igualdades <u>concretas</u>.

Entre las libertades concretas se puede decir que han adquirido carta de naturaleza en la mayoría de los Derechos vivos y actualmente vigentes, las siguientes: (1.) la libertad de conciencia y de pensamiento; (2.) la libertad de expresión o de manifestación del propio pensamiento; (3.) la libertad de asociación; (4.) la libertad de fijación del domicilio; (5.) la libertad de elección de estado civil y profesión.

Dentro de las igualdades concretas conviene distinguir dos grupos: el grupo de las igualdades civiles y el grupo de las igualdades políticas.

En el primer grupo se debe mencionar ante todo la igualdad de todos los ciudadanos ante la ley y en la ley. Este último inciso sirve para precisar que la formulación corriente de la igualdad ante la ley, no debe ser interpretada en aquel sentido restringido que la entiende como igualdad antes de o fuera de la ley, sino en aquel sentido estricto que implica una auténtica igualdad de trato dentro de la ley y que excluye todo tipo de discriminación por razón de sexo, de nacimiento -hijos legítimos o ilegítimos-, de raza, de religión, de lengua, de origen étnico, de nacionalidad, etc. En segundo lugar tenemos la igualdad social o en materia social, que excluye radicalmente toda división en castas o grupos sociales y exige unas mismas condiciones previas para subir en la escala social.

Dentro del grupo de las igualdades políticas se deben mencionar: (1.) La igualdad en el acceso a empleos públicos o al ejercicio de funciones públicas; (2.) la igualdad para exponer y difundir las propias opiniones políticas y para el uso de los medios de comunicación social; (3.) la igualdad ante el impuesto o la ausencia de todo privilegio o exención en materia tributaria; (4.) la igualdad en la exigencia de requisitos para el disfrute de bienes públicos, que excluye las denominadas listas de preferencia, de favor o de reserva. En todo caso se debe hacer la observación muy importante de que la problemática de las igualdades concretas se conecta inevitablemente con la problemática de las desigualdades concretas, dependientes de múltiples factores y que deben ser estimadas de acuerdo con un criterio de justicia.

II.

Una primera cuestión básica, dado este planteamiento inicial, es la de si la libertad y la igualdad, sobre todo entendidas como libertades concretas e igualdades concretas, constituyen derechos individuales o derechos sociales. Aquí se puede decir que se enfrentan radicalmente dos concepciones antagónicas: la concepción liberal-individualista y la concepción socialista y colectivista. Para la concepcion liberal-individualista, los derechos de libertad e igualdad, tanto en abstracto como en concreto, radican <u>en</u> y emanan <u>del</u> individuo, quien se constituye así en su única fuente y origen. Para la concepción socialista y colectivista, ambos derechos, máxime en sus formas concretas, proceden de la sociedad y solamente dentro de ella subsisten. Si tratámos de buscar un justo medio, la verdad es que ambos derechos, tanto la libertad como la igualdad, principalmente en sus formas manifestativas concretas, tienen una naturaleza bifronte: en la esfera <u>personal</u> se reconducen a la idea de <u>facultad</u> o de capacidad de exigir connatural a todo individuo; en la esfera <u>social</u> son propiamente un <u>poder</u> que se ejerce sólo <u>en</u> y <u>dentro</u> de una sociedad organizada.

III.

En todo caso, la cuestion anterior no elude otra de no menor importancia, cual es la de saber <u>cómo se forman</u> o surgen dentro de una sociedad dada <u>las ideas de libertad e igualdad</u>. En el plano individual esta nueva cuestión nos conduce necesariamente a la doctrina del conocimiento y dentro de ésta al apartado correspondiente a los condicionamientos sociales del conocer humano. En el plano social es preciso hacer referencia a los <u>valores sociales</u> y al modo cómo éstos se forman y se mantienen vigentes en una sociedad dada.

IV.

Sea cual sea la respuesta a las anteriores cuestiones, se debe constatar sin embargo que los derechos de libertad e igualdad, en cuanto derechos fundamentales, al ser formados y mantenidos en la consciencia colectiva, y posteriormente recogidos y plasmados en las legislaciones básicas de toda sociedad organizada, son simplemente <u>reconocidos</u> por ellas y no <u>constituídos</u>. Esto quiere decir que su propia y peculiar entidad normativa no depende radicalmente de su plasmación práctica, sino que precede a ella: su propia y peculiar entidad procede del ser humano en cuanto tal. Lo cual no excluye la posibilidad de error en la apreciación de su contenido normativo o de falsedad en su formulación concreta.

V.

En consonancia con el punto anterior, nos damos perfectamente cuenta que una comprension adecuada de los derechos de libertad e igualdad habrá de interrogar por su <u>fundamento</u>. Esta búsqueda de una base fundamentante es necesaria e ineluctable, y en ningún caso cabe juzgarla como inútil o vana. Ahora bien, la búsqueda del fundamento de las cosas es siempre sumamente problemática cuando se trata de aquéllas que por su misma naturaleza tienen una condición primaria. Es lo que ocurre con los derechos humanos. La solución aquí parece a primera vista más fácil, porque se piensa que, si son derechos <u>humanos</u>, su fundamento no puede ser otro que el <u>hombre</u>. Así se ha estimado al menos tradicionalmente en cuanto se ha señalado como fundamento inmediato de los derechos primarios, bien la <u>naturaleza</u> humana o bien el <u>ser</u> del hombre. Mientras la <u>doctrina</u> iusnaturalista, con sus múltiples variedades, prefería recurrir al concepto de <u>naturaleza</u>, las doctrinas modernas estiman con muy buen sentido que es preferible eliminar el concepto de naturaleza, esencialmente problemático, y hablar del <u>ser</u> del hombre sin más complicaciones.

Una prospección más profunda nos descubre, no obstante, que los derechos de libertad e igualdad, como derechos <u>humanos</u> básicos, son inherentes a la condición de <u>persona</u> que ostenta el ser del hombre, y que, como tales, constituyen parte sustantiva del <u>minimum</u> irreductible que la misma noción de persona humana requiere para que subsista.

Esta última consideración nos revela con toda claridad que los derechos de libertad y de igualdad son una consecuencia inmediata del principio ético-jurídico de la <u>dignidad de la persona humana</u>. En este principio clásico se enraiza la existencia de derechos humanos intangibles e inviolables en su núcleo esencial. Si se desconoce el principio de dignidad de la persona, pierden todo su sentido la libertad y la igualdad como derechos fundamentales. Con singular acierto la Declaración de derechos del hombre de 10 de diciembre de 1948 reconoce en su artículo primero que "<u>todos los seres humanos nacen libres e iguales en dignidad</u>..."

VI.

Supuesto que el principio de dignidad de la persona humana es el fundamento inmediato de los derechos humanos básicos, entre los que se cuentan la libertad y la igualdad, conviene preguntar: ¿La libertad y la igualdad son derechos <u>radicalmente primarios</u>? Pues bien, hay que decir que, la libertad y la igualdad, debidamente entendidas, no son derechos <u>radicalmente</u> primarios, pues presuponen otros derechos. ¿Cuáles son estos derechos? En primer lugar el <u>derecho a la propia vida</u>, presupuesto absolutamente necesario para poder ser sujeto existencial de otros derechos. Este derecho primarísimo incluye, además de la exigencia intangible e inviolable de vivir, que tiene todo ser humano desde que adquiere la condición de tal, <u>el derecho</u> a la subsistencia y a utilizar los medios necesarios para lograrla, así como el <u>derecho a la integridad de su cuerpo</u> y a que no se disponga del mismo con fines contrarios a la dignidad de ser humano. En segundo lugar, el <u>derecho a la personalidad</u>, que incluye su reconocimiento y proteccion tanto en el plano social como en el jurídico. En tal sentido tienen bastante razón aquellos autores iusprivatistas que estiman que el derecho a la personalidad es el fundamento de todos los demás derechos, una vez supuesto el derecho a la vida.

El derecho a la vida y el derecho a la personalidad son por tanto los derechos radicalmente primarios de todo ser humano, inmediatamente derivados del principio ético-jurídico de la dignidad de la persona humana. Supuestos estos derechos, les siguen los derechos <u>fundamentales</u> de libertad e igualdad. Esto no quiere decir que sean derechos secundarios, pues su condición de derechos <u>primarios</u> no puede ser discutida, aunque situados, tanto en un orden ontológico como en un order lógico, a continuación del derecho a la vida y el derecho a la personalidad.

VII.

Este carácter <u>primario</u> de los derechos de libertad e igualdad, en cuanto derechos <u>fundamentales</u>, no excluye cierta <u>relatividad histórica</u>, lo mismo en su comprensión teórica que en su formulación práctica. Tampoco parece que deba existir mayor dificultad en admitir que estos derechos, tanto considerados en abstracto como en concreto, constituyen una categoría lógico-conceptual históricamente variable. Pero esta misma variabilidad histórica revela que a través de su evolución progresiva o regresiva se observan una serie de <u>constantes</u> claramente diferenciadas, que configuran el núcleo esencial de estos derechos y el <u>minimum irreductible</u> sin el cual su misma noción desaparecería.

VIII.

Que existe hoy en día una <u>crisis</u> de libertad e igualdad, como derechos fundamentales, es un hecho innegable. Ciertamente que las modernas Declaraciones de derechos, tanto las que provienen de Convenciones u Organismos internacionales como las que se incluyen en las Constituciones o Leyes fundamentales de los Estados, contienen todas ellas mención expresa y clara de los derechos humanos de libertad e igualdad e incluso de ciertas libertades e igualdades concretas. Pero estas declaraciones, en el mejor de los casos, contienen normas meramente programáticas, cuya eficacia práctica alcanza muchas veces niveles mínimos, cuando no son totalmente nulas. Un estudio comparativo de los enunciados normativos de los textos constitucionales y su eficacia práctica en cada uno de los países, revelaría hasta qué punto la separación entre ambos polos es mucho mayor de lo que cabría pensar. Es más, si se considera atentamente el problema, la divergencia comienza ya entre las normas constitucionales de carácter programático y las respectivas leyes ordinarias que deberían desarrollar los principios contenidos en aquéllas. Así, por ejemplo, en las primeras se consagra el principio de que la libertad es un derecho inviolable de todo ciudadano, asi como el principio de la igualdad de todos los hombres--o de todos los ciudadanos--ante la ley. Pero en las leyes ordinarias, incluso en los Códigos civiles, se recogen una serie de disposiciones que desconocen o limitan las libertades concretas y que establecen desigualdades concretas por razón de sexo, de nacionalidad, de raza o color, de lengua, de religión, etc.

Si esto es así, es evidente que la libertad y la igualdad, más que derechos concretos y eficaces, son simples aspiraciones o desiderata, no siempre alcanzables en la vida social humana. O, en último término, son meras exigencias o demandas de algo que necesariamente debe ser, pero que, por diversas circunstancias, no es. En cuyo caso la afirmación de los derechos de libertad e igualdad, como derechos fundamentales, pertenece al reino ideal; en el mundo real las que existen son la carencia de libertad y la desigualdad.

This paper discusses the following doctrinal points:

It seems evident today that there is a very close relationship between the notions of liberty and equality. It is necessary to distinguish between abstract liberty and concrete liberties as well as between abstract equality and concrete equalities or concrete inequalities.

The first important question is whether the rights of liberty and equality, above all in their concrete formulations, are individual or social rights and what exactly is their juridical nature.

Another important question concerns the way in which the rights of liberty and equality <u>are shaped</u> in the individual and collective conscience.

Whatever may be the solution to these questions, what is clear is that the rights of liberty and equality, in so far as fundamental rights are concerned, are <u>recognized</u> and <u>not constituted</u> by the collective conscience.

This shows that an adequate comprehension of the rights of liberty and equality must be examined from its <u>basis</u>. This basis lies in the nature of the human person and, more concretely, in the ethical-juridical principle of the <u>dignity of the human person</u>.

Nevertheless, as is well understood, the rights of liberty and equality are not the most essential; the right to life and to personality has greater <u>priority</u>.

Furthermore, in its concrete formulation, it is necessary to admit a certain historical <u>relativity</u>; although the essential nucleus does not change, except for some errors and deviations of the historical human conscience.

That a <u>crisis</u> in the human rights of liberty and equality <u>exists</u> today is beyond dispute. In spite of the Declarations of Rights which are so fashionable and which, in fact are purely programmatic, the minimal standard content of both fundamental rights is declining.

BIBLIOGRAPHY

-LE FONDEMENT DES DROITS DE L'HOMME (Actes des entretiens de l'Aquila). Firenze, 1966.-L'EGALITÉ. Vol. I et IV. Travaux du Centre de Philosophie du Droit. Bruxelles, 1971 et 1975.-EQUITY IN THE WORLD'S LEGAL SYSTEMS: A COMPARATIVE STUDY. Brussels, 1973.- F. Elias de Tejada, LIBERTAD ABSTRACTA Y LIBERTADES CONCRETAS. Madrid, 1968.

ÜBER PERSÖNLICHE FREIHEIT UND SOZIALISTISCHES RECHT

Karl A. Mollnau

I.

Die Befreiung der Persönlichkeit beginnt mit der Befreiung der Gesellschaft. Der Übergang vom Kapitalismus zum Sozialismus ist der Weg in eine Gesellschaft, in der die Menschen ihre sozialen Lebensprozesse mit Sachkenntnis und in assoziierter Form gestalten und dabei ihre Fähigkeiten allseitig ausbilden und ausüben; d. h. persönliche Freiheit erlangen. Die persönliche Freiheit hat nicht im Recht ihre Geburtsstätte. Die Freiheit der Persönlichkeit ist das Produkt bestimmter gesellschaftlicher Entwicklungen; gleichwohl findet die Freiheit der Persönlichkeit auch im Recht ihren Ausdruck.

II.

Mit der Beseitigung des durch das kapitalistische Eigentum an Produktionsmitteln bedingten Gegensatzes zwischen kapitalistischer Aneignung und gesellschaftlicher Arbeit in der sozialistischen Revolution verschwindet der Zwangs- und Ausbeutungscharakter der Arbeit.

Zum ersten Mal nach Jahrhunderten bietet sich jetzt die Möglichkeit, für sich selbst zu arbeiten. Nunmehr wird der Gang der gesellschaftlichen Entwicklung die Realisation der menschlichen Arbeit selbst: der einzelne nimmt in seiner individuellen Tätigkeit mehr und mehr die produktiven Kräfte der Gesellschaft auf, wie umgekehrt seine individuellen Kräfte zu Kräften der Gesellschaft werden.

Unter diesen Umständen wird das Individuum immer stärker mit der Gesellschaft und ihrer Entwicklung verbunden. Die Lebensgrundlagen der Gesellschaft sind auch die des einzelnen Individuums. Das einzelne Individuum dient nur dann seinen Interessen, wenn es für die Gesellschaft arbeitet; umgekehrt kommt das, was der Gesellschaft nutzt, ihren Mitgliedern wieder zugute.

Diese Übereinstimmung zwischen den grundlegenden Interessen der Individuen und jener der Gesellschaft erwächst auf dem Boden des sozialistischen Eigentums an Produktionsmitteln und der Klassenstruktur, die entsteht, nachdem sich das sozialistische Eigentum an Produktionsmitteln durchgesetzt hat. Zwischen der regierenden Arbeiterklasse und den mit ihr freundschaftlich verbundenen Genossenschaftsbauern, der Intelligenz und den genossenschaftlich organisierten Handwerkern gibt es keine grundlegenden Interessendifferenzen. Fortschreitend identifizieren sich alle Werktätigen mit den Zielen, die die Arbeiterklasse verfolgt, um ihre Interessen zu verwirklichen. Gleichzeitig gibt es zwischen diesen Klassen und Schichten Unterschiede und differenzierte Interessen.

III.

Die Gemeinsamkeit der Grundinteressen ist eine Folge der gleichartigen Stellung jener Klassen und Schichten zu den Produktionsmitteln: alle sind Werktätige und verdienen durch ihre eigene Arbeit ihren Lebensunterhalt. Dass gleichzeitig differenzierte Interessen zwischen den Klassen und Schichten der sozialistischen Gesellschaft bestehen, ist die Folge ihrer noch nicht völlig gleichen Stellung zu den Produktionsmitteln, ihrer noch unterschiedlichen Rolle in der gesellschaftlichen Organisation der Arbeit und damit auch bestimmter Unterschiede in den Verteilungsverhältnissen. Hierauf basieren auch Unterschiede im Bewusstsein und in den Verhaltensweisen.

IV.

Die Gemeinsamkeit der Grundinteressen in der sozialistischen Gesellschaft ist eine dialektisch-widersprüchliche Interesseneinheit: die objektiv in der Eigentumsstruktur wurzelnde Gemeinsamkeit und Übereinstimmung zwischen den grundlegenden Interessen der Klassen und Schichten des Individuums und der Gesellschaft bedeutet nicht Abwesenheit des Widerspruchs überhaupt und schliesst auch nicht die Möglichkeit aus, die Einzelinteressen in einen Gegensatz zu den allgemeinen geraten lässt.

Die Ursachen für diese Widersprüche und Konflikte sind mehrschichtig. Sie sind einmal im Erbe des Kapitalismus sowie in Einflüssen des Klassengegners begründet, der von aussen in die sozialistische Gesellschaft hineinzuwirken versucht. Die Ursachen reduzieren sich jedoch nicht darauf; vielmehr gibt es auch im Sozialismus den

Widerspruch zwischen dem Alten und Neuen. So können beispielsweise Widersprüche entstehen, weil infolge gesteigerter Produktivkraftentwicklung bestimmte Seiten der Produktionsverhältnisse, die Wirtschaftsleitung veralten; einzelne Menschen aber, die diesen überalterten Strukturen verhaftet sind, an ihrem Fortbestand interessiert sind.

V.

Die gemeinsamen Grundinteressen sind weder der statistische Durchschnittswert aller Einzelinteressen, noch die Summe aller persönlichen Interessen; vielmehr repräsentieren sie die Interessen, die der real möglichen gesellschaftlichen Entwicklung entsprechen. Ihr Klasseninhalt wird durch die Interessen der Arbeiterklasse bestimmt. Sie bilden die Grundlage für die einheitliche Willensbildung des sozialistischen Staates, deren Ergebnisse vorrangig im Recht ihren Niederschlag finden. Der Inhalt des sich im sozialistischen Recht ausdrückenden Staatswillens wird durch die Interessen der Arbeiterklasse bestimmt, die im festen Bündnis mit den anderen Werktätigen, geführt von der Arbeiterklasse und ihrer Partei, die Macht ausübt. Im sozialistischen Recht erhält der Wille des sozialistischen Staates, der historischen Mission der Arbeiterklasse entsprechend zu handeln, seine Gestalt als besondere politische Entscheidung; der im sozialistischen Staat ausgedrückte Wille ist eine spezifische Erscheinung des sozialistischen Staates, den Zielen und Zwecken der Arbeiterklasse entsprechend auf die Gesellschaft einzuwirken. Ihr liegen die objektiven Erfordernisse der gesetzmässigen Entwicklung und die realen Bedürfnisse der Arbeiterklasse und ihrer Verbündeten zugrunde. Im sozialistischen Recht manifestiert die Arbeiterklasse ihre Entschlossenheit, die in den materiellen Lebensbedingungen der sozialistischen Gesellschaft liegenden jeweiligen realen Möglichkeiten auf dem Weg zum Kommunismus Wirklichkeit werden zu lassen.

VI.

Die Übereinstimmung zwischen den grundlegenden gesellschaftlichen und persönlichen Interessen ist die Basis, um in der sozialistischen Gesellschaft die persönliche Freiheit zu realisieren. Die persönliche Freiheit ist in den sozialistischen Gesellschaftsverhältnissen objektiv begründet und materiell gesichert. Sie ist die Freiheit der Werktätigen, ihre Fähigkeiten zu entfalten, ihre persönlichen Interessen zu befriedigen und damit einen individuellen Beitrag zur Realisierung des gesellschaftlichen Fortschritts zu leisten. Auch die persönliche Freiheit ist nur als erkannte und verwirklichte Notwendigkeit existent; wobei sich die Notwendigkeit in objektiven Erfordernissen der Gesellschaft verkörpert. Die persönliche Freiheit ist mit der gesellschaftlichen dialektisch verknüpft. Das bedeutet aber nicht, die persönliche Freiheit ginge völlig in der gesellschaftlichen auf.

Genau so wenig, wie sich die Übereinstimmung zwischen persönlichen und gesellschaftlichen Interessen spontan einstellt, so realisiert sich gesellschaftliche Freiheit mechanisch in persönlicher. Beides muss vielmehr bewusst herbeigeführt werden. Das sozialistische Recht hat dabei wesentliche Funktionen zu erfüllen. Die juristischen subjektiven Rechte und Pflichten sind in ihrer gesellschaftlichen Wirksamkeit ein Ausdruck der realen persönlichen Freiheit in der sozialistischen Gesellschaft.

Der sozialistische Staat wirkt auf die Beziehungen zwischen Individuum und Gesellschaft ein, indem er objektiven Erfordernissen entsprechende Rechte und Pflichten setzt und dafür sorgt, dass diese Rechte und Pflichten von den Mitgliedern der Gesellschaft verwirklicht werden. Dies ist wiederum nur über das willensbestimmte Handeln der einzelnen Gesellschaftsmitglieder möglich. Deshalb muss der in den einzelnen Rechtsnormen manifest gewordene Staatswille, dessen Inhalt in den materiellen Lebensbedingungen der sozialistischen Gesellschaft und ihrer führenden Arbeiterklasse wurzelt, in die verschiedenen Einzelwillen transformiert werden. Diese Transformation zielt darauf, die auf individuellen Bedürfnissen, Interessen und Meinungen fussenden Einstellungen und Handlungen der Bürger mit den im objektiven Recht zum Ausdruck gebrachten gesellschaftlichen Erfordernissen in Übereinstimmung zu bringen. Das geschieht weder in Gestalt pluraler Interessenkombinationen, noch in Form ausgleichender Interessenkompromisse. Die Interessen der Arbeiterklasse, die im sozialistischen Recht ihren normierten Ausdruck finden, sind vielmehr die unverrückbaren inhaltlichen Richtwerte, nach denen die Interessenübereinstimmung mit Hilfe des Rechts den objektiven Gesetzen gemäss herbeigeführt wird.

VII.

Für das Verständnis der Rolle, die das Recht spielt, um die Interessenübereinstimmung bewusst herbeizuführen, ist die dialektische Einheit von Rechten und Pflichten wesentlich.

Während sich im Kapitalismus der Mensch von der Gesellschaft absondert, sind in der sozialistischen Gesellschaft der wachsende Wohlstand der Bürger und ihre Rechte untrennbar mit den Erfolgen des sozialistischen Aufbaus verbunden.

Seine Persönlichkeit entfaltet jeder, indem er die Geschicke unserer Gesellschaft mitbestimmt und mitgestaltet. Es ist deshalb dem sozialistischen Staat nicht gleichgültig, wenn jemand zurückbleibt und grundlegende Rechte nicht ausübt. Wie mehrfach in der marxistisch-leninistischen rechtstheoretischen Literatur dargetan wurde, sind die Rechte der Bürger nicht Schranken und Fesseln, sondern Inhalt des sozialistischen Staates, Werkzeuge zur freien Persönlichkeitsentwicklung der Bürger.

Rechte und Pflichten sind im sozialistischen Recht beide Ausdruck objektiver Gesetze. Rechte und Pflichten sind im sozialistischen Recht zwar nicht im formal-logischen, aber im dialektischen Sinne identisch.

Die dialektische Beziehung zwischen Rechten und Pflichten tritt in verschiedenen Formen in Erscheinung. Einmal kann dem Recht einer Person die Pflicht einer anderen entsprechen, zum anderen können aber auch die an einem Rechtsverhältnis beteiligten Rechtssubjekte sowohl Rechte und Pflichten gleichzeitig gegenüber haben. Schliesslich gibt es auch die Fälle, in denen Recht und Pflicht zusammenfallen, eine Person also mit der Pflichterfüllung zugleich ihre Rechte wahrnimmt.

VIII.

In der sozialistischen Gesellschaft ist das subjektive Recht wie die Pflicht Element und Werkzeug, um die persönliche Freiheit zu realisieren.

Wie bereits betont, ist die persönliche Freiheit nur aus der Grundlage erkannter und verwirklichter gesellschaftlicher Notwendigkeiten realisierbar. Eben diese Notwendigkeiten liegen aber dem sozialistischen Recht insgesamt zugrunde; dem Bürger treten sie dann-- allerdings zumeist in vermittelten Formen--als individuelle Berechtigungen und Verpflichtungen in Erscheinung. Dem Grunde nach sind also Rechte wie Pflichten jeweils nur unterschiedliche juristische Ausdrucksformen ein und derselben Notwendigkeiten. Dass diese beiden juristischen Ausdrucksformen unentbehrlich sind, um das Tun oder Unterlassen der Bürger auf die Durchsetzung gesellschaftlicher Notwendigkeiten zu lenken, hat seine Ursache in dem dialektisch widersprüchlichen Charakter der

Einheit und Übereinstimmung der grundlegenden Interessen zwischen den Klassen sowie zwischen dem Individuum und der Gesellschaft.

Nur in der Verknüpfung von Rechten und Pflichten im rechtlichen Regelungsprozess wird es möglich, das Handeln des einzelnen an der Durchsetzung gesellschaftlicher Notwendigkeiten als in seinem persönlichen Interesse liegend zu orientieren. Subjektive Rechte wie Pflichten dienen also gleichermassen dem Bewusstwerden objektiver Notwendigkeiten und ihrer Durchsetzung durch organisiertes Handeln. Weiter: Die persönliche Freiheit ist gesellschaftlich bedingt. Der gesellschaftliche Charakter der persönlichen Freiheit kommt nicht zuletzt darin zum Ausdruck, dass sie nicht ausserhalb und neben den objektiven Gesetzen der gesellschaftlichen Entwicklung realisierbar ist. Die Durchsetzung gesellschaftlicher Gesetze ist aber im Sozialismus ein bewusst geleiteter und gestalteter gesellschaftlicher Vorgang. Überall dort, wo dies nicht sichergestellt ist, greifen individuelle Willkür und Spontaneität Platz. Die persönliche Freiheit kann sich deshalb im Sozialismus nur in vermittelter Form über die bewusste Leitung der gesellschaftlichen Prozesse realisieren. Die persönliche Freiheit ist deshalb wiederum mit der gesellschaftlichen Organisiertheit und Diszipliniertheit des einzelnen dialektisch verbunden. Persönliche Freiheit und Disziplin sind im Sozialismus keine Gegensätze, sondern bedingen einander.

Freedom of the individual does not originate from law. The amount of personal freedom is the product of a definite social development. Consequently the liberation of the individual cannot be conceived without the liberation of society.

The basis for realizing the freedom of the individual in socialist society is the correspondence of fundamental social and personal interests, gradually having been achieved by the emergence of socialist property in the course of socialist revolution. Freedom of the individual finds its objective basis and material guarantee in the socialist social order; its realization is a dialectical process in the course of which contradictions are developed, solved and newly set. It is understood as the freedom of the working people to develop their skills, to satisfy their individual interests and thus to contribute personally to the realization of social progress. Also freedom of the individual does only exist as recognized and realized necessity.

Socialist law constitutes an indispensable means of the state to mould, regulate and protect the process of realizing the freedom of the individual. Socialist law is just as indispensable for shaping the individual conditions, the subjective capacity of the individual citizen, and his ability to realize personal freedom.

While bourgeois legal thinking often defines subjective right as the only field of personal freedom, the socialist legal system includes subjective rights and subjective duties as forms of realization of the freedom of the individual.

EL HOMBRE, LA SOCIEDAD Y LA LIBERTAD

Domingo A. Labarca P.

El estudio del hombre es parte de los temas de interés en la hora actual. Ciencias como la psicología individual, la sociología y la psicología social, la antropología, criminología, etc., estudian la naturaleza humana, destacando cada una de ellas un perfil particular de dicha naturaleza. Es común, sin embargo, a todas, la búsqueda de las causas que determinan las reacciones diversas del ser humano. La manera particular de abordar los diversos modos de ser del hombre, es lo que fija la metodología propia de cada una de dichas ciencias. La psicología se encarga del estudio de la realidad psíquica del hombre, mientras que el objeto de la sociología es el enfoque del hombre en su relación con los demás, es decir, su vida en grupo y en sus interacciones sociales; para la psicología social, lo importante es el estudio del individuo en su relación con los demás, en cuanto los demás influyen en la realidad psíquica de ese individuo.

La separación de la temática objetal entre las diversas disciplinas a un nivel teórico es posible, como someramente lo hemos hecho. La referencia a un mismo objeto: el hombre, nos descubre la identificación, o mejor, la interacción entre las antes señaladas ramas del saber. La psicología nos habla de la percepción como un fenómeno complejo, resultado de un proceso en el cual participan nuestros sentidos, y en donde el sistema nervioso juega un papel de fundamental importancia. Sin embargo, un hecho como la percepción no es un fenómeno explicable a un puro nivel psicológico. "Por lo general--señala Simpson--resulta que fenómenos que se supuso claramente individuales, como la percepción y la memoria, han sido condicionados socialmente o aún socialmente determinados..."[1]

Existe en el "Homo Sapiens" una naturaleza biológica que también resulta de definitiva importancia considerar en el estudio del individuo y que determina el <u>ser hombre</u>, como resultante que es de los diferentes factores que se hacen presentes en el desenvolvimiento de su conducta.

El estudio de los instintos ha permitido una mejor comprensión de la naturaleza humana. La importancia de algunos de ellos determina la especificidad de la especie humana dentro del género animal.

Las antes señaladas ciencias humanas han recogido mediante la observación, de la conducta del hombre, datos que nos sirven de esquemas mediante los cuales es posible entender dicha conducta y conectarla con factores que la originan. Debemos destacar que la conducta humana es una resultante de diversas fuerzas que confluyen para determinarla. La psicología aborda el estudio de la misma, escudriña las experiencias íntimas, preocupándose por las vivencias de carácter interno que las provocan: causas mecánicas, orgánicas y vitales.

La teoría mecanicista en psicología destaca el nivel de los mecanismos físicos como fundamentos de la psique. El vitalismo entiende que el hombre es un ser de fines. "El principio teleológico es una propiedad adicional de la materia viva y escapa a la medida y a la predicción."[2] El organicismo destaca la realidad dinámica del individuo constituido por una naturaleza biológica que juega un carácter determinante en la conducta del individuo.

Independientemente de la teoría que se trate, es indiscutible que el hombre es un ser biológico. Ese ser biológico pertenece al reino animal. Tiene un sistema nervioso: el más desarrollado del de todos los seres vivos. Una corriente sanguínea recorre su cuerpo. Es depositario de glándulas que originan reacciones determinadas en los individuos. Estas constituyen las fábricas químicas del cuerpo humano. Dichos elementos químicos son repartidos por todo el continente del hombre a través de los ríos de sangre que recorren el cuerpo. El funcionamiento de las glándulas determina la personalidad del individuo.[3]

El hombre es un ser complejo, como muy bien lo expresa Wolff: "Lo psíquico no es, pues, el resultado exclusivo de la función cerebral sino que está integrado en todo el organismo. Alfredo Adler fue el primero que señaló el hecho de que la gente no piensa solamente con su cerebro, sino también con sus glándulas, su estómago, su corazón, sus pulmones y otros órganos. A este fenómeno lo llamó Adler 'el dialecto de los órganos'. A esta interrelación entre los fenómenos psíquicos y los somáticos se le llama ahora psicosomática... Es decir, que una tendencia psíquica puede provocar una tensión de los órganos e incluso un hiper o hipodesarrollo de la función glandular, lo cual repercute sobre el organismo. En esta forma se crea un ciclo biopsíquico entre los estímulos psicológicos (p) y los biológicos (b): $p \longrightarrow b \longrightarrow p \longrightarrow$

b ⟶ p, etc".[4]

Otro factor que la psicología no puede descartar es el sistema nervioso. Esta complicada red de cables que se extienden por todo el cuerpo humano son los conductores de la electricidad que imprime dinamismo al hombre. Ellos llevan al cerebro la "información" que recogen en el organismo humano y transmiten las respuestas por todas las partes del mismo. Son tres las grandes formaciones de esa armonía de filamentos: El sistema nervioso autónomo, el central y el periférico.

Mediante el sistema nervioso autónomo y el central se determinan los músculos involuntarios y las glándulas, y la dirección consciente del organismo, respectivamente. Hablamos del organismo humano. En efecto, el cuerpo humano es un todo armónico. Existe una distribución de funciones que forman ese conjunto que es el hombre como realidad biológica. Los sentidos son parte de ese todo. El hombre capta la luz por un sentido: la vista; sabe de la dureza de los cuerpos por el sentido del tacto, diferencia los olores a través del sentido del olfato, etc. Sin los sentidos el hombre estaría perdido. No tendría noción del tiempo ni del espacio. Se confundiría. Viviría como parte de una totalidad, sin vivencia de su individualidad.[5]

La psicología comienza con el estudio de la naturaleza humana de índole biológica. El hombre es, sin embargo, una naturaleza social. Aristóteles en LA POLITICA afirma el carácter instintivamente social del hombre. Hablar de psicología social, es aceptar esa peculiaridad humana. La presencia de la antes señalada disciplina surge de la demanda que la propia realidad plantea al psicólogo y al sociólogo. Se observa que una unilateral consideración del hombre desde un punto de vista individual, resulta una explicación científica incompleta. En este sentido señala Simpson: "La psicología social no surgió sólo de la fisiología-biología y de la doctrina evolucionista. Proviene de la necesidad que sintieron los primeros sociólogos de explicar la interacción de los individuos en los grupos sociales y en las comunidades, y de los intentos de los primeros antropólogos de analizar y explicar las actuaciones de la mente del hombre primitivo..."[6]

Los sociólogos norteamericanos captan la importancia del estudio del hombre a nivel social. A través de la consideración de los instintos y mente de grupo, se emprende el estudio de la vida social. "Los instintos se usaban para analizar las maneras en que el individuo se socializa, o sea, se vuelca consciente de sus obligaciones y responsabilidades y participa como miembro de un grupo; y el así llamado instinto gregario era usado por algunos para explicar la vida social en general..."7

El antes mencionado método, tiene sus antecedentes muy remotos en los propios griegos. Aristóteles estructura toda la teoría del Estado partiendo de la consideración del instinto gregario. La consideración tomista del hombre, homo hominis amicus et familiaris, se mantiene dentro de la misma línea aristotélica. Descubriendo la naturaleza humana, concluimos el instinto social y político del hombre: naturaliter animal sociale et politicum. De lo anterior resulta que el Estado o la sociedad responden a la naturaleza humana; no son productos del pecado.

La psicología social ha recurrido a los métodos experimentales a objeto de elaborar esquemas que permitan una comprensión de las relaciones interindividuales. Las estadísticas han jugado un papel primordial en el establecimiento de cuadros en los cuales se busca reflejar el desenvolvimiento de la conducta humana social. Mediante los estudios comparativos se han tratado de establecer constantes en la vida social de los diferentes grupos y aplicar los logros de investigaciones en diversos grupos sociales, a otros grupos.

La antropología ha visto, por lo menos, la teoría funcionalista de Bronislaw Malinowski, que es posible determinar relaciones constantes en la vida social de todos los pueblos. Lo mismo en el hombre primitivo que en el hombre moderno: "El hombre primitivo, de acuerdo con sus actitudes científicas, debe aislar los elementos pertinentes del conjunto, de los proporcionados por el medio de las casuales adaptaciones y de su experiencia, e incorporarlos en sistemas de relaciones y factores determinantes. El motivo final de todo esto es principalmente la supervivencia biológica. La llama era necesaria para calentar y cocinar, para seguridad e iluminación... Todas estas provechosas actividades tecnológicas estuvieron basadas en una teoría, en la cual fueron aislados los factores decisivos, en la que se apreció el valor de la penetración teórica y en lo que la previsión del resultado se basó en experiencias anteriores cuidadosamente formuladas." Lo anterior es aplicable a las más modernas investigaciones científicas. En el fondo se encuentra una constante: la supervivencia biológica del individuo.8

Malinowski considera que es posible elaborar una teoría de la conducta organizada. El hombre en su desenvolvimiento cultural tiene que organizar su conducta. El hombre se encuentra rodeado por un medio geográfico, instrumentos de trabajo y otros artefactos, al igual que ciertos bienes que le pertenecen. Su conducta, el uso de los instrumentos de trabajo y demás artefactos que le son necesarios para su subsistencia y la de todos los otros bienes que le pertenecen, se realiza a través de reglas técnicas y de normas de conducta, bien jurídicas, o morales. Esas diversas actividades que el hombre realiza desde que se levanta hasta cuando se acuesta, son actividades organizadas en torno a fines determinados, fines que son definibles, que se encuentran entrelazados; así como también, es determinable el efecto que dichas actividades producen en la comunidad o sociedad, consideradas como un todo: "Comprobaremos--dice Malinowski-- una vez más que dondequiera y en todo acto concreto el individuo puede satisfacer sus intereses o necesidades y llevar a cabo cualquier acción sólo dentro de grupos organizados y por medio de la organización de las actividades."[9]

En los planteamientos anteriores vemos que se trata de generalizar mediante el trabajo de campo o el trabajo experimental. Es decir, se trata de establecer las constantes en el desenvolvimiento de la conducta humana social. A la metodología utilizada en las formas de trabajo señalado se le objeta, el no tomar en cuenta una serie de variantes que se producen en cada grupo social. "Se han puesto en tela de juicio los intentos de cuantificación por medio de métodos estadísticos, sobre la base de que las mediciones estadísticas tienden a ocultar los motivos, actitudes y valores subyacentes, que se pretende que sean no-mensurables. Estos debates son paralelos con debates similares en sociología, y a veces coinciden con ellos, e implican también fenómenos de motivos, actitudes y valores..."[10]

En todo caso, lo que no se discute, incluso por parte de las teorías más individualistas, es que la sociedad inscribe en el individuo una serie de caracteres que determinan su existencia.[11]

Los estudios modernos acerca de la influencia de los medios de comunicación social han servido para evidenciar la gravitación en el individuo de formas y modos de pensar que determinan sus reacciones y sus preferencias. Esto ha hecho que hoy en día se considere el problema de la comunicación social como uno de los aspectos de mayor importancia en la actualidad. Los grandes medios de comunicación social, que permiten lanzar las ideas de un individuo o un grupo de personas a millones de individuos, representan uno de los temas contrales de la psicología social y de quienes se ocupan de la educación social. Es indudable que la moda, o las necesidades de consumo, incluso; el tono del lenguaje están estrechamente ligados a la comunicación. El hombre moderno vive poco tiempo en diálogo. El individuo que trabaja en la

fábrica tiene poco tiempo para la conversación. La vida
actual es un monólogo en el cual hablan solamente los medios de comunicación social: prensa-radio-televisión-ci-
ne, incluso el teatro, porque en este último también el
hombre es un mero espectador de tesis con las cuales él
no dialoga. En este sentido dice Simpson: "En el estudio de la opinión pública y propaganda, los psicólogos
sociales se han unido a los sociólogos y a los científi-
cos de la política para mostrar cuán difícil ha sido para
el público identificarse bajo condiciones de sociedad de
masa y cuán fácilmente puede lograrse que la propaganda
aparezca como verdadera. La propaganda ha sido estudia-
da no sólo como mecanismo político--una forma de manipu-
lar la opinión pública sobre problemas de política inte-
rior y exterior--sino también, como técnica económica para ganar un mercado para bienes, para mantenerlo y para
expandirlo. Aquí hay un campo abierto a la psicología
social de la publicidad, que incluye análisis de la forma en que se usa la 'libre elección' en el mercado como
clisé para ocultar lo que importa para la elección con-
trolada, debido al hecho de que los consumidores están
inundados de extravagante verborragia en pro de los productos."[12]

Nos encontramos evidentemente ante una situación sumamente grave: ¿qué queda del individuo? Se nos podría decir que nuestro planteamiento implica una interferencia
en la libertad de otros individuos y que nuestro análisis se fundamenta en un criterio que pretende ser impuesto, con lo cual a su vez se niega la libertad de la persona. Se nos podría argumentar por otro lado, que el individuo es libre de elegir aquello que se le presenta como bueno o de rechazarlo considerándolo como malo.

En realidad, no se trata de imponer un determinado criterio, se trata dentro de lo posible, de establecer los
mecanismos, que permitan que el hombre desenvuelva su capacidad creadora y que no se le utilice como instrumento.
Es indudable que el argumento que señala una supuesta libertad en el hombre de decidirse a aceptar o rechazar un
objeto que se le pretende imponer mediante la propaganda,
es totalmente falso. La verdad es que en el hombre se
logran fomentar tendencias y hábitos que forman parte de
lo que algunos han denominado muy bien una segunda naturaleza. No parecería absurdo, de acuerdo con eso, que
un árabe prefiera morir antes que comer carne de cerdo o
que se enferme y sufra un ataque de vómito cuando sepa,
que la ha comido, ignorándolo.

No se trata tampoco, de pretender destruir creencias y
costumbres. Se trata de racionalizar la vida del hombre.
En dicha tarea las ciencias tienen que desempeñar un papel fundamental en la superación de creencias y prejui-

cios deshumanizantes.

Permaneciendo en la referencia al problema de la influencia de las ideas imperantes en una sociedad, en los individuos que la forman, podemos citar el caso de la India. En este inmenso país del Asia, observamos que en medio del hambre de un pueblo, se mueven doscientos cincuenta millones de cabezas de ganado vacuno, fuentes de ricas proteínas que se desperdician y que podrían servir al propio pueblo de la India, y del mundo entero, para superar o mitigar en gran parte el problema del hambre.

Muchos hombres, destacados intelectuales y hombres comunes, no se explican dentro y fuera de Alemania, cómo fue posible una guerra como la conducida por los nazis. La destrucción en masa de 6 millones de judíos es una realidad que nos impone pensar sobre el problema de la utilización del individuo, mediante la propaganda y la elaboración de estructuras irracionales. La psicología social se ha ocupado de la consideración de esta cuestión: "Como parte del estudio de actitudes, se ha desarrollado en la psicología social una ciencia subsidiaria del prejuicio que trabaja en términos del concepto de esteriotipo (producto de estudios de opinión pública) y conceptos tales como proyección, racionalización y desplazamiento para explicar los comportamientos que implican prejuicio. El estudio del prejuicio ha vinculado necesariamente el estudio de la psicología social de grupos minoritarios por medio de la investigación de actitudes engendradas en miembros de estos grupos y de las mantenidas por el grupo mayoritario. Aquí han sido muy reveladores los estudios sobre antisemitismo, prejuicios antinegros, y los mecanismos que suscitan esas actitudes y sentimientos."[13]

De acuerdo con todo el desarrollo de nuestro análisis queremos señalar la siguiente conclusión: El hombre se nos presenta como una realidad biológica. Lo psíquico está determinado por los factores de orden biológico y también por los factores de carácter social. Existe un ciclo biopsíquico resultante de la relación entre lo biológico y lo psicológico del individuo, pero una comprensión correcta del individuo no es posible si la limitamos puramente a dicha relación. El ejemplo señalado de un árabe, que comiendo carne de cerdo sufre un ataque de vómito, nos pone de relieve, la importancia de lo social en las reacciones de los individuos. Existe pues, un condicionamiento de la libertad del hombre que no se queda en una determinación por parte de su naturaleza biológica. Existe un condicionamiento del hombre por parte de lo social, condicionamiento que se basa incluso en la propia realidad biológica, y puede afectarla. De manera que si queremos representar con un esquema esa relación diríamos que el individuo es como un ángulo en el cual se unen los factores sociales y biológicos que puntualizan su realidad psíquica.

La presencia de lo social en el hombre nos impulsa a pensar en la importancia de una contemplación multidisciplinaria del ser humano. Horkheimer, expresa la idea del condicionamiento social del ser humano en estas palabras: "Tanto ayer como hoy cuanto los hombres saben y el modo en que lo saben, desde acerca de sus autopistas, sus poblamientos y sus talleres, hasta sobre su amor y su miedo, está condicionado por su vida en común y por la organización del trabajo..." Horkheimer nos habla de un "tipo humano que produce este mundo..."[14] Cuando decimos que el mundo cambia, estamos señalando que el hombre cambia con él, como parte del proceso total de una realidad de la cual él forma parte.

La criminología como ciencia causal explicativa del delito ha penetrado la problemática del hombre y sabe de la naturaleza compleja del mismo. El filósofo y criminólogo argentino José Ingenieros, critica a la psicología (pensamos que se refiere el sabio argentino a la psicología individual) cuando señala, que ésta trata al hombre desde una sola perspectiva: "Los psicólogos suelen estudiar las condiciones intrínsecas del carácter, sin tomar en cuenta sus condiciones sociales..." Más adelante escribe: "Las anormalidades del carácter pueden ser congénitas o adquiridas, ora producto de una mala constitución bio-psíquica hereditaria, ora de una mala influencia educativa del medio social. Se puede nacer antisocial, por temperamento; se puede perder un buen temperamento por la mala educación, en cuyo caso, la degeneración del carácter es adquirida..."[15]

Lo expuesto por Ingenieros, plantea la necesidad científica de tratar al hombre desde una perspectiva dual. Su determinación por parte de lo social, nos impone un análisis filosófico que tome en cuenta, que el hombre no es propietario de una libertad absoluta. La coincidencia de Horkheimer con el filósofo latinoamericano es fundamental. En efecto, dice Horkheimer: "La filosofía era una meditación acerca del sujeto. Mientras éste pareció ser exclusivamente el yo individual, la psicología era la ciencia que estaba más ligada a las intenciones filosóficas; una vez que, en el idealismo alemán, el sujeto aprendió a concebirse ya no únicamente como individual, sino al mismo tiempo como la fuerza de los hombres, que, activos, enlazados mutuamente, están arrastrados y, sin embargo, hacen su propia historia, como sociedad, la sociología se ha convertido en la disciplina filosófica en sentido eminente..."[16]

La determinación por parte de lo social en la realidad vital del individuo ha conducido al determinismo social.[17] El determinismo social ha presentado una problemática que como lo señala Kwant no puede dejar de ser atendida: "Quien no conoce la fundamentación del determinismo social está expuesto al peligro de afirmar la libertad del hombre de una manera irresponsable."[18]

No se trata de adherirse en forma absoluta al determinismo social, se trata de la necesidad científica de atender esta dimensión del hombre para lograr la mejor comprensión del mismo. El tema de la libertad del hombre se encuentra inseparablemente ligado a su ser biológico y a su ser social. El hombre está determinado por ambas realidades.

FOOTNOTES

1. George Simpson, EL HOMBRE EN LA SOCIEDAD 80 (Trad. de Elizabeth Gelin), Paidos, Buenos Aires, 1970.

2. Werner Wolff, INTRODUCTION A LA PSICOLOGIA 8 (Trad. de Federico Pascal del Roncal), Fondo de Cultura Económica, México, D.F., 1963.

3. Ibid., pp. 27 y ss.

4. Ibid., p. 32.

5. Ibid., pp. 41 y ss.

6. Simpson, op. cit., p. 84.

7. Ibid., p. 85.

8. Bronislaw Malinowski, UNA TEORIA CIENTIFICA DE LA CULTURA 20 y 21 (Trad. de A.R. Cortazar), Sudamericana, Buenos Aires, 1970.

9. Ibid., pp. 59 y 60.

10. Simpson, op. cit., p. 86.

11. Ibid., pp. 86 y 87.

12. Ibid., pp. 88 y 89.

13. Simpson, op. cit., p. 89.

14. T. Adorno y M. Horkheimer, SOCIOLOGICA 18 y 19 (Trad. de Víctor Sánchez de Zavala), Taurus, Madrid, 1971.

15. José Ingeieros, CRIMINOLOGIA 80 y 81, Elmer, Buenos Aires, 1957.

16. Horkheimer, op. cit., p. 17.

17. Remy Kwant, FILOSOFIA SOCIAL 96, Carlos Lohlé, Buenos Aires, 1969.

18. Ibid., p. 96.

Man presents himself to us as a biological entity, he has instincts, senses, and a nervous system. His psychological aspects are determined by both biological and social factors; there exists a psychosomatic cycle which is the interplay between the psychological and biological aspects.

But man is, nevertheless, a social creature. Social psychology tries to explain the interaction of individuals in social groups. The basic constant common to all peoples is the biological survival of the individual. A further aspect of society, however, is that it inscribes onto the individual a series of characteristics that determine his existence. The communications media have served to inculcate modes of thinking that have determined his reactions and preferences. The tendencies and habits frequently become that which we describe as "second nature."

Man must be treated from a dual perspective. His social determination imposes on us a philosophical analysis that takes into account the fact that man does not have absolute liberty. Social determination of the reality of the individual has led to social determinism. The theme of liberty of man is inseparably connected with his biological being and his social being. Man is determined by both realities.

EQUAL WORTH OF PERSONS*

Robert Ginsberg

Equality is one of the great beacons for the understanding of social and legal systems. Ambiguities and variants in definition make much of the discussion pompous, trivial, or puzzling. Not all conceptions of equality are of equal worth. Sophisticated analysis can easily miss the <u>sense</u> of equality which men feel and act upon. Sketched here are guidelines for further thought on the concept: (1.) Equality is to be viewed qualitatively not quantitatively; it is a commitment to worth, not a measure of something observable. (2.) But the exact worth of people is open-ended; it cannot be definitively settled. (3.) While we must practically judge better and worse among persons, we must also operate with a presumption of their equal worth. (4.) Equality affirms the importance of human differentiation and laments uniformity. (5.) Equality as principle is goal-directed; it is to be realized in action rather than fully articulated in discourse. (6.) It cannot be analyzed adequately as an independent concept but must be seen in the context of associated open-ended concepts, including person, liberty, and happiness. (7.) Equality as commitment requires appropriate modes of action or "treatment," sometimes quantitatively equal, sometimes quantitatively unequal.

The equality of human beings is their identical worth as persons. But what is the exact worth of a person? <u>Inestimable.</u> That means that persons are worth more than anything (and more than any thing). No price can be set on a person, though a person has incalculable value, as Kant understood.[1] To recognize the equality of human beings is to recognize the preeminence of persons in a world of things; it is to draw the line between the human and the non-human. To insist on the equality of man is to affirm humanity.

All men are equal is another way of saying all men are men. "Men" should be understood in this context to apply to the human community, regardless of sex, age, or race. What good is equality if it is reducible to humanity and results in a tautology? The significance is that equality does not exclude anybody. We are all human. Women and children too. No one is left out of the valuable status of persons. Equality is membership. To be equally members of the human race is to have all the rights and privileges that appertain thereto. The members differ individually, they lack uniformity, and there is little observable exterior similarity among them. Since equality is a non-excluding principle, then it follows that the right to recognition as a human is not subject to forfeit. Equality is the inalienability and undeniability of being human. In drawing its line between the human and non-human, equality draws us all within its circle, and this draws our attention to one another.

As we examine one another we make different value judgments, for people are different. No person is the same as another person. Indeed, that is one of the things to be insisted upon in the proper sense of person: Individual uniqueness. Moreover, we are likely to judge our own character or accomplishments more favorably than others even if these were to be identical, for the simple reason that we have a more intimate knowledge of ourselves. Hobbes analyzes this as the problem of diffidence in the condition of equality that marks the State of Nature and that leads to conflict.[2] It now appears that if all men are human, then to be human is to differ in fact and in judgment from everyone else. Equality perforce would be impossible among human persons.

Yet each is valuable _as_ different and the final value of each cannot be fixed. Equality, then, is a status of value without limit. One may have more wealth than another but that doesn't change one's worth as a person. Master and servant, governor and subject, policeman and criminal are not equal in their functions or social contributions, but they are equal in their persons.

A man may be a better lawyer or logician than another. So in every field of endeavor we judge excellence and favor it. That is not to say that the better lawyer is a better person. But may we not also determine who is a better person? Is not the virtuous victim better than his cruel assassin? For practical purposes we do much such judgments. Identifying good persons and evil ones provides guidelines for emulating the one kind and avoiding the other.

It seems essential to an harmonious society and an ethical life to abandon equality in judging persons. If some men are not better than others, than I cannot be expected to better myself. If no men are worse than others then I can do no wrong. The commitment to the belief <u>some people are better than others</u> may be a prerequisite for human responsibility because of these implications for the self.

But what is the exact worth of another human? Or even of oneself? We cannot say. There are four inhibitions to accurate judgment of the inequality of persons. (1.) We never have all the information. For the most part we are on the outside looking at the deeds and words of others. The biographer and the family member may see the inner life but not all of it. No person can be known fully by another. We are even puzzled by and unsure of the one person we know best by daily inner experience. (2.) People may improve. Great sinners have repented and become great saints. Part of what we recognize to constitute a person is this capacity for reversal and radical change. The books are not closed on anyone alive. (3.) Judging the person by his deeds may not be fair. An evildoer need not be an evil person. Context and intention must be known for deeds to be understood. Socrates warns that no one willingly does evil.[3] (4.) The source of our judgments of persons is a person. We are engaged in being what we are engaged in judging. We do not hold to equal standards, though there may be general agreement about what is good or bad. If persons differ so do their standards of evaluating persons. As there are reversals in life so there may be reversals in judgment.

These inhibitions do not suffice to renounce moral judgment of persons but they render all such determinations tentative. Though for practical purposes we have to approve or condemn, the final worth of a human being remains an open question. Some people may be worth more than others in the long run, but everyone is equal in that their worth is inestimable. If all men are unequal, it would only be in God's eyes.

Hence the assassin is equal to the victim as person, and should be treated appropriately with due process and a respect for his life. The punishment inflicted upon him may constrain the person but it should not otherwise deny him use of his capabilities for a full life. Once we treat some men as beasts, say by executing them, excluding them from human worth, then we are confirming their own deeds which we deplore.

Though we must judge men for their actions and consequently treat them unequally at times, yet we should allow the equal possibility of improvement and fulfillment. Equality is a principle of faith in the potentiality of the lowliest, of doubt in the supremacy of the mightiest, of tolerance for the eccentricity of others, of hope for the improvement of ourselves. According to this principle people are not to be treated below the level of humans (nor above it). But how exactly are humans to be treated?

We are in the process of finding out, and that requires a certain amount of trying out. A fairly reliable list of ways not to treat humans has been compiled: Don't kill, maim, torture, injure, etc. The positive treatment is roughly: Assist human beings so they may develop and fulfill their capabilities as persons (this is what "equality of opportunity" comes to). Again these capabilities differ, whence inequality. But again everyone has such capabilities, whence equality. There is a circularity to the stipulation for positive treatment, for it assumes we know what the distinctive capabilities of a person are. We come to know more of what persons can be as we treat all men as persons. This process of approximation and discovery is the adventure of becoming fully human. Equality, then, should be understood as the final goal toward which social organization aims. It is not a simple principle definable in advance and mechanically applied to social interaction, but it points to the optimum outcome of all human endeavor.

Seen in this light, equality blends with other grand concepts of human purpose, notably liberty and happiness. Equality calls for the maximization of the liberty of each individual to develop themselves within an association of individuals. Everyone is equally to be free, though their freedoms will differ as they themselves differ as persons. This is a far cry from the deprivation of individual liberty by the imposition of a thoroughgoing quantitative equality. Equal does not mean uniform. It does mean differences may have the same worth. If the fulfillment of the individual is happiness, then freedom is indispensable for happiness. Equality affirms that everyone may be happy. Each is worthy of fulfillment as a human being.

Liberty is often thought of merely as freedom from restraint, equality as quantitative sameness, and happiness as euphoria. These senses wreak havoc with a theory of purposive society. What counts is freedom to become oneself as a person, equality of personal

worth, and happiness as fulfillment of person. The
three notions overlap. Equality is only one way of
talking about what we are aiming at by other ideas.
There is no profound concept of equality independent
of these other concerns of person and purpose.

Equality has proved most troublesome when interpreted
as a quantitative measure rather than as a qualitative
value. This leads to hasty dismissal of it as a principle, for in measurable ways people are quite different. It also leads to superficial application of
equality as mode of treatment, because giving people
the same salary for the same work might not be treating them as equal persons. Marx points out that this
is a bourgeois notion of the right to equality, whereas
the truly equal treatment of the more needy worker requires a greater salary than the co-worker.[4]

Equality understood as qualitative principle pointing
to worth obliges us to adopt appropriate actions. Concept and treatment are intertwined. "Equal treatment"
is remarkably ambiguous. It may simply mean quantitatively identical treating regardless of individuals.
In this sense it can be a method of action rather than
a principle about persons. "Equal treatment" may also
mean treating people as equal in worth, which often
requires quantitatively different applications in order
to respect the special differences of persons. Bus
fares are at a standard level for the general populace,
but special rates apply for the elderly, children, the
handicapped, students, etc.

The appropriate way of treating people as equals is
often by equal quantitative methods, closing our eyes
in certain areas, especially in the criminal law, to
the special status of any persons. It would be an
enormous error to be perpetually blind to such considerations and to enact equality primarily in this
uniform manner. The fundamental sense of equal treatment should be the respect of the individual person,
and we must be willing to open our eyes to the
individual.

It is important to see how these two procedures of
treatment balance one another in the operations of law.
What initially may be set down as general rules regardless of individuals yet comes to be softened by judges,
juries, and arbitrators in terms of the specific persons involved. We are inclined to say that here justice is tempered by mercy, but it would be better to
conceive of this tempering as part of the justice.
The root of law is the person.

There is another important way of speaking of equal treatment: Treatment of people as _if_ they were equal. This approach avoids the value commitment to persons while at the same time sets a strategy for acting toward them in terms of their worth. "Persons may not be equal in worth, but let us act on the supposition that they are." This stipulation is useful for those skeptical of the more fundamental commitment to equality. But the difference in formulation leads to no difference in practice, and therefore they pragmatically are the same. Treating people as if they were qualitatively equal is really treating people as qualitatively equal.

Natural rights theories, as well as other brands of social thought, have celebrated equality, liberty, and happiness as the central purposive notions, usually termed rights, for social organization and legal systems. No more eloquent expression of them can be found than this passage, now entering its two-hundredth year and still worthy of our reflection and respect:

> We hold these Truths to be self-evident, that all Men are created equal, that they are endowed by their Creator with certain unalienable Rights, that among these are Life, Liberty, and the Pursuit of Happiness.

FOOTNOTES

* My study of equality has been assisted by a Fellowship from the National Endowment for the Humanities and a Grant from the Liberal Arts Central Fund for Research, The Pennsylvania State University.

1. Immanuel Kant, GRUNDLEGUNG ZUR METAPHYSIK DER SITTEN (1785), Section II.

2. Thomas Hobbes, LEVIATHAN (1651), Pt. I, ch. xiii.

3. Cf. Aristotle, NICOMACHEAN ETHICS, Bk. VII, 1145^b.

4. Karl Marx, RANDGLOSSEN ZUM PROGRAMM DER DEUTSCHEN ARBEITERPARTEI (1875), Pt. I.

THE EXHAUSTION OF THE IDEALS OF FREEDOM AND EQUALITY IN THE UNITED STATES

Lester J. Mazor

The ideals of freedom and equality, in the particular form they have taken in the thought and culture of the United States, were born in tension with each other and with their historical context. As the contradictions inherent in each of these ideals and in their relation to each other increasingly have become apparent, desperation more and more appears as the dominant tone in the voices which speak of them. What in the character of the initial statement of these ideals and in their historical unfolding has brought political thought in the United States to the brink of the abyss? What capacity is there to create a bridge to the future by giving new expression to the meaning of freedom and equality?

The limits inherent in these democratic ideals were powerfully stated at an early date by Alexis de Tocqueville. He saw that the concept of civil liberty, as the embodiment of the freedom ideal in this culture, was bounded by a sharp separation drawn between political and social existence, between status as a legal subject and in the daily relations of work and domestic life. Thus, from the standpoint of civil liberty, one was free while living under stern social disfavor, prejudice and persecution, so long as these were not the immediate consequence of legal and, therefore, direct governmental sanction. The conceptual separation of the public and private spheres served both as a cherished guarantee of the realm of freedom and, simultaneously, as the device through which cultural ghettoes and prisons of the spirit were erected. So long as in legal form one was protected in the right to own property, and those who did were perceived to have a measure of control over their own fate and, especially, over their subordinates, this version of the ideal of freedom appeared intact.

The notion that liberty is freedom was common to other lands. The extent of the acceptance of the ideal of equality, however, was in Tocqueville's view the unique contribution of this portion of North America to the liberal ethos. In this formulation, equality was defined by the absence of an hereditary nobility and, more generally, as equality of opportunity, the chance to achieve the liberty of a property owner in a fluid social structure under the protection of general legal norms. The weakness in this conception he saw in the tendency of a culture so founded to permit centralization of power and ultimately to welcome stratification by status. This potential was inherent in the limited capacity of this version of equality to overcome the barriers of class, race, ethnicity, wealth, sex, age and other factors given discriminatory weight in the society.

In the historical experience of the people of the United States, these limits frequently have been approached. No sooner had a national government been formed than the people found it necessary to demand a Bill of Rights as a limitation upon it. But this document of civil liberty proved incapable of restraining the wave of political repression of aliens and critics of the government which followed as those in power sought to consolidate their position. Nor as the first century of the country's history proceeded were the provisions of the state and federal constitutions sufficient to prevent severe deprivations—in the genocide of the native tribes; in the continuation, expansion, intensification and extension of involuntary servitude; in the imposition of forced service in the military upon those who could not purchase exemption; and in the denial of political rights to the majority of the population. What protection was civil liberty against the exploitation of ethnic minorities, the imposition of inhumane working conditions upon adult and child alike, the restriction of political action and expression to the confines of the established political parties, and from the responsibility for financing through taxes filibustering expeditions, military invasions and the extension of political and economic hegemony over other peoples? Even the ameliorating measures adopted only after much struggle and bloodshed brought in their train greater bureaucracy and new heights of taxation, at least for those of modest earnings.

In addition to the contradictions within the ideal of civil liberty and equality of opportunity, there was also the deep constraint which followed from conceiving freedom and equality as necessarily in opposition. Any extension of the domain of liberty was perceived as a further limitation of the possibility of realizing equality. Since freedom in its liberal version (liberty) implies possession and exclusivity, it entailed at least a partial denial of access by others. Even more powerfully did it seem clear that an increase in the degree of equality would have to come at the expense of a measure of freedom. Equality could only be seen as an enforced homogeniety obtained at the expense of room for diversity and personal fulfillment.

For more than a century the contradictions within and between the democratic ideals of civil liberty and equality of opportunity were disguised by a variety of masks. Frequent ceremonies of public worship of their names and forms were offered as a substitute for their substance. Denials of their fulfillment were excused as imperfections to be reformed, as grist for the eternally slow wheels of justice, as the subject for continuing progressive effort, or justified as the appropriate chastisement to be visited upon the undeserving, because they did not share in the common heritage, speak the common tongue or willingly accept the portion allotted to them. When promises of future performance grew dry in the mouths of orators, the cry of patriotism could be drawn upon as the ultimate reserve. Higher levels of apology were, as usual in such matters, entrusted to the priesthood, whose secularization in the universities did nothing to impede the fulfillment of their historic mission. Official philosophy endorsed the system of power, warning that liberty should not be construed to permit license and that equality should not be corrupted into that levelling tendency which had destroyed civic virtue elsewhere. The more prominent the examples of an expansive notion of human freedom and equality in other parts of the world, the more shrill was defense of the limits given to these ideals by the tradition of the country. The loss of their vitality was signalled by the fact that the point of reference grew ever more remote as time passed. Freedom in its name of liberty became a fusty phrase in the mouth of a colonial patriot; equality a word in the antique rhetoric of a bearded president whose every feature of origin, of frame and of mien marked him as a mythical figure of a golden age long passed.

A new freedom might be depreciated into a new deal, poverty made the subject of an internal war, yet the conceptual constellation remains fundamentally intact. Only a few, and they quickly branded as deviants--

radicals or reactionaries beyond the pale of thought to
be given serious consideration--might suggest that the
ideals were not only inadequately fulfilled but were
insufficient in their very formulation. As a second
centennial approached, the leaders of opinion instead
announced the end of basic controversy, asserted that
the nation's past was a tale of consensus, and defined
both society and the political order as areas of general agreement.

   Yet a turbulent decade stirred anxiety. Against the
background of a period of social disquiet and profound
questioning there was a reawakening of interest in
fundamental issues of the social, political and economic order. Books and articles of a kind which had
long been thought to be appropriate only to a less
sophisticated time, an early period of inchoate consciousness, suddenly appeared. Their authors appear
to have been inspired to return to the discussion of
fundamental questions by the confusion evident in the
argument over central political issues; their critics
welcomed the effort, even when in disagreement, for
the opportunity it represented to revive the ancient
dialogue. In these recent works the fate of the
liberal ideals of liberty and equality is revealed.

   Foremost, without doubt, of the works of this late
phase, is the book by John Rawls, A THEORY OF JUSTICE.
Despite the author's patient disclaimer of novelty,
this book immediately became an idol, dominating discussion even among those who lacked the patience to
read it. Thoroughly within the tradition, Rawls places
freedom and equality at odds with each other. Having
asserted this antithesis, he follows the conventional
view by placing freedom in ascendancy. Freedom for
him is liberty, whose definition, when it is not merely
vague, is confined to the kind of political and civil
rights central to the seventeenth century political
movements which the eighteenth century spokesmen from
which Rawls' position is derived thought important.
The concept of liberty represents a minimalist view,
a fear that its priority would threaten to destroy the
equality principle were its definition more expansive.
Equality, specifically, distributive justice, is the
real subject of the book. In the guise of a theory of
justice we are given instead a theory of injustice, a
defense of inequality. The task which the book sets
out to accomplish and almost fulfills is to provide an
explanation to those who must receive less from their
lives why they should willingly accept the fact that
others receive more. The necessity that there be a
differential is an almost unspoken assumption,

revealed only in the suggestion that inequality is given in nature. Thus is the class basis of society reified.

The significance of this book is not in its content, for there is little that is new in it and its main contribution consists in its stark confession of the assumptions on which such a theory would have to be founded, a kind of reductio ad absurdum of the social contract doctrine. The significance of the book lies instead in its reception. The enthusiasm with which it was welcomed in both scholarly and popular press, the continued attention which it has received, the inclination to make it the subject of glosses, all betoken some deep crisis of faith in the doctrine it proclaims. The version of liberty it embraces, the claim it asserts to the priority of liberty against equality, the endorsement of equality of opportunity--these come together as an apology for the welfare state, at precisely the moment when the welfare state as an attempt to forestall a more basic restructuring of social relations has foundered.

Rawls' book, however, is at least a valiant attempt to maintain the validity and primacy of the values of freedom and equality. Despite its failure to transcend the contradictions in their traditional liberal formulation, it is only a symptom of the failure of nerve. In other authors who share the same time and place, the game has been abandoned altogether. Robert Nozick, for instance, rejects human experience even more radically than does Rawls. Where Rawls' theory is merely ahistorical, Nozick conjures a world which is in no way connected to human experience. This is a world in which there is property without the state, in which people have entitlements without explanation of their source.

Whereas Rawls pushes at the limits of the capacity of the liberal version of the freedom and equality ideals in a vain effort to shore up a tottering political and economic structure, Nozick regresses to a more primitive state of liberal thought, in which it is possible only to assert that people have rights and that social justice is a matter solely of the rules for the acquisition, transfer and recapture of entitlements, his euphemism for property. In his effort to escape the weaknesses of the contract position exposed by Rawls' stark exposition, Nozick accepts uncritically a view of human beings as possessive individuals, each a composite of impenetrable desires and rational calculation. Since historicity is only a formal criterion for Nozick, he never encounters the actual political experience of humanity; by the same token, his empty concept of utopia is incapable of pointing a direction toward the good society. If Rawls at bottom is an apologist for the welfare state, Nozick offers ammunition to those whose strategy in the face of the collapse of the legitimacy of the liberal order is to cling to what they have, on the bare claim of prior possession.

Both freedom and equality have little content in Nozick's account. Although his book purports to embrace a philosophy of freedom, in particular, libertarianism, his description of this freedom never achieves concreteness; it remains on the level of such abstractions as "rights" and "entitlements." His treatment of equality is even more severe, however. He finds the concept of distributive justice essentially meaningless, ignoring injustice in the pattern of the distribution of goods. Once a title has been examined and found to be without flaw, he rests his case; there are no further questions. The effective abandonment of equality as an ideal through this maneuver suits well the call of other conservative social theorists, reacting to the evident strain upon modern democratic ideals. Robert Nisbet in a recent article urges that we draw back from the pursuit of equality. In an argument that can only be read as a final surrender of the claim of modern political thought to universal benefit, Nisbet sees that each attempt to rectify inequality leads to the recognition of further levels of inequality. Horrified at the prospect of a thoroughly unstratified society, in which domination would lose its place and hierarchy be dissolved, and seeing no logical stopping place for the egalitarian tendency, he longs for the restoration of traditional society.

While Nisbet prepares to jettison equality, his compatriot Robert Heilbroner has embraced authoritarianism for similar reasons. Scouting the future, Heilbroner, who had long been optimistic that the rest of the world might be developed in the image of the United States, now sees instead the prospect of a forcible redistribution that might undo the entitlements of the rich Western nations. Rather than accede to such a fate, he is willing, as de Tocqueville forsaw those like him would be, to cast off democratic ideals in their entirety.

But it has been left to a lawyer utterly to reach bottom. Reviewing three epochs of law in the United States, Grant Gilmore moves remorselessly to the conclusion that the dream of creating a just society of freedom and equality through law has been a false hope.

Not content to stop with a refutation of the capacity of law to produce the good life, however, Gilmore goes on to assert that what has been revealed as empty is the enlightenment belief that it is possible to find pattern in social existence, to have a science of man. Intensely aware of the shambles of contemporary legal thought, sensitive to the exhaustion of the concepts of liberty and legal equality, yet apparently restrained by parochialism from a deep enough awareness of other perspectives to see any way out, Gilmore proclaims his nihilism. In saluting a predecessor for having the nihilistic courage to give up writing, he implies that not only his literary life but that of the whole world has ended--that the rest of us have nothing to say either.

The odor of approaching death which this body of thought exudes indeed would be suffocating were it not for the breath of vitality coming from other directions. For in movements of political action, in the cause of blacks, women, children, homosexuals, the aged and all those who experience the sense of injustice, a new version of the concepts of freedom and equality is emerging. Under the name of liberation freedom is finding a more spacious home, and one which welcomes equality as a needed and desired companion, rather than as a competitor. Equality too, as seen in the frame of liberation, demands not mere equality of opportunity, but an equality of condition which does not accept the colorless homogenity and oppressive conformity that Nisbet fears.

The philosophic exploration of the concept of liberation is only barely begun. The dimensions of freedom, founded not in possession and exclusivity, but in openness and connection are yet in an exploratory stage. The means of sharing that freedom in a global ecology are only beginning to be articulated. But it ought to be clear enough that the philosophic challenge is not to be found at the graveside of the dying concepts of civil liberty and formal equality.

BIBLIOGRAPHY

1. John Rawls, A THEORY OF JUSTICE, 1971.

2. Robert Nozick, ANARCHY, STATE, AND UTOPIA, 1974.

3. Robert Nisbet, "The New Despotism," in Commentary, June, 1975.

4. Robert Heilbroner, AN INQUIRY INTO THE HUMAN PROSPECT, 1974.

5. Grant Gilmore, "The Age of Anxiety," in Yale Law Journal, April, 1975.

6. Herbert Marcuse, AN ESSAY ON LIBERATION, 1969.

BROWN v. BOARD OF EDUCATION AND EDUCATIONAL EQUALITY

Klaus H. Heberle

The establishment and maintenance of a democratic society in a racially heterogeneous population of which one part were at one time slaves has posed the problem of equality in a particular form for the United States. Our experience is of interest because the ideological commitment to human equality has always had to compromise with the social and political reality of racial prejudice--a compromise made necessary in part by our understanding of the principles of free limited democratic government. (Our experience, incidentally, is that "merely social discriminations" arising out of racial differences do give rise to "conduct denying or restricting human rights" sometimes with legal sanction attached. See the paper of Luis Recaséns-Siches in this volume.)

In this paper I want to touch on some of the problems arising out of the Supreme Court decision in Brown v. Board of Education[1] and subsequent attempts to define and actualize equality in American education. The problems are instructive because (1) They indicate the limits of modern empirical social science in guiding and evaluating public policy designed to achieve equality and (2) They indicate that the real democratic tension is not between liberty and equality but over the meaning of equality.

In Brown v. Board of Education the Supreme Court held that state imposed segregation by race in public schools denies equal protection of the laws in violation of the Fourteenth Amendment of the Constitution. One year later, in Brown II[2] the Court directed school districts operating racially segregated schools to move "with all deliberate speed" to free their schools of racial discrimination. In the intervening twenty years the courts and school districts have struggled to define and implement what was required of them by the Brown decisions. The task was not made easier by Chief Justice Warren's "laconic"[3] opinion.

On one level it appeared that the state might not by legislation or administration classify its population along racial lines. On that basis the Court in the decade following Brown held invalid every form of state compelled racial segregation to come before it, usually in unexplained per curiam decisions citing Brown. But the prohibition against racial classification has caused some trouble in cases where the state has attempted to confer a benefit on Negroes--usually in an attempt to undo what were perceived to be the effects of past discriminatory practices--and whites have complained of denial of equal protection of the laws.[4] The problem was posed in classic form (but not solved) in the De Funis[5] case involving preferred entrance quotas for Negroes in a state supported law school. The problem is briefly canvassed by D. D. Raphael in this volume and was the subject of several papers and lengthy discussion at the Third Annual meeting of the American Section of this Association in the fall of 1974.

I would only comment here that Raphael's formulation that "Equal opportunity means ignoring artificial inequalities but following natural inequalities"[6] is adequate as a starting point but leaves us with the difficult task of disentangling the artificial from the natural inequalities. A brief example may illustrate the problem. Take two children, one born into a middle class family, one born into a poor family. The former receives proper physical and mental nurture and enters school knowing numbers, colors and the alphabet. The latter is not properly nurtured and enters school without knowing numbers, colors or the alphabet and because of improper nourishment is in poor physical health, has a short attention span, etc. Their respective I.Q.s are 110 and 90. They are clearly unequal in terms of their ability to perform the tasks expected of first graders--just as the applicants in the De Funis case were unequal in their capacity to perform the tasks expected of entering law students. But how do you disentangle the artificial and the natural inequalities between the two children so as to be able to ignore the former and follow the latter? Analysis of this problem and the debates it has engendered lead to the proposition that the real tension in democratic societies is not between liberty and equality but over the question of defining and separating natural from artificial inequalities among human beings.

The other element in the Brown decision was the Court's concern not with race relations in general but with equality in education.

> In these days, it is doubtful that any child may
> reasonably be expected to succeed in life if he
> is denied the opportunity of an education. Such
> an opportunity, when the state has undertaken to
> provide it, is a right which must be made available to all on equal terms.

But racially segregated education is inherently unequal because

> Segregation of white and colored children in public schools has a detrimental effect upon the colored children. The impact is greater when it has the sanction of law; for the policy of separating the races is usually interpreted as denoting the inferiority of the Negro group. A sense of inferiority affects the motivation of a child to learn. Segregation with the sanction of law, therefore, has a tendency to (retard) the educational and mental development of Negro children and to deprive them of some of the benefits they would receive in a racially integrated school system.[7]

The analysis is based on an assessment of the effect of Negro children's perception of racial separation on their self-concept and their motivation to learn. The effect is said to be "greater" if the separation has the sanction of law. It is merely a matter of degree whether the separation is caused by state law or by other factors such as residential separation. It follows that equality in public education can be provided only by establishing and maintaining racially heterogeneous schools.

This analysis rested on the findings of research in social and educational psychology.[8] The intervening twenty years have provided an instructive example of the interaction of law and social science in the attempt to define and deal with the problems arising out of implementation of the 1954 decision.

The <u>Brown</u> decision and its implementation were and are matters of great public controversy, leading to attempts to evaluate. Evaluation requires standards. The decision suggests standards. If racial separation fosters feelings of inferiority which in turn hinder learning, then racial integration should lessen the feelings of inferiority--improve the children's self-concept--and learning should improve. The inequality between achievement of white and Negro children in the schools should disapper. That was the expectation--the standard. Innumerable studies[9] have been done to determine whether or not the <u>Brown</u> rationale is valid and whether expectations based on it are being met. The results are mixed. They neither support nor refute the anlysis in <u>Brown</u>. Nor can it be said without equivocation that the standard implicit in <u>Brown</u> is or is not being achieved by the policies that have been adopted.

The reasons for the equivocal character of the results of this massive research effort are instructive for the concerns of this conference. They illustrate the inherent limits of empirical social science research as a means of evaluating policy designed to implement equality They indicate that such evaluation presupposes a resolution of the question of the meaning of equality, a task that requires the definition and disentangling of natural from artificial inequalities, a task that has not been achieved.

POINT ONE

The <u>Brown</u> decision rested on the assumption of a causal relationship between segregation, low self-concept, low motivation and low rate of academic ahievement. Much research effort has been devoted to determining what kinds of school situations will improve the self-concept of children.[10] It has become apparent from these efforts that there is little agreement in the "scientific community" on how to measure self-concept and for that matter what self-concept is. Ruth Wylie, after an exhaustive review of the work in this field, points out that there is no generally accepted operationally defined conception of self or of self-concept and that the studies which attempt to measure self-concept have used instruments "which have been used only once or a few times and are completely unvalidated for their purpose".[11]

POINT TWO

The Coleman Report[12] presented as its major finding that the attributes of fellow students derived from their social class status are a major determinant of educational achievement. In other words the dominant social class structure of a school has more to do with the educational achievement of all students in the school than the racial mix of the school. But as Weinberg points out, "no (statistical) technique to isolate the effects of social class on performance in interracial schools" has been developed.[13] Because most Negroes are poorer than most whites, we do not know whether or to what extent improvement in performance, when it occurs, is due to racial integration (the <u>Brown</u> rationale) or to socio-economic integration and only incidentally to racial integration.

POINT THREE

The Brown decision provided no standard in terms of which to judge whether or not a school system had been freed of racial discrimination. The resulting debate over implementation led to the distinction between desegregation and integration. "Desegregation is achieved by simply ending segregation and bringing blacks and whites together; it implies nothing about the quality of the interracial interaction. Integration involves... positive intergroup contacts, cross-racial acceptance, and equal dignity and access to resources for both racial groups."[14] The research that has been done by and large does not pay attention to this distinction--largely, one suspects, because the latter conditions are in fact too rarely achieved to supply a data base sufficient for drawing any conclusions about the effects of such conditions on achievement.

POINT FOUR

Underlying the Brown analysis is the assumption of natural equality of whites and Negroes with respect to educability. Arthur Jensen's work at least raises the suspicion that that assumption is not well founded.[15] He concludes that "the most tenable hypotheses...is that genetic (natural) as well as environmental (artificial) differences are involved in the average disparity between American Negroes and whites in intelligence and educability...All the major facts would seem to be comprehended quite well by the hypotheses that something between one half and three-fourths of the average I.Q. difference between American Negroes and whites is attributable to genetic factors and the remainder to environmental factors and their interaction with genetic differences."[16] Jensen's work has of course not stood unchallenged,[17] but the root of the difference between the parties to that debate is the problem of defining and separating natural and artificial differences.

POINT FIVE

The standard which has been accepted for the evaluation of the policy of racial integration is equality of results, not for individuals but for the racial groups. The apparent failure to achieve that result through integration of schools has led to the proposition that the expectations of the effect of schools was too high-- that children of minority groups come to school with serious deprivations and to expect the integrated school to rectify the deprivation is expecting too much. True educational equality calls for society to intervene earlier in the child's rearing and nurture to insure that children come to the schools equally equipped--to insure that artificial differences due to deficient early childhood environments not mask the natural abilities of the children. Generally accepted theories of child development place the crucial stage for mental development in the period between 6 months and three years.[18] Following this line of analysis we do encounter a tension between liberty and equality. But it is not between the liberty and equality of the children. It is between the liberty of parents to raise their children free of community intervention and policy to insure equality for the children.[19] In this form the drive for equality of education comes into tension with the traditions of American society which regard the family as the basic social unit--a tension that was given classic expression in President Nixon's veto of the 1971 Office of Economic Opportunity continuation bill which contained provisions for a comprehensive program for child development.

Other problems are involved but lack of space prevents their elucidation here. What is clear is that the social sciences are still very limited in the kinds of assertions they can make with any degree of certainty about matters of disputed public policy as it applies to education. It is probably that this limit is inherent in empirical social scientific investigation. Points One, Two and Three, above, are merely illustrative. Gordon Foster has summarized the state of the art in the field of our concern as follows: "Research on the cognitive and affective results of desegregation will and should continue, but it may never really tell us much except what we want to hear."[20]

Points Four and Five indicate that the real problem is the philosophic problem of the meaning of equality and the definition and disentangling of natural and artificial inequalities. Most fundamentally we are faced with competing theories of the nature of man and his relationship to society,[21] a realm of analysis in which empirically limited social sciences are of little help.

FOOTNOTES

1. 347 U.S. 483, 1954.

2. 349 U.S. 294, 1955.

3. Alexander Bickel, THE SUPREME COURT AND THE IDEA OF PROGRESS 118.

4. Ibid., pp. 118-119.

5. De Funis v. Odegaard, 416 U.S. 312, 1974.

6. D. D. Raphael, "Tensions Between The Goals of Equality and Freedom," this volume.

7. Brown v. Board of Education, 347 U.S. 483, 1954.

8. Ibid., footnote 11.

9. Meyer Weinberg, DESEGREGATION RESEARCH, 2nd ed. summarizes some fifty major projects completed prior to 1969.

10. Ibid., ch. 3.

11. Ruth C. Wylie, THE SELF-CONCEPT, rev. ed. ch. 7 and p. 324.

12. James S. Coleman et al., EQUALITY OF EDUCATIONAL OPPORTUNITY.

13. Weinberg, op. cit., p. 27.

14. Thomas F. Pettigrew et al., "Busing: A Review of the Evidence," in Nicholaus Mills, ed., THE GREAT SCHOOL BUS CONTROVERSY 129.

15. Arthur Jensen, EDUCABILITY AND GROUP DIFFERENCES.

16. Ibid., p. 363.

17. See critiques in "Environment, Heredity and Intelligence," Harvard Educational Review.

18. Benjamin S. Bloom, STABILITY AND CHANGE IN HUMAN CHARACTERISTICS; Jean Piaget, THE ORIGIN OF INTELLIGENCE IN CHILDREN.

19. Pierce v. Society of Sisters, 268 U.S. 510.

20. Gordon Foster, "Desegregating Urban Schools: A Review of Techniques," 43 Harvard Educational Review 5, 30-31.

21. E. G. John Locke, SECOND TREATISE and J. J. Rousseau, A DISCOURSE ON THE ORIGIN OF INEQUALITY.

LOS DERECHOS HUMANOS Y EL BIEN COMUN

Jorge Iván Hübner G.

Nos proponemos poner de manifiesto en esta ponencia el hecho de que la plena aplicación práctica de los derechos humanos, tal como hoy los concibe la doctrina y como se enuncian en la Declaración Universal de las Naciones Unidas, enmarcados dentro de una norma social superior de carácter espiritual y moral, coincide con la realización de lo que clásicamente se ha llamado el Bien Común, o sea, con el bienestar general de la sociedad. Más aún, podemos afirmar que el mensaje de estas prerrogativas fundamentales del hombre, que ha ido madurando a través de los tiempos, se identifica, substancialmente, con el ideario de toda doctrina de redención social.

Debemos comenzar afirmando con optimismo que estamos viviendo en el siglo de los derechos humanos. A primera vista, esta aseveración podría parecer extraña, desconcertante o inexacta, si nos atenemos a la situación real que se observa notoriamente en esta materia en la mayor parte del orbe. La geografía del despotismo, como sistema, sigue siendo mucho más amplia y extensa que la "geografía de la libertad".[1] Pero, es necesario aclarar que, al formular aquella tesis, no nos referimos a la observancia efectiva de estos atributos, sino a la viva y creciente toma de conciencia de la ineludible necesidad de los derechos humanos, que se advierte, con mayor o menor intensidad, en el mundo entero.

La historia de la humanidad en este campo ha constituido un notable testimonio del progreso de la conciencia moral. Este ejemplo demuestra palmariamente que no sólo progresan las ciencias naturales, sino también las ciencias morales, en el descubrimiento de las leyes que están en la naturaleza misma de las cosas. Si comparamos sólo la situación que existe actualmente en el mundo, en lo que respecta a estos derechos, con la que imperaba hace cien años, y, más aun, hace uno o dos milenios, veremos que, en general, el adelanto ha sido inmenso. Pero, mucho más notorio resulta este contraste si, después de leer algunas de las grandes declaraciones actuales, nos remontamos a la antigüedad greco-latino o a las más brillantes civilizaciones de los pueblos del Oriente, en las que nos encontraremos con que ni siquiera se formuló el concepto de los derechos humanos. Fue necesaria la enseñanza del cristianismo y una evolución varias veces secular, para que el hombre tomara conciencia de sus propias prerrogativas, que no vino a inventar en nuestro tiempo, sino a descubrirlas dentro ese ámbito tan próximo, y a veces tan lejano, que es su propia conciencia. Allí estaban presentes, como lejanas constelaciones, que esperaron millares de años en el fondo del espacio que el ojo humano, con el progreso de los tiempos, llegara a detectarlas, proyectando una nueva luz en el camino que conduce al Bien Común de la Sociedad.

En este proceso histórico, que aun no ha llegado a su plena culminación, podemos distinguir cuatro grandes etapas:

(1) <u>Primera etapa.</u> <u>La aparición del concepto de los derechos fundamentales de la persona humana.</u>

La concepción de estos atributos es un acontecimiento nuevo en la larga trayectoria del hombre sobre la tierra. Es cierto que la semilla de los derechos humanos surge en tiempos remotos, en el Libro de los Libros, cuando el relato bíblico nos revela que Dios creó al hombre a su imagen y semejanza. Un ser de tan eminente naturaleza, dotado de un alma que es reflejo del propio Ser Supremo, tiene una dignidad que exige el respeto de sus congéneres, de donde se siguen una serie de importantes consecuencias para la convivencia social. Pero, el hombre no estaba todavía preparado, por la dureza de su corazón, para comprender todo el alcance de las palabras inspiradas. Tampoco tomó conciencia de que el precepto de No Matar, contenido en el Decálogo, no sólo incluía la prohibición de quitar la vida a otro, sino que también debía significar no atentar en ningún aspecto contra la integridad física, espiritual y moral del prójimo. No matar involucra respetar el derecho a la vida; y este derecho, tratándose de seres humanos, implica la prerrogativa de vivir como tales y no como bestias, de donde puede deducirse, además, la amplia gama de los modernos derechos sociales (justo salario, adecuado nivel de vida, seguridad social, etc.). Pero, la tierra no estaba abonada en aquella época para que germinara la semilla. Ni los antiguos pueblos del Oriente, ni los griegos, ni los romanos, sin excluir a sus más preclaros filósofos, conocieron nada que pudiera compararse a los derechos humanos. Tampoco la organización social, uno de cuyos pilares era el régimen de la esclavitud, permitía el establecimiento de condiciones que aseguraran el Bien Común. Fue necesario el advenimiento del cristianismo, con su sublime doctrina de la filiación divina, del amor "Amarás a tu prójimo como a ti mismo" y de la fraternidad universal, para que comenzara a perfilarse en la conciencia, aun en forma intuitiva, la imagen de los derechos humanos.

(2) <u>Segunda etapa. Consagración de estas prerrogativas en declaraciones de creciente amplitud</u>.

Bajo la influencia del cristianismo, el reconocimiento de determinados derechos humanos surge en las costumbres políticas y en las declaraciones medievales antes que se formule una concepción doctrinaria sobre la materia (lo que solo ha de ocurrir siglos más tarde). La práctica se anticipa a la teoría. Con los antiguos fueros españoles, y con ciertas garantías individuales que se reconocen en el Imperio Carolingio y en el Imperio Germano (siglos VII a XI), se inicia un proceso evolutivo que culmina, en la Edad Media, con la Carta Magna (1215). Este proceso se caracteriza, hasta nuestros días, por un movimiento de doble y creciente expansión: por una parte, los derechos humanos crecen en número y contenido (compárese por ejemplo, los breves y balbuceantes derechos de los fueros primitivos, o incluso las garantías individuales de las primeras declaraciones, con la vasta y completa nómina de la Declaración de las Naciones Unidas); por otra parte, los derechos humanos se extienden en su ámbito de aplicación personal y territorial, pasando de determinados estamentos a la población entera y de una vigencia puramente regional al campo nacional y, actualmente, internacional.

En esta fecunda evolución, muy vinculada también a la fundamentación del régimen democrático, desempeñan un papel importante la Escuela Española del Derecho Natural, en los siglos XVI y XVII, con Suárez y Vitoria a la cabeza; la corriente iusracionalista, en la que se destacan especialmente Hugo Grocio y John Locke y algunos de los heraldos de la Revolución Francesa Montesquieu y Rousseau. Los grandes jalones históricos de este proceso, después de la Carta Magna, son la Petición of Rights (1628); el Acta del Habeas Corpus (1679); el Bill of Rights (1689); el Acta de la Independencia de los Estados Unidos (1776), seguida por la Constitución norteamericana y sus diez primeras enmiendas; y la Declaración de los Derechos del Hombre y del Ciudadano (1789) de la Revolución francesa. Estos importantes textos abren el camino al constitucionalismo liberal-democrático del siglo XIX. Pero, todavía los derechos humanos constituyen más una expresión prográmatica que una efectiva garantía jurídica.

(3) <u>Tercera etapa. Estableciemiento de instituciones y procedimientos legales de resguardo de los derechos humanos, incluso a escala internacional</u>.

El siglo XX presencia el esfuerzo por conseguir que los derechos humanos no sean sólo una declaración romántica, sino una efectiva realidad social. Con este objeto, se organizan instituciones y se dictan normas substantivas y de procedimiento, destinadas al resguardo jurídico de estas prerrogativas. La protección de los derechos humanos se extiende al plano internacional, a través de tres grandes sistemas de reconocimiento y amparo, que señalamos en orden cronológico: (1) El sistema de la Organización de Estados Americanos (1948); (2) El sistema de las Naciones Unidas (1948) y (3) El sistema de la convención Europea de los Derechos del Hombre o Convención de Roma (1950). Este último es, sin duda, el que ha representado el más importante avance, como una significativa conquista de nuestra época, porque comprendió la creación de un efectivo mecanismo jurisdiccional, con facultades para imponer sus resoluciones en un plano supranacional, a través de la Comisión y la Corte Europeas de los Derechos Humanos, además de la intervención que le corresponde, en ciertos casos, al Comité de Ministros del Consejo de Europa.

(4) Cuarta etapa. La real y efectiva observancia de los derechos humanos en el mundo.

No es necesario abundar en mayores consideraciones para poner de relieve que aun no se ha llegado a esta etapa. Si contemplamos el panorama que se nos presenta hoy en día en esta materia, podemos llegar muy pronto a la conclusión, como lo anotábamos anteriormente, de que sólo en escasas regiones del planeta existen sistemas que garantizan de una manera más o menos estable y completa la vigencia de estas prerrogativas, sin perjuicio de las restricciones temporales que puedan producirse por circunstancias de emergencia. Desde los tiempos antiguos hasta nuestra época se han dado, en verdad, pasos de gigante. El progreso de los derechos humanos, a través de las diversas etapas que hemos reseñado, ha sido inmenso. Este progreso ha sido particularmente notable en nuestro siglo. Pero, nadie podría negar que queda aun mucho camino por recorrer para que la aplicación práctica de estos atributos se extienda por la faz de la tierra. Postulamos que ese camino es, fundamentalmente, el mismo que conduce a la realización del Bien Común, como puede domostrarse en pocas líneas.

¿Qué es el Bien Común? ¿Cómo se ajustan en armónica ecuación, los dos grandes polos de la convivencia colectiva, que son la persona y el Estado? ¿Cómo se integran en el Bien Común los derechos individuales y los derechos sociales, sin que la realización de aquéllos vaya en desmedro de éstos o vice-versa? No es posible, dentro de los márgenes de este trabajo entrar a analizar y debatir estos polémicos interrogantes, que son enfocados con distintos criterios por las grandes concepciones filosófico-juridicas y políticas.

En nuestro concepto, el Bien Común es el bien general de la comunidad, en su doble aspecto, en espiritual y material. S.T. Delos anota que es "el conjunto organizado de las condiciones sociales gracias a las cuales la persona humana puede cumplir su destino natural y espiritual".

De estas y otras definiciones se desprende que todo concepto del Bien Común implica, esencialmente, la realización del bien personal de los integrantes de la comunidad, de la plenitud y perfección humana, ya que como tanto se ha dicho y repetido, no se ha hecho el hombre para servir a la sociedad, sino la sociedad para servir al hombre. Lo dicho se entiende sin perjuicio de las justas limitaciones que el interés social exija imponer al ejercicio de los derechos de las personas, ya sea en forma permanente o temporal por circunstancias especiales, pero sin menoscabar jamás, la intangible dignidad del hombre. Rechazamos, por lo tanto, como aberrantes, tanto el individualismo como el colectivismo. Pues bien, la plenitud y perfección de la vida personal, que es carácter esencial de la realización del Bien Común, está condicionada por la plena vigencia de los derechos humanos, en términos que la aplicación práctica de estas prerrogativas, rectamente entendidas y dentro de una norma superior de respeto a la Verdad, configuran, globalmente el bienestar general de la Sociedad.

Se realiza el Bien Común en efecto, en toda Sociedad en la que, por una parte, sus integrantes gozan de los derechos a la vida, a la igualdad, a la libertad, de reunión, de asociación, de propiedad; y en la que, por otra parte, se ha dado un conjunto de condiciones sociales que hacen posible que cada persona pueda tener amplio acceso a los bienes espirituales y culturales de la comunidad, recibir una justa remuneración de su trabajo, mantener un nivel de vida adecuado, satisfacer sus necesidades de alimentación, vivienda y vestuario, ser amparada en el desempleo, la enfermedad, la vejez y otras vicisitudes, por los servicios de seguridad social, etc. En suma, si se cumplieran íntegramente en una sociedad los derechos individuales y sociales enunciados en la Carta de las Naciones Unidas, integrados dentro de una unidad superior de carácter espiritual y moral, se darían las condiciones necesarias para considerar que esa sociedad estaría realizando el Bien Común. ¿Cómo lograr este objetivo? ¿Cómo conseguir que los derechos humanos se extiendan y se observen en regiones cada vez más mayores, hasta hacerse efectivos en el mundo entero? Tarea de titanes dada la imperfección del hombre, éste es sin embargo, el camino, de avatares, avances y retrocesos, que ha venido siguiendo la humanidad desde sus orígenes.

Mucho más allá de las posibilidades de un sistema jurídico que contemple en forma eficaz el reconocimiento y la protección de estos atributos,--condición necesaria, pero no suficiente, para su observancia--, la realización de los derechos humanos implica la concurrencia de múltiples complejos factores de carácter religioso, moral, cultural, cívico, social, económico. En síntesis, pensamos que el avance en esta materia se identifica con el desarrollo y progreso de los pueblos. Esto explica, en nuestro concepto, que,--salvo algunas excepciones cuyas razones históricas sería demasiado extenso analizar aquí--, las naciones en que mejor se han realizado en nuestra época los derechos fundamentales de la persona humana son, precisamente, aquellas que han alcanzado un más alto grado de cultura, de civilización y de progreso, en todos los órdenes.

El esfuerzo en favor de la observancia y universalización de estas prerrogativas debe encauzarse, como lo señalamos en nuestro libro PANORAMA DE LOS DERECHOS HUMANOS, en un doble sentido: luchar por difundir en todo el género humano los grandes valores religiosos, espirituales y morales, conducentes a promover una conducta de auténtica fraternidad humana; e "impulsar una gigantesca cruzada internacional en favor del progreso social, económico y técnico de las regiones subdesarrolladas". "Dentro de un cuadro de suficiente adelanto cultural y material,--agregábamos más adelante--, podrán generarse y perfeccionarse en todos los pueblos sistemas políticos que garanticen la libertad y ordenamientos jurídicos que no sólo reconozcan y consagren en forma abstracta los derechos humanos, sino que contemplen también mecanismos y procedimientos verdaderamente eficaces para hacerlos respetar en los casos en que sean atropellados o menoscabados." Estos medios de organización institucional podrían culminar en organismos jurisdiccionales supranacionales "y, tal vez, en una etapa final, en un Tribunal Mundial del ramo".[2]

Esta vasta e ímproba tarea en favor de los derechos humanos y, por ende del Bien Común, no es patrimonio sólo de los Gobiernos, de las instituciones sociales, de los partidos políticos o de otros grupos societarios, sino que debe ser labor constante y tesonera de todos los hombres de buena voluntad, cualesquiera que sean los lugares en que estén situados o en los ámbitos en que ejerzan sus actividades.

En esta campaña por el mejoramiento de la humanidad, no debe haber decepción ni desfallecimiento por el hecho de que nuestra obra tarde años o siglos en fructificar. Lo fundamental es seguir adelante con la conciencia de que estamos cumpliendo con un deber ineludible y de que, aunque no veamos de inmediato sus resultados, nuestro esfuerzo no será estéril.

Como dice Juan Enrique Rodó, "mientras la muchedumbre pasa, yo observo que, aunque ella no mira al cielo, el cielo la mira. Sobre su masa indiferente y obscura, como tierra del surco, algo desciende de lo alto. La vibración de las estrellas se parece al movimiento de unas manos de sembrador."[3]

FOOTNOTES

1. Cfr., Louis de Villefosse, GEOGRAPHIE DE LA LIBERTE. LES DROITS DE L'HOMME DANS LE MONDE (1953-1964), Robert Laffont, ed., París, 1965. 404 págs.

2. Cfr., Jorge Iván Hübner Gallo, PANORAMA DE LOS DERECHOS HUMANOS 153, Editorial Andrés Bello, Santiago de Chile, 1973.

3. Juan Enrique Rodó, ARIEL 210 en OBRAS COMPLETAS, Ediciones Antonio Zamora, Buenos Aires, 1956.

In this paper we try to show that the full practical application of the human rights, within spiritual and moral superior standards, coincides with the fulfilment of Common Good and all doctrines of social redemption.

It is said that we are living in the century of human rights, not from the point of view of its practical realizations, but from the conscience of humanity about its importance. The history of human rights is a living testimony of the progress of man's conscience. Not only the progress of the Natural sciences, but also the sciences of man, discovering the law that is within the nature of things. Nowadays the situation is much better than before. In early times, those rights were not even conceived. Christian teaching and an evolution of centuries was necessary before man took cognizance of them. In this historical process we can distinguish four stages:

(1) The birth and improvement of the concept of these rights. Its base can be found in Genesis which states that God created man in His image and after His likeness, but man did not understand in his time all its scope.

With the arrival of Christianity the image of such rights began to be outlined.

(2) Its consecration in declarations of increasing broadness. Here are mentioned the principal declarations. It is observed that a movement of double expansion of these atributes exists, as to its quantity and contents and as to its territorial extension which today has international coverage.

(3) Establishment of institutions and procedures of legal caution. This has been the main task of the twentieth century. Working to cause the declarations to become a reality instead of remaining a mere programme.

(4) Real and effective observance of human rights in the world. This stage has not yet been reached in spite of the long way already passed. This way is the same that leads to the fulfilment of Common Good, that the author conceives as the general good of the community, in its spiritual and material aspects. The fulfilment of Common Good, of the plenitude and improvement of human life, is conditioned by the application of human rights--individual and social--integrated in a superior rule of regard for the truth.

This objective surpasses the possibilities of law, and depends on many complex factors of religious, moral, cultural, social, political and economic factors, which identify themselves with the development and progress of the people. The task in favor of the actuality and universalization of human rights and of progress and Common Good, must be accomplished by all men of good will, without weakness. Even if we shall never see the results, our efforts will not be fruitless.

SOME QUESTIONS CONCERNING THE CORRELATION BETWEEN PERSONALITY AND LAW

Vilmos Peschka

I.

Value and importance of being and becoming a personality can hardly be summed up more beautifully and concisely than by Goethe in the unforgettable lines of his Westöstlicher Divan:

> Volk und Knecht und Überwinder
> Sie gestehn zu jeder Zeit:
> Höchstes Glück der Erdenkinder
> Sei nur die Persönlichkeit.
> Jedes Leben sei zu führen,
> Wenn man nicht sich selbst vermißt.
> Alles könne man verlieren,
> Wenn man bliebe, was man ist.

The substantial element in these lines is the idea of evolution, the processes through which man in the course of his life becomes personality, evolves into person. Law, as a social objectivation, is doubtless one means of the unfolding of man's personality, of the potential that man be able to display and preserve his personality in a given society. And this very society is actually the creator of the human personality. If we interpret personality as the unity of man's attitudes, man's acts, that of the various motivations resulting into such acts, or, in other terms if man is understood as a rational and moral being, we must be impelled to draw a paradoxical conclusion as to the correlation between law and personality: While law is usually one means of the unfolding and the protection of the human personality, in the world of law persons, i.e., subjects at law usually do not emerge as the above-defined moral beings who rationally take into account their conduct, their actions and the motives thereto, and assume responsibility therefor. What appears is rather the man as a particular individual, born with certain properties, capabilities.

The history of law, quite naturally, has made us acquainted with a number of legal systems which have defined special and strict preconditions of being a subject at law. In the circumstances of modern commodity production, however, in highly developed legal systems the usual situation is that man becomes a subject at law, a person from the legal point of view, already at birth--in some legal systems even from the date of conception. All this indicates that, for the purpose of law, it is not an absolute criterion of personality to be a rational and moral being, e.g., a child or an insane person may hardly be regarded as such.

As a matter of fact, the condition of being a subject at law, a personality from the legal point of view, i.e., to have the status that a person <u>can be</u> the subject of rights and obligations, does not require any rational comprehension or moral sense of responsibility on his part. This means that law, in order to co-ordinate the circumstances of social life, disregard man's certain properties and faculties which develop with age, education, participation in the life of society, etc., and which render man a conscious, rational and moral being having the faculty of comprehension and assuming responsibility, and that law qualifies man as a person, as a subject at law, merely as a natural being and as a particular individual of society.

The co-ordinated settling of social conditions, their keeping within defined channels and limits, the smooth circulation of commodities, undisturbed transfer of ownership, etc., are evidently pre-conditions of unfolding and preserving the personality of man as a rational and moral being. But law while realizing this applies among others, rather paradoxically, such ways and means which from certain aspects and facets of legal regulation disregard these criteria of the personality, and degrade the person from a rational and moral being into a merely particular individual. Law uses no fiction in this respect, does not employ the method of regarding, e.g., a child or an insane person <u>as if</u> they were rational and moral beings. This is out of the question, but is not necessary either for defining the status of being a subject at law, i.e., for defining the sphere of persons who can be bearers of rights and obligations.

The situation is different as concerns the active acquisition of such rights, the active establishment of such obligations, the exercise of such rights or the performance of such obligations. It goes without saying that such activities and processes can be displayed or controlled only by persons who are able to

size up the true importance, meaning and effects of
such activities, by persons, in short, who are <u>rational
beings</u>, who are accountable, i.e., have disposing capacity. This means that certain acts and processes
occurring in the world of law presuppose the person to
be a human being who acts rationally. The law, therefore, usually defines the sphere of persons having
disposing capacity, can act rationally, as a special
category, and as a rule makes this capacity conditional on age and mental state. The consequence is that
persons without the requisite rational ability may not
perform certain legal acts, or may perform them only
with the aid of others, as in the case of children.
There is no need of further discussion. It is evident
that acts producing certain legal effects can only be
performed by beings possessing the required rational
ability.

A more problematic question is whether the criterion
to be a moral being is really indispensable in respect
of the person playing a part in the world of law. Or,
in other terms, whether only moral beings are actually
regarded by the law as persons? At first sight it
would seem that while constructing a particular civil
or criminal system of responsibility, evaluating the
conduct and acts of persons, calling them to account,
the law relies on moral consciousness, moral responsibility and acceptance of such responsibility. Suffice it here to mention some elements and categories
of legal responsibility, such as guilt, culpability,
wilfulness and negligence of conduct to be judged
legally, which elements have, beyond doubt, entered
the legal sphere from the world of morals through the
medium of ethics.

The Kantian distinction assuming an entirely abstract, objective law and merely subjective, prescinded
morals is therefore an exaggerated generalization too
rigid for real conditions. Still, the history and
function of law tells us that the Kantian distinction
between merely subjective morals and entirely objective
law is not void of a rational nucleus and a real foundation. No need to consider more than the historical
development of liability for damages, the objective
system of responsibility in primitive law, or the
placing on objective bases of civil legal liability in
modern legal development, e.g., in cases of damage
caused by works, involving increasing danger.

Moreover, the objectiveness of the validity of law,
the fact that the validity of any and each legal act
is, as a matter of necessity, independent of the consciousness and arbitrariness of the given individual,

also indicates that what is decisive for the existence, the direct operation of law, is human conduct, action and their results, and not the motivation, subjective side, wilfulness or negligent nature of the relevant activity, not the sense for, and assumption of, responsibility on the part of the involved person. The importance of subjective elements is secondary at best, if not indifferent for the mechanism, for the direct functioning of law.

It evidently follows from all this that the quality to be a <u>moral</u> being is not considered as an absolute criterion for persons to play a role and act in the world of law, although these same persons are actually, and usually, moral beings. As concerns human action and the responsibility for it, law contents itself with the recognition and observance of rules and standards, and may function very well without the subjective conscience, sense of duty and guilt of the acting persons. Moreover, law establishes legal obligations also toward persons who are not moral beings. All this, of course, does not mean that morals and ethics should play no part in law, should have no effect thereon. What is involved here is merely the direct existence, operation and mechanism of law, and the persons taking part in all this. Naturally, the moral views and convictions of people are of extreme importance in the genesis, development and extinguishment of law. But this sociological fact does not affect the independence of the direct operation of law from the moral consciousness and convictions of the individual; nor does it affect the circumstance that a person proceeding and acting in the world of law is not necessarily a moral, but only a rational being.

II.

The problem of the personality appears in philosophical literature and in the literature on the philosophy of law practically, and usually, inseparably from the subject-matter of man's natural rights, of the so-called human rights. It is a stereotyped, general conclusion that natural, human rights are due to men, just because they are persons. The first question of all arising in this connection is whether there exists, or not, another law, the so-called natural law or human right, in addition to positive law.

From the aspect of the Marxian philosophy of law, a consistent approach to law as social objectivation leads to the theoretical result that we cannot speak of natural law, of human right as a real, valid law that precedes positive law and stands above it. There exists only one type of law in any society -- the valid, positive law issued by a given state. But this denial of natural law, of human right, on the part of the Marxian philosophy of law by no means excludes the category, the concept of natural law, of human right as a mental phenomenon reemerging incessantly during the evolution of mankind over many thousand years. We cannot emphasize it enough: The Marxian philosophy of law denies natural law only as law; but it recognizes its existence as an ideology which, as appears from the history of law, may influence considerably the trends, the development of law, without being law in itself. It is only a legal view, notion, desire produced consciously, be it by a false consciousness.

Natural law, human right as an ideology always has expressed real, objectively existing social-legal problems, has been aimed at their solution; yet as a mental phenomenon and desire, it reflected the objective social-legal relations through a false consciousness and not with the adequacy of scientific cognition. We are perhaps not mistaken if we say that the thesis in the paper by Professor Plamenatz according to which "human rights refer not to claims that are actually made and recognized (acknowledged to be just), but to claims which the speaker thinks it desirable should be made and recognized," expresses the same idea, namely that human rights do not mean real law but only a claim, a desire raised in respect of law. Human rights become part of real law if and when a given state enacts them, expresses them in its positive law. It is in this sense that we may speak also of real human rights enacted into positive law, meaning by this the generalization of the so-called human rights laid down in the different systems of positive law.

But the decisive question remains this: Where do these human rights come from, if only as a claim, as an ideology, or expressed also in a positive legal system? The answer of the Plamenatz paper is that they come from man's personality, from the fact that man exists as a person, i.e., is a rational and moral being. This seems to be consequent to the abstract, general formulation of human rights, according to which these mean some personal rights in general. Yet, in reality, this abstract and general expression of human rights always expresses very real and concrete social conditions and relations. Human rights demand freedom of speech, assembly and combination, etc., not because this would be an indispensable precondition of the existence of a rational and moral being, of the person.

It was always the concrete historical, socio-political situation that determined the concrete contents of the category of human rights. The source of human rights, therefore, is not to be found in man as a personality but in the society and in its economic, political and legal system. Both the concept of human rights and its deduction from the personality is an abstraction, prescinded from concrete socio-historical conditions. Usually they express in an abstract, general manner the circumstance that the existence of a given social and legal system has become questionable, problematic. The adherents and demanders of a new social and legal system, the "revolutionizing" class, to use the term of Marx, usually aspiring for power, inevitably and always speak in the name of the entire society, and present their very concrete and particular social and legal interests as the human rights of the said entire society, as those of all its members, moreover as those of the person in general.

The person, and the human rights arising from it is but one outward form of this inevitable social abstraction and generalization. The ideological nature and role of human rights becomes manifest exactly in this dual concealment of the real situation, of real conditions; on the one hand, by presenting human rights as part of law and, on the other, by presenting them as the person's general rights in an abstract manner, deprived of their concrete social contents.

LA RECEPTION ET LE DEVELOPPEMENT DES IDEES OCCIDENTALES DE LA LIBERTE ET DE L'EGALITE AU JAPON MODERN

Yukio Uehara

I. PREAMBULE

L'histoire humaine montre que les idées de la liberté et de l'égalité sont été soutenues par les hommes depuis longtemps. Et aujourd'hui, ces idées, aussi bien que celle de la dignité humaine, sont exprimées nettement dans "La déclaration universelle des droits de l'homme (1948)". Nous y trouvons le consensus commun, tout au moins, sur les valeurs sociales fondamentales. Cette situation actuelle nous permet de présupposer ces valeurs comme des axiomes et de chercher à trouver le moyen de les réaliser, sans nous préoccuper de les justifier philosophiquement.

C'est pourquoi je me restreins ici à esquisser le trait caractéristique de la réception et du développement des idées occidentales de la liberté et de l'égalité au Japon modern et à en obtenir, s'il est possible, quelques suggestions en vue de la réalisation plus parfaite de ces idées. Une telle tentative ne serait pas, je crois, sans importance, parce que l'histoire humaine montre aussi que quelques obstacles ont empêché les hommes de réaliser ces idées malgré leur aspiration.

Prof. Luis Recaséns-Siches, dans son article "Dignity, Liberty, and Equality", [supra] cherche à justifier philosophiquement la dignité humaine. "It seems to me," écrit-il, "that through the ways of purely philosophic meditation it is possible to establish the principle of the dignity of the human individual." Et il me semble que la base de sa théorie se trouve dans le passage suivant. "Although philosophic idealism has been superceded by the philosophy of human life or existence, it has preserved as a firm truth that my consciousness constitutes the center, support, and proof of all the other realities. The consciousness is inescapably and necessarily the born center of the Universe." Mais, si la dignité humaine se fonde sur la conscience de l'individu, un bébé, un fou, ou surtout un homme qui a perdu la conscience tout à fait (nous l'appelons "homme végétal"), n'a-t-il pas de dignité humaine? Cette question ne signifie jamais que je nie la dignité humaine. Au contraire, je pense que même si l'on ne réussit pas à la justifier philosophiquement, on doit la reconnaître pour tous les hommes. Il vaudrait mieux, je crois, trouver le moyen de l'assurer ou construire en détail son système axiomatique.

II. L'OCCIDENTALISATION AU JAPON--LA LIBERTE ET L'EGALITE

(A.) Avant La Deuxième Guerre Mondiale

Au milieu du XIXème siècle, le Japon s'ouvrit au monde extérieur, donné l'occasion par le Commodore Perry qui se présenta avec ses bateaux noirs au larges des côtes japonaises. Dès lors commença l'occidentalisation du Japon, d'abord dans le domaine des sciences techniques conformément à la maxime "la morale orientale et l'art occidentale" qui avait été déjà établie. Cependant, aussi les idées sociales ne purent ne pas s'introduire et de plus en plus attirèrent quelques Japonais lucides. Et au début de la Restauration de Meiji se traduisirent ON LIBERTY de J.S. Mill (1872), DE L'ESPRIT DES LOIS de Montesquieu (1876) et DU CONTRAT SOCIAL de J.J. Rousseau (1877, 1882). Ainsi, les idées des droits de l'homme et de l'isonomie devinrent la base du Mémoire sur l'établissement du parlement démocratique adressé au Conseil Politique en 1874.

Cependant ce Mémoire ne réussirent pas à établir la démocratie au Japon. La Constitution impériale de Meiji promulguée en 1889 établit, au contraire, l'absolutisme impérial du Japon. Ainsi, la liberté et l'égalité furent dégradées du titre de droit naturel et admises seulement sous réserve des lois qui visèrent la richesse et la puissance militaire de l'Etat sous la direction de l'Empereur deifié.

Or, qu'est-ce qui permit d'étouffer le mouvement démocratique déjà assez répandu et d'établir l'absolutisme impérial? Sans doute on peut énumerer, à cet égard, quelques facteurs, comme les sentiments traditionnels des Japonais de l'exclusivisme et de l'unité nationale, le confucianisme, surtout celui de Sorai Ogiu qui attache l'importance à l'autorité du roi ou du saint, l'étude philologique et philosophique des classiques japonais (kokugaku), plus directement les doctrines allemandes de R. Gneist et de L. Stein, etc. Mais, ici, notamment les deux facteurs suivants devraient être remarqués.

En premier lieu, la conception japonaise du droit.

"Pour nous, occidentaux," écrit Prof. Jacques Robert, "le mot droit évoque avant tout l'idée de la justice absolue appliquée comme principe dirigeant aux actions humaines. On l'oppose ainsi facilement à la force, à l'injustice, à l'illégalité. Il suppose une société relativement égalitaire qui entend être soumise aux mêmes règles librement voulues et acceptées par tous. Il n'existe que pour protéger l'homme contre les abus éventuels du pouvoir ou des autres hommes et, au besoins, à le garder contre ses propres égarements. Il trouve sa raison d'être dans l'effort qu'il poursuit pour concilier--le mieux possible--les impératifs de la cohésion sociale et la nécessaire liberté de chacun."[1] Pour les Japonais, une telle notion du droit était, ou est étrangère. La notion du droit, s'il existe, n'évoque point chez eux l'idée de la justice absolue, et partant le droit ne s'applique guère comme principe qui doit être observé à tout prix. Le droit se recule aisément devant la raison d'Etat quelconque et la tension, s'il existe, entre le droit et la réalité, c'est-à-dire entre ce qui doit être et ce qui est, se relâche presque toujours au profit de celle-ci. Ce fut ce qui détermina la réception des idées occidentales au Japon et qui dénatura la notion du droit naturel sur laquelle fut été basées la liberté et l'égalité.

En outre il existait, en deuxième lieu, l'oppression des grandes puissances, les Etats-Unis, l'Angleterre, la Russie, la France et les Pays-Bas, qui imposèrent au Japon des "traités inégaux" au milieu du XIXème siècle. Aussi, dès le commencement de son ouverture, se chargea le Japon d'une lourde tâche: S'il voulait maintenir son indépendance, il lui fallut avant tout de la richesse et de la puissance militaire. On pourrait y trouver le leitmotif qui suscitera le nationalisme pour l'extérieur et l'absolutisme pour l'intérieur, au prix de la liberté et de l'égalité des individus. En fait, dans cette circonstance, les libérals japonais renoncèrent, de gré ou de force, à faire aboutir leur revendications. Même Yukichi Fukuzawa, "libertin" le plus éminent, écrivit en 1875, "L'indépendance du Japon, c'est le but, la civilisation du peuple, c'est le moyen de l'atteindre." Ainsi, l'oppression des puissances occidentales finit par aider, paradoxalement, à dénaturer les idées occidentales de la liberté et de l'égalité et à rédiger la Constitution de Meiji.

Il se peut que la réflexion sur la conception japonaise du droit mentionnée ci-dessus ne serte qu'à comprendre le réalisme des Japonais qui donne la priorité à la politique plutôt qu'au droit. Mais la deuxième serait significative en vue de la réalisation universelle de la liberté et de l'égalité. Parce que l'on peut y trouver un exemple de la tension des relations entre les pays évolués et les pays arriérés qui peut donner la naissance à l'absolutisme dans ceux-ci au détriment des droits de l'homme. "The great error committed by antihumanism or transpersonalism," dit Prof. Luis Recaséns-Siches avec raison, "is the following: It does not take into account that the collectivity has no substantive reality, that it does not have a being for itself independent of the being of the individuals who compose it."[supra] Mais, si l'histoire humaine montre que l'on a souvent commis ou passé une telle erreur si apparente, il nous faudrait aussi en connaître la cause empiriquement.

(B.) Après La Deuxième Guerre Mondiale

La défaite de 1945 apporta au Japon une nouvelle constitution qui fut rédigée en vue de sa démocratisation et promulguée en 1946. L'empereur n'a plus de pouvoirs substantiels et tout individu est censé libre et égal aux autres. Voilà, à coup sûr, une des constitutions les plus démocratiques. Mais il faut ajouter que cela ne veut pas nécessairement dire que la société japonaise soit réellement démocratisée. Comme le dit Prof. Yoshiyuki Noda, "La question demeure entière de savoir dans quelle mesure, derrière cette façade occidentale, le Japon a subi en profondeur une transformation et accueilli l'idée de justice et de droit telle qu'on la connaît en Occident."[2]

A cet égard, on devrait tenir le plus grand compte du fonctionnement faussé au Japon du régime parlementaire à l'anglaise. On s'y souvient que Montesquieu a prévenu du danger d'une sorte du régime parlementaire. "Lorsque dans la même personne ou dans le même corps de magistrature, la puissance législative est réunie à la puissance exécutrice, il n'y a point de liberté; parce qu'on peut craindre que le même monarque ou le même sénat ne fasse des lois tyranniques pour les exécuter tyranniquement."

Non que le parti gouvernemental du Japon qui, sous le régime parlementaire, pratiquement détient la puissance exécutrice et la puissance législative soit expressément tyrannique. Mais, au Japon, depuis plus de vingt ans le parti conservateur demeure au pouvoir sans alternance des partis à la faveur de l'indifférence ou l'inertie du peuple pour l'élection. Il en résulte que la politique conservatrice est poursuivie obstinément, parfois au mépris des droits de l'homme. Au "bonum commune" se substitue secrètement l'idéologie conservatiste, totalitariste ou capitaliste.

Par ailleurs, c'est à peine si fonctionne le principe du contrôle judiciaire de la constitutionnalité des lois emprunté aux Etats-Unis. On ne trouve que deux arrêts

de la Cour suprême qui exercent positivement ce principe
--concernant l'égalité devant la loi (1973) et la liberté
du commerce (1975).

Dans ces conditions où la liberté et l'égalité sur le
plan des principes, bien qu'elles ne soient niées expli-
citement, sont parfois restreintes ou laissées en suspens,
on pourrait dire que "les structures sociales et l'esprit
libéral que supposaient les codes d'inspiration occidentale ne se retrouvent qu'à un faible degré dans la réalité japonaise".[3] Alors doit-on renoncer à développer la
démocratie parlementaire au Japon? Certes la situation
ne donne pas une perspective rassurante et certains vi-
sent en effet au socialisme ou à la démocratie populaire.
Mais, d'autre part, les idées de la liberté et de l'égalité vont incontestablement s'amplifier et les campagnes
politiques commencent à s'augmenter parmi les Japonais
qui se trouvent être maîtres de leur destinée. Voilà le
développement réel et l'avenir de la démocratie au Japon.
Il se peut trouver des gens qui se soucient d'une confusion sociale ou d'une anarchie dans le processus de ce
développement. Je suis convaincu toutefois avec Tocqueville que "l'anarchie n'est pas le mal principal que les
siècles démocratiques doivent craindre, mais le moindre".

Quelles sont les causes du développement, dans ces der-
nières années, de la démocratie au Japon? On peut réfé-
rer à l'alternance des générations, par exemple, ou à la
corruption pécuniaire du parti gouvernemental et à la dé-
gradation de l'environnement par les déchets industriels
qui éveillent le peuple. Mais la plus essentielle serait
la liberté de l'opinion dont l'assurance est relativement
solide même au Japon depuis la défaite. Et c'est cette
liberté de l'opinion qu'on doit respecter le mieux dans
l'application du principe "No freedom against freedom"
qu'établit Prof. Luis Recaséns-Siches [supra] afin que
l'autorité n'en abuse pour opprimer la contrepartie.
Parce que, si ce principe exerce les fonctions du gardien,
la réponse à la question, "quis custodiet ipsos custodes",
ne peut être, me semble-t-il, autre chose que l'opinion
publique.

FOOTNOTES

1. "La pensée juridique française au Japon" dans Revue
de Collaboration culturelle Franco-Japonaise, No. 24,
Mars 1969, p. 28.

2. "Le droit japonais" dans LES GRANDS SYSTEMES DE
DROIT CONTEMPORAINS 556, éd. par René David, Dalloz,
6ème éd. 1974.

3. Y. Noda, déjà cité, p. 557.

Now that the Universal Declaration of Human Rights has been proclaimed, perhaps the ideas of freedom and equality can be accepted as axioms and we can devote ourselves to finding means to their realization. This is why I try to draw some lessons from the reception of these western ideas in modern Japan. Prof. Luis Recaséns-Siches tries to justify human dignity through purely philosophic meditation, but I think human dignity must be recognized even if it cannot be justified philosophically.

The Meiji Constitution was established in conformity with political absolutism in spite of the western ideas of freedom and equality introduced into Japan before or after the Meiji Restoration. In this respect, two factors were important. (1.) The realistic conception of law in Japan, which relativized the western idea of natural right and made it yield to raison d'Etat. (2.) Pressure of great powers, which gave a pretext for the suppression of liberalistic movement in Japan. The complicated relations between liberalism, nationalism and absolutism of those days suggest why antihumanism, in spite of its error pointed out by Prof. Luis Recaséns-Siches, occasionally has great influence.

In order to democratize Japan, a new constitution was established under the influence of Anglo-American law. In Japanese social and mental structures, however, there have been some difficulties in giving satisfactory effect to those principles such as parliamentary government and judicial review, which have permitted the enforcement of conservative policies for more than twenty years in disregard, if necessary, of human rights. But, on the other hand, freedom of speech has been, comparatively speaking, firmly assured after the War. And it is in this noteworthy fact that we can find great possibilities of further development of the parliamentary democracy in Japan. Prof. Luis Recaséns-Siches presents an interesting principle "no freedom against freedom". Then, who will distinguish freedom against freedom from freedom? If the people or public opinion are to be the judge, freedom of speech must be respected at any price.

EQUALITY OF BASIC RIGHTS

Gene G. James

If all human beings are similar in certain fundamental respects, does it follow that they are entitled to the same basic rights, whereas if they are not similar in those respects, they are not entitled to the same basic rights? Thinkers such as Locke and Jefferson who held that all men are entitled to the same basic rights, and those such as Plato, Aristotle and Nietzsche who denied this, seem to have assumed the truth of the foregoing principle. However, since one can always find both similarities and differences between human beings, no one seems to have maintained that all similarities or differences are relevant in deciding the rights to which people are entitled. Why should the presence or absence of one characteristic be relevant to this issue and another not? What characteristic or characteristics do we, and should we, take as essential in determining if a being has rights? Is any such characteristic shared by all human beings? If so, is it shared with any other beings?

I shall argue that: Only those beings which I call interested beings have rights; All interested beings have rights; Statements ascribing rights to individuals are irreducibly normative and justificatory; There are two fundamental kinds of rights, passive rights and active rights; The individual to have passive rights, need not do anything whatever; The passive right against infliction of needless suffering is shared by both humans and animals; This is a natural right if any is, and is basic to concepts of morality which have evolved in both the Western and non-Western world.

I.

Imagine a universe in which there are objects coming into being and passing away, planets revolving around suns, earthquakes, tidal waves and other events. Imagine also that there are beings in this universe who are aware of these occurrences but who are indifferent to them, having no preferences regarding their outcomes. Such a universe would be a place where differences made no difference. It would be a universe devoid of value. Now imagine a universe in which there are beings who are not only aware of changes going on about them, but who have preferences regarding those changes. Imagine that those changes produce pleasure or pain, suffering or enjoyment for these beings. Events would no longer be mere occurrences, but would be differences that make a difference. Only a universe which contains beings of this sort can be said to contain value, and only beings of this sort can be said to experience value. This would be the case even if the beings in question were incapable of acting to influence their environment. It is not the capacity to act which is essential to value; it is the capacity to suffer or rejoice.

I shall speak of beings with the capacity to suffer or rejoice, and with preferences regarding the outcome of events, as interested beings. States of affairs which produce pleasure or satisfaction for such a being are in that being's interest, and those which produce pain or frustration are not. Only those states of affairs compatible with a being's long-range interests and goals are totally in its interest. Conditions which bring about a state of affairs in a being's interest may be said to fulfill its interest while conditions which prevent or hinder such a state of affairs do not fulfill but frustrate that interest.

Only interested beings may be said to have needs because to say of anything that it is needed is to say that it is required to fulfill some interest. Furthermore, since only animate beings seem to have interests, only they have needs. For example, although machines require oil to function, because they are incapable of experiencing satisfaction or frustration and are indifferent to whether or not they function, they can only be said to need oil in a metaphorical sense. Nor can oil be said to be in their interest or "good for" machines, except in a metaphorical sense. The concept of a necessary condition is all that is required to explain the sense in which machines "need" oil. It is only because people have interests in the functioning of machines that we speak of them as having needs or say that things are good or bad for them. Similar remarks apply to inanimate objects in general.

Only beings with needs have rights. To say of an individual that it has a right is to say either that other individuals would be unjustified in acting in certain ways toward that individual, or that the individual would be justified in acting in certain ways toward other individuals. Statements ascribing rights to individuals are, therefore, statements prescribing limits to conduct, and are irreducibly normative and justificatory in nature. The reason that only beings with needs have rights, then, is that it makes sense to talk of justified or unjustified conduct with regard to an individual only if there are things which can be said to be, or not to be, in that individual's interest. Beings which are indifferent to their treatment cannot be treated justly or unjustly.

Not all needs are of equal importance. If some interests are not fulfilled, the individual will perish or suffer irreparable harm. Fulfillment of these interests may, therefore, be said to be basic needs. For example, protection from extremes of temperature and an adequate supply of nutrients are basic needs of all animate beings. Although individuals have rights only if they have needs, since not all needs are of equal importance and one individual's needs may conflict with another's, individuals do not necessarily have the right to have all of their needs fulfilled. However, an individual always has at least a prima facie right to have its basic needs fulfilled. And, as I shall argue below, there is at least one basic need which individuals always have the right to have fulfilled.

II.

When other individuals would not be justified in acting a certain way toward an individual, I shall say that he has a passive right. When an individual would be justified in acting a certain way toward other individuals, I shall say that he has an active right. Only individuals capable of acting toward others can have active rights, but both individuals capable of acting toward others and those not capable of such action, may have passive rights. To have a passive right is to be entitled to receive a certain type of treatment from others. Thus, D. D. Raphael has used the term 'right of recipience' in much the same way I use the term 'passive right.'[1] The latter term is preferable for my purposes here, because it draws attention to the fact that to have some rights it is not necessary to do anything at all. It is not necessary, for example, to be self-directing, to be able to follow rules, play societal roles, enter into contracts or agreements, have duties toward others, or to make claims of any sort; it is necessary only that one have the capacity to suffer or rejoice. This is not to deny, however, that passive rights may come about as a result of the foregoing activities.

Since the most basic or fundamental need of all interested beings is to avoid pain and suffering, the most basic or fundamental right is to not have needless or avoidable suffering inflicted on one. This is the most basic or fundamental right because if a being did not have this right it could not have any other rights, and it is the only right shared by all interested beings, both humans and animals.

III.

Many people would dispute the last statement because they deny that animals have rights at all. There seem to be two primary reasons for this denial. First, there is the belief that one or more of the activities mentioned above is a necessary condition for having rights. For example, Kant held that only self-directing beings capable of legislating duties for themselves have rights, and T. H. Green held that only individuals who play a societal role have rights.

Since for one individual to have a right is for some other individual to have an obligation, Kant was correct in thinking that there is a correlation between rights and obligations. In fact, the foregoing claim, which is part of the so-called doctrine of the logical correlativity of rights and obligations, follows from an analysis of what it means to say that an individual has a right. But, while it is true that for one individual to have a right, another must have an obligation, it is neither logically necessary nor in accordance with our practices of assigning rights, that for an individual to have rights he must also have obligations. If this were the case, then not only animals but also infants and severely handicapped people would have no rights. But because we do recognize infants and severely handicapped people as having rights, even though they are incapable of having obligations or playing societal roles, positions such as Kant's and Green's are mistaken. Similar arguments hold against positions which maintain that other of the activites mentioned above are necessary for one to have rights.

The second reason people are reluctant to acknowledge that animals have rights is the fear that this will lead to the conclusion that animals ought to be treated exactly the same way as people, and it would no longer be justifiable to do such things as perform medical

experiments on them or slaughter them for food. Peter
Singer recently advanced an argument very similar to
this.[2] But, while no rational or moral justification
can be given for preferring the well-being of one interested being to that of another, considered simply as
interested beings, justification can be given for preferring the well-being of one type of interested being
to that of another, e.g., for preferring the well-being
of human infants to that of even the most highly developed other animals. For, although it is true by definition that all interested beings have the capacity to
suffer, it is by no means apparent that all types have
equal capacity to suffer. In fact, evidence based on
examination of nervous systems, etc., indicates that
they do not. Furthermore, the capacity to suffer is
not the only relevant criterion in deciding what rights
individuals have. Any of the activities mentioned above
in the discussion of active and passive rights, such as
entering into contracts and agreements, may also give
rise to rights. Treating beings with different capacities differently, therefore, is not a prejudice as
Singer maintains, but is merely conformity to the second half of the equality principle as formulated by
Aristotle. The fact that people may misuse this principle by citing differences which are not relevant in
determining the rights to which individuals are entitled does not mean that the principle does not have
legitimate uses. Singer is thus correct in arguing that
animals have rights and that human behavior toward them
is frequently unjustified, but he is wrong in suggesting that remedying this requires treating animals exactly the same way we treat people.

IV.

Because there have been numerous attacks on the doctrine of natural rights from Bentham to the present,
and because they reject the epistemological and metaphysical theories on which the doctrine was based, most
recent thinkers sympathetic to the moral claims involved in the doctrine have abandoned the term 'natural
right,' replacing it with the term 'human right.' This
is unfortunate because, since no new epistemological or
metaphysical arguments have been given in support of
human rights, the change of terminology has served to
relegate epistemological and metaphysical problems to
the background without coming to grips with them, and
it has tended to obscure the fact that humans and animals have at least one right in common--the right to be
free from infliction of needless or avoidable suffering.
The latter is important not only because it draws

attention to our obligations to animals, but also because of the light it throws on such issues as the rights of fetuses, children and severely handicapped people.

Although I do not have the space to defend the concept of a natural right, and like most contemporary philosophers, reject the epistemological and metaphysical assumptions on which the traditional doctrine was based, I would like to point out that the right not to be subjected to needless or avoidable suffering has most or all of the characteristics traditionally attributed to natural rights and, therefore, may be called a natural right if any right is. It is a universal right which belongs to all interested beings by virtue of the kind of being they are. It is in no way dependent on an individual's playing societal roles or entering into contracts or agreements and is, therefore, "natural" in the sense that the natural is opposed to the conventional. Since it is an entitlement to a particular type of treatment by virtue of the kind of being one is, and is in no way dependent on social or conventional arrangements, it is inalienable in the sense that there does not seem to be any way to talk meaningfully of its being transferred, waived or voided. Even if one were to state that he no longer wished to receive this type of treatment, that would not take away his entitlement to receive it. Because it is not the right to receive some specific mode of treatment, but only the right to be free from infliction of needless or unjustified suffering, it is absolute in the sense that there does not seem to be any way that it could conflict with other rights. It is a moral rather than a customary or legal right, because any justification for it would have to be based ultimately on a moral appeal rather than an appeal to the customary or legal. Although the claim that no being should have unjustified suffering inflicted on it is not "self-evident" in the sense that it cannot be meaningfully questioned, it is self-evident in the sense that it is fundamental to the dominant conceptions of morality which have evolved in both the Western and non-Western worlds. This is clearer in non-Western moral codes which incorporate ahimsa or non-violence as a way of life than in Western moral codes, but can be seen with regard to the latter also to the extent that cruelty to animals is forbidden in almost all Western countries. However, use of the term 'self-evidence' in even this restricted sense is misleading because it suggests that the issue is one of truth or falsity when it is not. If pressed by the skeptic to justify the claim that all beings are equal in the sense that they ought not have needless or unnecessary suffering inflicted on them, all one can do is to state the idea as clearly as possible, show that it is possible to incorporate it within a consistent morality, and insist that one is rationally justified in adopting it as a normative principle.

I am aware in concluding this paper that, because of the complexity of the topic, the limited space available, and my own limitations, many questions have been left unanswered and many assumptions undefended. Much more thought, e.g., needs to be given to stating precisely the relations which hold between interests, needs and rights, and to the ways in which human rights differ from animal rights. The various types of human rights also need to be specified in greater detail. Nevertheless, I hope that at least the outline of a systematic theory of rights has emerged.

FOOTNOTES

1. D. D. Raphael, PROBLEMS OF POLITICAL PHILOSOPHY 68, Praeger Publishers, New York, 1970.

2. Peter Singer, "Animal Liberation," The New York Review of Books, April 5, 1973, reprinted in James Rachels, MORAL PROBLEMS, 2nd ed. 163-77, Harper and Row, New York, 1975; and "All Animals are Equal," I Philosophic Exchange 103-16, Summer, 1974.

PERSONS AND PUNISHMENT

Wade L. Robison

It is a common assumption that a person being punished must be pained in some way. Yet numerous examples of legal punishment are difficult to account for on this assumption. What is one to do with the masochistic convict? Or the occasional convict who, faced with parole, prefers staying behind bars?

These sorts of examples do not force one to deny that punishment involves pain, for the assumption can be protected in two ways to meet the counter-examples. S. I. Benn makes both protective moves when he says that "characteristically, punishment is unpleasant."[1] With the addition of the "characteristically," one can admit punishment without pain, for it becomes atypical. After all, the masochistic convict is an unlikely sort, and few willingly forego parole. It is asserted that there is punishment in such cases, because the concept is the same. It is just not a normal case of punishment, for that involves pain.

The other protective move is to slide from punishment involving pain to its involving something less or different--"unpleasantness," to quote Benn, "distress of some sort,"[2] "the frustration of a person's desires,"[3] "suffering,"[4] or the "infliction of hardship" or having "the aim...to hurt."[5]

These descriptions are typical and clearly differ. To frustrate someone's desires may not hurt him. If he desires to hurt himself, frustrating his desires will not hurt him. Again, one need not be distressed if caused an unpleasantness. One may be nasty and wish to be caused an unpleasantness, and to have that known, to distress one's host, not oneself.

This disagreement ought to make one suspicious about what punishment really involves. That the differences are unmarked ought to make one wonder if there is not some underlying basic agreement that makes the differences seem minor to those not marking them. And there is.

For whatever differences there are between suffering and hardship, distress and pain, unpleasantness and frustration, they are in the same conceptual plane. They are all ways in which a person can be, let us say, hurt. The underlying agreement is thus that the subject of punishment is a living being capable of sensation, emotion, and so on. But this assumption is false for legal punishment. Corporations are punished, for example, but it would be a ludicrous analogical extension to call one pained by the experience. And it would hardly help clarify the nature of the extension to call the case atypical. But to make my position clear, let alone plausible, I must distinguish between legal and non-legal conceptions.

There is a tendency to confuse them. If I buy a piece of property, a house, we all say, without a blush, that I own a house. But one must distinguish between the physical entity, the house, and one's rights over that physical entity:

> In a strict legal sense, land is not 'property,' but the subject of property. The term 'property,' although in common parlance frequently applied to a tract of land or a chattel, in its legal signification 'means only the rights of the owner in relation to it.' 'It denotes a right over a determinate thing.' 'Property is the right of any person to possess, use, enjoy, and dispose of a thing.' Selden, J., in Wynehamer v. People, 13 N. Y. 378, 433; 1 Blackstone's Com., 138; 2 Austin's Jurisprudence, 3rd ed., 817, 818...[6]

In short, one must distinguish between property and its subject.

Usually these two go together. If I go to Sears and buy a table saw, I get a saw and a table top, but no legs. That is what Sears calls a table saw. They could mislead more subtly by selling me a table saw and telling me, upon my query about the legs, that they sold me rights to a table saw and that though some of those rights had subjects, some did not. Given what usually happens and the presumption that creates, this would be outrageous, but conceptually it is acceptable. For property and its subject are different kinds of entities.

A particular house, for example, occupies a particular spatial location for a particular temporal duration.

It begins to exist because concrete is poured, bricks are laid, and so on, and it ceases to exist because it burns, or collapses, or deteriorates. A person's property, his ownership of that house, is his legal right to reside there, his legal right to prevent others from entering it, and so on, and these rights can cease to exist because the legal system under which they are operative has ceased to exist, because relevant laws have changed, and so on. One's property can thus cease to exist for reasons having nothing at all to do with the existence of the house, and vice versa. For the house may cease to exist without affecting the owner's property rights. An insurance company could not refuse to pay him by claiming that since it burned down, he no longer owns the house he insured.

This distinction between legal and physical entities can be made in regard to persons. In one sense of persons, "We ascribe to ourselves <u>actions</u> and <u>intentions</u>. (I am doing, did, shall do this); <u>sensation</u> (I am warm, in pain); <u>thoughts</u> and <u>feelings</u> (I think, wonder, want this, am angry, disappointed, contented); <u>perceptions</u> and <u>memories</u> (I see this, hear the other, remember that)."[7] But someone can be a person in this sense and not be, as it were, a legal person. The man without a country had sensations, intentions, and so on, but if he was truly without a country, had no legal status at all. He had no rights, obligations, powers, immunities, or privileges. He could not make contracts, sue, or vote. And someone may be a legal person and not be a person in the sense of having sensations and intentions. We may say of a dead person that certain of his rights and even obligations still exist. We need not say this, for we may also say that his estate has these rights. But Hume's analogy of a person with a republic can fit here. For we can look upon a legal person in the way in which Hume asks us to look upon a person--as a bundle of rights, powers, privileges, immunities, and obligations created by law and held together by legal ties. The subject of that legal person, the living being with sensations and intentions, can change and even die without the legal entity ceasing to exist.

I do not mean to create a new world of entities, legal shadows of physical entities. It will not do to say of a legal person what one judge said of property, viz. that "...these things (like houses), ...though the subjects of property, are, when coupled with possession, but the <u>indicia</u>, the visible manifestation of invisible rights, 'the evidence of things not seen.'[8]" One needs conceptual mastery, not legal mystery. I must leave unclear for now the ontological status of legal entities, but it is clear enough, I think, that one's legal status can be affected without affecting one's physical status. And that is enough for me to proceed.

Punishment involves a change in a person's legal status, a change, that is, in his legal relations. Obviously not every change is punishment. One changes one's legal relations by marrying, but only as a joke is that called legal punishment. But some changes in legal relations are clearly instances of punishment. Given that it is done for the commission of a crime, done properly, and done by the proper authorities, to deprive a citizen of the right to vote is to punish him; he is being prevented from exercising all the rights of a citizen. To require a citizen to pay a fine is to create an obligation a citizen not fined does not have. To incarcerate someone is to punish him; he is being deprived of the right to determine for himself where he shall sleep, where he shall walk. What particular forms of punishment a legal system adopts is a matter to be determined by utilitarian considerations, by its commitment to some rights over others, and so forth. In our system, for instance, we punish a thief by depriving him of rights regarding his freedom of movement. In other systems a thief may be punished by creating an obligation for him to pay the person robbed. But none of these variations affect the point that legal punishment involves changes in a person's legal relations.

The status of this point needs to be made clear. I am not denying that what we call legal punishment in fact often involves pain. All of us are cognizant of how much pain is involved in incarceration, though "pain" and its variants hardly seem of sufficient diversity to cover all that can go wrong, such as crowded jails, loss of privacy, loss of a sense of dignity, knifings, fear of bodily harm, etc. It would be foolish to deny that these occur. I want rather to emphasize that they do. For my claim is not that they do not occur, but that they <u>cannot</u> occur and be legal punishment. This is not to make a moral claim. I am not claiming that they ought not to occur, though that is true too. I am making an ontological point, one about the status of legal entities and about what is appropriately ascribed to them.

The analogy with property and land is helpful. Legal authority can create rights in regard to land, remove them, change them; it can create powers in regard to trespass on land; it can create an immunity from taxation. But none of these changes change the physical properties of the land. They change its legal status.

In the same way, the only way a legal system can affect a person is to change his or her legal relations. It can grant a right to vote or remove it; it can create powers to marry and divorce or remove them; it can create obligations to pay taxes. All of these affect only the person's legal status.

This is not to deny that some changes in legal relations have causal consequences for the subjects of those changes. If a person is incarcerated for a period of time, the incarceration may cause loss of sleep and weight, loss of muscle tone, and so on. But these are causal consequences of a change in legal relations which are contingent. The assumption with which I disagree is that punishment <u>must</u> involve pain. It clearly need not if pain is only a contingent causal consequence of a change in legal relations.

I am also not denying that some forms of punishment directly involve pain. If the captain of a ship sentences a sailor to twenty lashes, then the sailor will not be punished if he fails to get his twenty. But such punishment is not legal punishment. Again, the point is not a moral one. I am not suggesting that such punishment is somehow illegitimate and therefore not legal. The point is an ontological one. I am saying that such punishment essentially involves more than a change in legal relations, and a legal system can <u>as a legal system</u> affect only legal relations.

Thus the only way a legal system can do something that might involve paining a person is to deprive that person of a right not to be harmed. But if this is punishment, it is still the deprivation of a right, and it involves that deprivation primarily and only secondarily, at best, paining. For in depriving a person of a right not to be harmed, one is not thereby harming him, but putting him in a position where he can be harmed without legal recourse. One is guaranteeing that the state will not enforce any laws against anyone who harms him. So the deprivation of a right not to be harmed neither is, nor causes, pain.

If punishment involves essentially only a change in legal relations, and not pain, then a number of consequences of practical importance follow. It is to two of these that I wish to turn briefly.

Persons are immensely hurt by the present system of punishment in ways that are in no way related to their being punished. They are subjected to indignities, pained, suffer lack of privacy, etc. If punishment is a change in one's legal relations, then none of these other harms are required by punishment. More important, the most familiar case of punishment is a deprivation of one's right to liberty. But if that is the only change in legal relations mandated by the sentence, then one's other rights remain intact. In particular, one maintains one's right not to be harmed--either by other persons or by the state. The consequence of this fact for the present system ought to be clear. If the state maintains such conditions that those persons being deprived of their rights to liberty are harmed or put into a position to be harmed without legal recourse, then it must justify what it does in some other way than by saying that it is punishing those persons.

So one consequence of clarity that punishment is simply a change in one's legal relations is that the pain we normally associate with punishment is not required. Another consequence is that some things we do to persons is essentially legal punishment, though we do not call it that. In New York City in 1974, 80 percent of the felony cases were settled by plea bargaining. That is no aberration, for plea bargaining is widespread. What makes it effective is the present bail system, for that guarantees that those accused of committing a crime are kept in jail until they go before a judge to make their initial plea of innocent or guilty. Given the delays built into the system, and some that are not, oftentimes a person has been held for trial quite a number of months before he first sees a judge. At that time he faces an unpalatable alternative. He can leave jail that day by pleading guilty or he can plead innocent and return to jail to await trial. Such a system is not just, for it extorts pleas of guilty from the accused. An essential element in that extortion is that the accused are held in jail waiting trial because they cannot afford bail. But the only difference between holding someone in jail awaiting trial and putting him in jail after conviction is that in the former he is being deprived of his liberty without a trial. He is being incarcerated, in short, not for conviction of a crime, but for suspicion of the commission of a crime. The system recognizes this fact by using the time awaiting trial as time served in jail--if a guilty plea is forthcoming and the person has been on good behavior. We ought to recognize the essential similarity between such incarceration and punishment, and, recognizing it, we ought to combine prudence with conceptual clarity and, as best we can, prevent such incarceration from occurring before conviction.[9]

FOOTNOTES

1. "Punishment," in Paul Edwards, ed., VII ENCYCLOPAEDIA OF PHILOSOPHY 29, Macmillan, New York, 1967. Antony Flew makes the same moves, saying that "'punishment,' in the primary sense, ... must be an evil, an unpleasantness, to the victim" ("The Justification of Punishment," in H.B. Acton, ed., THE PHILOSOPHY OF PUNISHMENT 85, Macmillan, London, 1969).

2. K.G. Armstrong, "The Retributivist Hits Back," in Acton, op. cit., p. 41.

3. C.W.K. Mundle, "Punishment and Desert," in Acton, op. cit., p. 68.

4. Anthony M. Quinton, "On Punishment," in Acton, op. cit., p. 61.

5. K.E. Baier, "Is Punishment Retributive?", in Acton, op. cit., pp. 130, 132.

6. Quoted in Wesley Newcomb Hohfeld, FUNDAMENTAL LEGAL CONCEPTIONS 28-29, Yale Univ. Press, New Haven, 1919.

7. P.F. Strawson, INDIVIDUALS 89, Methuen, London, 1959.

8. Quoted in Hohfeld, op. cit., p. 29.

9. I wish to thank Michael Pritchard and Richard Pulaski especially for their objections to earlier drafts of this paper.

HUMAN NATURE AND THE STATE

Edward Sankowski

This paper will be about issues connected with what might be called "the argument from the evil or defects in human nature." This argument can be regarded as the germ of a position attempting to justify the necessity of a state; that all things considered it is better for there to be a state of some sort or other. Those who accept the argument from evil can be construed as committed, explicitly or implicitly, to the defense of certain values which prominently include a kind of freedom and a kind of equality. Thus it is often held that the flourishing of some sorts of especially valuable freedom, freedom to engage in justifiable activities, requires the coercive curtailment by the state of other sorts of freedom, freedom to harm others unjustifiably. And the coercive legal mechanisms of the state are also often held to be necessary to protect some valuable sorts of equality. The law, on such a view, can, when operating properly, equalize relations between the more and the less powerful, for example.

It is in the protection of such values, according to defenders of the argument from evil, that the justification of the thesis of the necessity of the state is to be found. This thesis is supported by claims about the evil or defects in common human nature, at least in any sort of environment which could reasonably be expected to exist. If the argument from evil fails, as I think it does in the form in which it is usually advanced, that will not of course demonstrate that a state is unnecessary. If it can be shown that the contention in question fails, however, one common argument for the necessity of the state will be shown, at the very least, to need much subtler development and elaboration if it is to work at all. Moreover, corresponding to comparatively general questions about the necessity of the state are more specific issues of what to do about particular social problems. Should the remedy be primarily through the action of the state in legislation, (and if so what sort of legislation), or should it not be legislative at all? Political positions may be classified usefully by the range of social problems which they seek to solve through the implementation of legislation attaching penalties to behavior. Although I do not here essay any very detailed study of a particular social problem, the following reflections should also suggest questions worth asking about the reflex for seeking certain sorts of legislative remedies for social problems.

The existence of a range of human propensities for anti-social behavior poses a problem on two levels, practical solutions and philosophical theories of social organization. One way to handle such propensities is to design legislation which will deter the behavior by affixing penalties to it. If this solution is defended for a substantial proportion of problem cases, the way is clear for a familiar line of argument for the necessity of the state. Whatever the variations about details, in its most general form the argument from the evil or defects in human nature is of the following skeletal form:

(1.) Ordinary human nature is such that human beings commonly have anti-social impulses which it would be wrong to gratify. The existence of a state, and only the existence of a state can control the gratification of some of the impulses mentioned by affixing penalties to the undesirable behavior.

(2.) Some of the impulses mentioned in (1.) which only the existence of a state can control, must be controlled (because, e.g., their gratification would prevent the realization of personal and social goals which ought to be realized).

(3.) Therefore, on balance, a state is desirable.

It is natural to take this pattern of argument as invoking the necessity for limiting some sorts of freedom through coercive law to protect other sorts of freedom. The argument can also be construed, however, as involving some sort of equality, e.g., the equal right of all in similar positions to protection by the law, as we shall soon see in more detail. Indeed in many contexts the relevant forms of freedom or equality seem to be two aspects of the same value. Critics of the argument from evil also depend explicitly or implicitly on an appeal to freedom and equality. The crucial difference between defenders and critics of the argument from evil, as I hope to show, is not that one side depends on freedom and equality while the other does not. The crucial difference is in the kind of freedom and equality valued and in the conception of how best to defend what is valued.

Among the classical political philosophers, Hobbes, for example, clearly expresses views with the purport above. LEVIATHAN is littered with comments like these: "Men have no pleasure, but on the contrary a great deal of grief, in keeping company, where there is no power able to over-awe them all ... <u>Out of civil states, there is always war of every one against every one</u>."[1]

Rather than exploring for my purposes various classical philosophical arguments for the necessity of the state, however, I would prefer to document some of the <u>actualité</u> in the position I am examining by considering relevant contemporary views, of which there is no shortage. For example, Robert Nozick's argument for a minimal state in his ANARCHY, STATE AND UTOPIA[2] could plausibly be fitted into the framework above. Also, although the terms of Nozick's essay would need to be translated, he could be construed as defending the state when and only when it protects favored varieties of freedom and equality: The sort of freedom associated with the exercise of certain specific individual rights, the sort of equality manifested in all having similar sorts of rights, despite the fact that Nozick attacks some prominent kinds of egalitarianism. Showing all this in detail would, however, require considerable rearrangement of Nozick's argument.

I mention Nozick's book because if it presents a variant of the argument from the evil or defects in human nature, this suggests all the more forcefully how widespread is the acceptance of this style of argument for the necessity of the state in some form or other, and how much acceptance cuts across many political viewpoints. But let us turn to a clearer contemporary example. Richard Taylor argues that there are some rather restricted circumstances under which anarchy (by which he means roughly the absence of government) is a viable social condition. He writes:

> A state of anarchy or absolute liberty is ... possible, given the world and men precisely as they are, but only on a minute scale. Within a family one may leave his belongings lying about without fear of theft and turn his back to the others without the remotest fear of assault, but it is far otherwise when he ventures forth into those larger bodies politic which we call the city, the province, the state. Here one can by no means be sure that the man behind his back intends him no harm, that his goods will not be

coveted and, at the first opportunity, seized, for
the love that bathes the home life cannot be assumed
to exist outside in any sufficient force to abolish
all concern a man may have for his safety. In the
larger society, in which the love of man is weaker,
something else is needed in its stead, and this
something else can of course be nothing but laws of
restraint. There is needed, in short, the coercive
force of government. Where love is wanting, freedom must be compromised in the sheer interest of
safety. Abstractly considered, it is possible that
the day might come when the utopia that actually
exists in the ideal and sometimes real family circle will become sufficiently comprehensive to include much larger societies, but there is not the
slightest sign that such a day is near, and there
are perpetual reminders that the greed, vice, and
callousness of our race are as durable as the race
itself. This is enough to insure that the age of
anarchy, if it is to be anything better than pure
savagery, will never come. The deprivation of freedom, the corruption of the anarchical state, is of
course a great evil, being the deprivation of an
unqualified good; but it is a necessary one, for
the threat to body, life and property that otherwise flourishes in every quarter is a greater evil
still, being the threat of frustration of _every_ aim
and purpose.3

Now Taylor's assumptions do seem to fall into the pattern set out above. By implication his mentioning the
perpetual reminders that "the greed, vice and callousness of our race are as durable as the race itself" as
well as the reference to "laws of restraint" correspond
to the first step in the argument, the references to
"body, life and property" and "the threat of frustration of every aim and purpose" correspond to the second
step, and the conclusion is clearly that "the coercive
force of government" is desirable. This fits the general form of the arguments from the evil or defects in
human nature. In his book Taylor is explicit in invoking the coercive force of government as necessary to
protect freedom of a sort by limiting freedom, presumably of another sort. By implication, the law could
also be construed on Taylor's view as equalizing relations between those more powerful and those less powerful in certain respects, and between those more scrupulous and those less scrupulous in honoring the rights
of others.

A variety of social philosophical positions either
explicitly or implicitly would cast doubt upon the
argument from evil in different ways and to different

degrees. The most thoroughgoing repudiation is to be
found in certain "anarchists". But there are relevant
views defying conventional classifications which emphasize the application of a technology of behavior, reject the concept of personal responsibility, and in
particular downplay punishment as a method of behavioral control, all on supposed scientific psychological
grounds (e.g., B.F. Skinner and others). There are
some "liberals" who may favor fairly extensive legislative action to cope with anti-social behavior, but
much of it indirect action which does not affix penalties to undesirable behavior, rather aiming to change
circumstances held to generate the undesirable behavior. There are some sorts of "conservatives" who are
skeptical about how much good any government can do.
There are a few among the "Marxists" who question the
value of the state, and there are various others whose
positions are more difficult to label.

Defenders of some version of the argument from evil
probably represent and always have represented the generally accepted view against the critics. Yet defense
often takes the form of a misunderstanding of the criticism. This can be illustrated if we consider replies
by defenders to those anarchists who criticize and reject the argument. Such critics of the argument from
evil are accused of having an unrealistic belief in the
goodness of human nature. Most often it is assumed by
the defenders that such critics of the argument from
evil flatly deny that the state should exist, and that
they challenge the force of the argument from evil by
denying that human beings have serious anti-social tendencies. Taylor defends the argument from evil on this
basis. Such a defense, however, misconstrues much of
the opposition.

To begin with, although some anarchists repudiate the
rightness of the existence of any state under any circumstances, many allow that under certain circumstances
the existence of a particular sort of state is justifiable, and some are even uninterested in maintaining
that under the best of circumstances the state should
not exist in any form. (Paul Goodman and Robert Paul
Wolff maintain the last sort of view). But even those
anarchists who are more persistently and directly opposed to the state as such, far from denying the evil
or defects in human beings, often explicitly acknowledge and emphasize for their own purposes the anti-
social aspects of human nature. This can easily be
documented with thinkers like Godwin, Proudhon, Tolstoy,
Kropotkin and others. Often, so seriously do such
critics of the argument from evil take the dangers in
human nature (whatever the sources of the dangers) that

they argue that it is dangerous and unwise to centralize power in the ways demanded by the state. Their recognition that human nature has its anti-social aspects thus links up with their advocacy of decentralization of power. Proposals for decentralization are usually grounded in specific ideals of both freedom and equality. Very often, the favored sorts of freedom and equality are conceived as fitting together. Enhanced freedom for more persons and a greater relative equality in decision making power are generally thought by such theorists to be valuable consequences of the application of decentralist proposals.

There are, however, as I have mentioned, theorists of various other perspectives whose positions would cast doubt upon the argument from evil. Broadly, positions critical of the argument from evil challenge in various ways one style of social organizational strategy--legislation and implementation of a law by which penalties are attached to the unjustifiable behavior. Of course there are considerable differences in worth of the criticisms and alternatives presented by the variety of positions mentioned. For example, while some theorists who stress behavioral technology appeal in an interesting way to evidence of the unnecessary damage that can be done by punitive means of social control, such theorists are often insensitive to other sorts of issues such as the possibility of highly centralized and manipulative though non-punitive organizational styles. Similarly, while extreme individualistic anarchists and conservatives sometimes raise genuine issues about the capacity of government to resolve certain kinds of social problems, such positions are often both flawed by a very confined account of the sorts of abuses which characterize non-governmental social organizational forms of various kinds and also flawed by a correspondingly confined conception of viable organizational alternatives which might provide fresh sorts of remedies for social problems, in accord with humanistic values. Threats to freedom and equality emanating from the state are only some of the threats to freedom and equality which need pondering; but comprehending all the significant relevant dangers demands reflection on the varieties of freedom and equality, and what cancels them, within many sorts of governmental and non-governmental organizational forms.

Many of those whose positions are opposed to the spirit of the argument from evil share a suspicion of concentrated power. On such views, concentrated power tends to curtail the most estimable sorts of freedom and equality. Such freedom and equality are often conceived as logically consistent with and indeed mutually demanding one another. Admittedly critics of the argument from evil differ among themselves significantly in the range of concessions they make to the necessity for centralization and in the sort of decentralization espoused. Still, those whose positions are opposed to the argument from evil tend to converge in a sympathy for decentralization of power. As we have seen, often this is partly supported by evidence of the evil or defects in human nature to which proponents of the argument from evil appeal. Explicit critics of the argument from evil often see centralization as risky since those who wield disproportionate power in and through the state presumably share the defects imputed to human beings. Indeed, some such critics claim that many of those who wield power through the state are likely to have graver anti-social characteristics than most persons, either because of the tendency of positions of power to attract anti-social, manipulative types or the deleterious moral effects of holding power, or both.

The problem for defenders of the argument from evil is exacerbated if any or all of three propositions are accepted by proponents of the argument. First, the difficulty is intensified if no radical difference is acknowledged in favor of those directly involved in government (and implementation of the law) over the governed with respect to anti-social propensities. Thus a kind of descriptive egalitarianism (that is, the idea that persons are in fact roughly equal with respect to anti-social propensities) which is psychologically plausible tends to conflict with the rational acceptance of the argument from evil. An elitist political philosopher like Plato evades this particular kind of internal tension, while remaining open to other objections. With the spread of democratic and egalitarian ideas, as well as some growth of a connected tendency to demythologize the personal qualities of holders of political power, however, apologists for the relevant sort of elitism face severe difficulties. Second, the problem for defenders of the argument from evil is intensified if human nature is condemned as anti-social at its very root in an extreme and sweeping way. Such condemnation only encourages all the more questions about why government itself is not also (perhaps even more disastrously) infected by a corresponding weakness. Third, the difficulty for defenders of the argument from evil is of course intensified if they adopt a plausible and widely held view that legal systems generally share the defects of the persons who legislate and implement the law. Strengthening their defense would require them to explain how this could not be so. Otherwise the objection would go unanswered that the potential loss of freedom to coercive law, and the inequality between those exercising influence on and through the law, are necessarily more ponderable than the freedom and equality protected by the law. Ancient views on which a superstitious reverence for the laws might be rooted in a belief in the divine or quasi-divine origin of the laws no longer seem acceptable. The difficulty of finding some plausible demythologized substitute is considerable.

Criticism of the argument from evil challenges the capacity of defenders of the argument to deal adequately with the evil or defects in human nature. In part this criticism can make a case without itself offering alternative social organizational solutions to the problems presented by human anti-social propensities. Plumbing the depth of the criticism at a more profound level, however, seems to demand a further development of the theory of social organization. For example, although we can make a tentative judgment, we cannot yet with complete confidence say what sorts of freedom and equality are more to be favored and how this bears on the defenders and critics of the argument from evil.

Maintaining the views advanced in this essay suggests the need for a reconsideration of the relations between philosophy of law and its close relative, political philosophy, on the one hand, and social philosophy, on the other hand. A change from the perspective of many philosophers seems desirable. Philosophical thinking about social organization has too often been dominated to an excessive degree by the philosophy of government and law. Punitive legislation, and indeed legislation of any sort, need to be seen as constituting only some and by no means all the important methods for solving social problems. Without scanting the importance of the philosophy of government and law, a thinking through is needed of the sorts of non-governmental social organizational forms which could perhaps be encouraged by legislation and other means, and which might remedy social problems generated at least partly by human anti-social propensities better than punitive legal methods. This requires extensive philosophical reflection on all sorts of social organizational forms—school, union, business, and so on. Such reflection must take account of internal structure and external relations, including, among the latter, the degree and quality of dependence and influence on governmental and legal systems. Only in the light of a much more developed philosophy of social organization than exists at present could all the issues involved with the argument from evil be examined in a fully satisfactory way.

It is somewhat ironic that on the level of much current theory, we can find a verbal commitment to a philosophy skeptical of the extensive uses of state power expressed by discussing almost exclusively the state and its proper limits, comparatively little worthwhile being said in detail about non-governmental organizational forms. Explaining the proper limits of the uses of state power is certainly important; but then so is extensive critical and imaginative discussion of non-governmental organizational forms. It is all the more to be expected that any philosophy skeptical of ultimately satisfactory solutions to social problems by extensive uses of state power should explain what modes of social organization are preferable and why they are preferable. Philosophers suspicious of governmental solutions to human problems must beware of becoming so fascinated by what they are criticizing that their expressions of their own views come to be shaped more and more by what they oppose. Nonetheless, such theorists may usefully emphasize how those who readily and unremorsefully have recourse to exercises of state power to promote a particular pattern of personal or communal freedom and equality, too often overlook the possible loss of valuable freedom and equality to the state which is empowered to carry out the right policy. Until a more adequate, broadly-ranging philosophy of social organization is developed, those with positions explicitly and implicitly opposed to the argument from evil are surely helpful in reminding us of the difficulties in advancing punitive legal remedies for the weaknesses in human nature.

FOOTNOTES

1. Chapter 13.

2. Basic Books, New York, 1974.

3. FREEDOM, ANARCHY, AND THE LAW, 44-5, Prentice-Hall, Englewood Cliffs, 1973.

THE PARADOXES OF EQUALITY

Alejo de Cervera

A frequent contention is that we should always strive after equality until equality is reached, or, more sophisticatedly, irrevocably reaching higher and higher degrees of equality. Others contend that we should strive towards differentiation or hierarchy. The relationship between the former contention and the latter is thought of as opposition. The corresponding drives are also thought of as opposed drives. There are those who think of the possibility and desirability of the definitive triumph of one of the two drives. To promote the former drive, the "causes" of inequality are investigated, while on the other side justifications are given of hierarchies.

In spite of their conventionality, or perhaps because of it, something must be deceptive in the preceding statements because, from the beginning, loaded words are used. In effect, the statements above give the former attitude the word "equality," always inseparable from the feeling of justice; it could hardly be otherwise because the former attitude has the monopoly of the word, a monopoly perhaps claimed and defended in good faith; defenders of this attitude proclaim themselves "egalitarians." By the same token, the word is denied to the latter attitude, its low-keyed claims to the word being largely neglected, either because of lack of grounds, or because of incapacity to correctly articulate the claim.

The word "inequality" (as suspect as "equality" is ingratiating) is routinely ascribed to the latter attitude, an ascription that advocates of the latter attitude try (unsuccessfully) to avoid by using the words "differentiation" and "hierarchy." Of course, advocates of the former attitude would rather use every occasion to stigmatize the latter one with the word "inequality," as when investigating the "causes of inequality;" and on many occasions, the defenders of the latter attitude seem resigned to such an allocation.

Consequently, there must be more than meets the eye concerning equality. If we are to avoid the distorting effect of loaded words we must make an investigation which would attempt to disclose what is involved in the alluded to attitudes and drives. Only then would it be possible to make a rational judgment concerning attitudes and drives.

A first step is to realize what experience teaches if only we open our eyes, namely, that the drive towards equality is always circumscribed to a group, tending to the equalization of all individuals included in such a group. Unless it is a question of the human race, there always will be individuals left outside. It is unlikely that the group will be humanity as a whole. Therefore, we may well proceed without any further concern for such an eventual role for humanity. This is also the case with respect to the drive towards hierarchy.

Egalitarians tend to use the national group within which to promote the egalitarian drive, if only because of the multiple means at their disposal to promote equality (welfare, taxes, etc.). To think of some other group smaller than the national one is rather unusual in "egalitarians." When reminded of their emphasis on the national group, "egalitarians" feel uncomfortable and begin to think of some larger group, at times humanity itself. They immediately realize, however, that there are hardly any means for achieving the desired equality beyond the national group. The most important drive towards hierarchy remains also circumscribed to the national group.

A second step is to realize that there are many groups, and that all of them determine and include an equalizing drive of all the individuals integrating the group. This is common experience. The very fact of the elaboration of the group of professors of law, stockbrokers, nurses, dockers, tends to the elaboration of equal conditions of living, and, consequently, of the individuals themselves. It is only natural because any group entails an emphasis on all that is common, and a neglect of all that is singular. A deliberate drive towards equality inside such a group emerges very soon, and any tendency towards hierarchy is frowned upon. We will see, however, that any group also determines and includes a drive towards hierarchy.

Both drives are ingredients in the elaboration of groups. On the one hand, the perception of common traits, with or without awareness, raises a feeling of equality, which provides an immediate spring to the elaboration of groups, so that they all strive after equality, and cannot but strive after equality. On the other hand, the perception of differences supports the aspiration towards hierarchy; also a necessary drive as will be further explained later. However, the necessity of both drives does not say anything concerning the groups to which the drives will be circumscribed. In fact, we can select the necessary group. Here is where the problem begins. Some would be satisfied with equality inside a club; some would be satisfied with hierarchy inside the army.

To proceed to a third step, we would do well if we consider the articulation of groups, and distinguish between vertically articulated groups (those groups in the relationship of the-one-inside-the-other) and horizontally articulated groups (those groups in the relationship of the-one-outside-the-other).

IMPACT OF DRIVES IN VERTICALLY ARTICULATED GROUPS

In vertically articulated groups, every drive towards equality gives rise of necessity to the drive towards hierarchy, which, in turn, gives rise to another drive towards equality, which, in turn, gives rise to another drive towards hierarchy, and so on in an unending chain. The presence of both drives, and their necessity, may be seen in the march towards smaller groups and in the march towards larger groups. Correlatively, both marches must prove to be also necessary.

The march towards smaller groups is unleashed of necessity by singularities. No matter which group we have in mind, the individuals have singularities to which the group pays no attention. As soon as the singularities are shared, and some surely will be, an effort is made to form a smaller group whose concern will be such singularities. The equality drive will also be immanent in the new smaller groups, but new singularities will also emerge which will prompt another smaller group, and the process may repeat itself threatening atomization. In this context, the trend towards hierarchy is prompted by the fact that many interests (in singularities) are classified according to notions of importance: The trend towards smaller groups implies thereby a trend towards hierarchy inside the larger group.

The march towards larger groups is unleashed also of necessity by those traits held in common beyond the group. As no group can provide the commensurate frame of reference for those interests held in common beyond the group, the very presence of those interests encourages the elaboration of a larger group. The drive towards equality inside the larger group is bound to make the drive towards equality inside the smaller group a drive towards hierarchy. Other interests will emerge which would claim for another still larger group, promoting both equality and hierarchy, and the process may repeat itself as though in an attempt eventually to encompass all of humanity.

The two marches might be thought to correspond to two processes. It would give a better picture, however, to hold that they integrate one and the same process moving in two directions. Accordingly, the drive to equality and the drive to hierarchy are no longer seen as being in opposition. A drive towards hierarchy means a drive towards equality elsewhere, and vice versa. More accurately stated, both drives are really one and the same, splitting into two contrary directions. All such social drives are equalizing and hierarchizing by the very fact of the drive (not to use the expression "at the same time," which inaptly involves considerations of time). In addition, from the fact of the necessity of both drives the conclusion is legitimate that the drive towards equality cannot suppress the drive towards hierarchy, nor vice versa. Definitive triumph of one of the two drives is unattainable.

We may further realize that every equalizing drive consolidates and fortifies the group to which the drive is circumscribed, thus postulating a better hierarchical position for the group. My drive towards equality inside a club is apt to be considered by others as a drive towards hierarchy. But to exercise power the group needs hierarchical organization. Therefore, hierarchy inside a group also tends toward the consolidation of the group, up to a point. If pressed beyond that point hierarchy tends to destroy the group. The drive towards equality must again take preeminence at some point, and so on again in an unending chain, or the group is destroyed. But, of course, some other group must take over, with the same inside automatism. We can understand now why any tendency towards hierarchy (necessarily inside a group) is frowned upon. It does tend to the elaboration of minor groups inside the larger one, apt to fight the larger one, with special interests, thus to the detriment of the larger one, whose interests must suffer.

Hence, there is after all some opposition. But it is a special opposition because both drives also merge, while opposing each other, in a variety of combinations: (a.) The drive towards equality helps the drive towards hierarchy in all the larger groups. (b.) The drive towards hierarchy helps the drive towards equality inside the groups whose hierarchical position is promoted. (c.) The drive towards equality inside a group goes against the drive towards equality of any other more or less concentric group. (d.) The drive towards hierarchy inside a group goes against any drive towards hierarchy inside the groups whose hierarchical articulation is promoted.

In personal terms, my drive towards equality, necessarily inside a group, means an attack against the hierarchy in which the inside groups are articulated, and thus against the corresponding drives towards equality, tending to destroy them, and means also an effort to increase the inequality inside any larger group. My drive towards hierarchy inside a group means a helping hand to equality inside the resulting groups, but may become a threat to the group inside which we are attempting the equalizing drive.

If we now try to elaborate a complete statement, we might say: 1. There is tension, as described, between the drive towards equality and the drive towards hierarchy. 2. There are tensions between the same drive--located in different groups; when equality (or hierarchy) is losing ground inside a group, it is winning ground inside some other group. 3. Either of the two drives is helped by the other drive if located elsewhere. 4. The articulation of drives and different groups shows the impossibility of the definitive triumph of either of the two drives.

IMPACT OF DRIVES IN HORIZONTALLY ARTICULATED GROUPS

The drive towards equality has a special impact on external groups, even though the eventual presence of larger encompassing groups always end by imposing some influence bound to render applicable even here the considerations made when studying the impact of drives in vertically articulated groups. But if only horizontally articulated groups are considered, the drive towards equality inside any group necessarily tends to establish a pattern of differentiation with respect to outsiders. A mere collection of individuals is apt to register a full spectrum of degrees of differentiation. But if some of the individuals elaborate a group, a tendency towards equality will arise which automatically will impose itself as differentiation with respect to those on the outside. Automatically also, the individuals located outside will tend to feel as aggressive the drive towards equality inside the group which makes them outsiders. As a means of defense, they will tend to elaborate a group of their own, or many groups, with the same drive towards equality inside each one.

The two eventually attained degrees of equality will be located, however, at different levels depending upon the groups, so that a tendency will result towards a separation of the two groups ever more complete, and towards a classification of individuals more uniformly unequal. Consequently, any drive towards equality, necessarily inside a group, tends to the elaboration and consolidation of external groups, and thus, promotes other efforts towards equality, and imposes uniform degrees of differentiation.

NATIONAL GROUPS

National groups are the key groups because, while the national group is prevented by its own convenience from destroying the inside groups (it needs them), it may not tolerate serious challenge by any other group, external or internal, larger or smaller. Of all tensions and interplay of drives, those apposite to the national group must therefore be the most relevant.

If only the national group is considered, there is only a drive towards equality, and another drive towards hierarchy. Their articulation may be described as follows: Any step towards equality means the same step away from hierarchy, and vice versa. If many steps can be taken in any of the two directions, it can be done only by compressing the drive in the opposite direction; as the trend is apt to be corrected by the other drive, and the more compressed the latter the more apt will it be to introduce the correction. There must eventually arrive a moment of the sudden reversal of the trend, eventually heading to another reversal, and so on indefinitely. There must be, and there is, thus, automatism between the trend towards equality and the trend towards hierarchy. If not, if the march towards equality could be pushed to reach absolute equality, it would be a march towards death, as also would be, in such a case, the march towards absolute hierarchy.

As both drives are indeed necessary, necessity arises for a compromise at some middle point. Some points will be better than others, that is, giving better social results. These will be located at times nearer complete equality, at times, nearer rigid hierarchy, in constant movement.

In connection with other national groups, there is no automatism because of the lack of a commanding group encompassing the two sovereign groups. The drive towards equality resulting in the ever more complete separation of the two groups and the classification of individuals ever more uniformly unequal, is normally irreversible. The more complete is the equalization of the American group, the more clear cut would be its separation from the Mexican group, and the differentiation of their individuals. Of course, the drive towards equality can be reversed through the relinquishment of degrees of sovereignty. Though difficult, such relinquishment is not impossible, and history proves beyond a doubt the fact of the constant transfer of sovereignty from certain groups to other groups. In final result, the fact of sovereignty cannot ultimately prevent the elaboration of encompassing groups, and thus the interplay of drives corresponding to vertically articulated groups.

If the preceding considerations are correct, to be simply "egalitarian" is an illusion, and a dangerous one at that. Equality means hierarchy elsewhere. Defenders of hierarchy are actually defending equality in other settings. Equality brings about inequality regarding others. When Aristotle tells us that equality is not to treat all equally but to treat equally what is equal, so that the treatment can appear unequal if what is treated is unequal,[1] he has in mind the paradoxes of equality.

FOOTNOTES

1. Aristotle, NICOMACHEAN ETHICS, Book V., Ch.3. "If they [the persons] are not equal, they will not have what is equal, but this is the origin of quarrels and complaints--when either equals have and are awarded unequal shares, or unequals equal shares."

THE FULL DEVELOPMENT OF PERSONALITY IN SOCIALIST SOCIETIES

Peter Popoff

The concept of "person" is not primary, as non-Marxist social and legal philosophers assume, in understanding the relations of society and the state. Such a primary concept is an illusion or an intentional myth of the bourgeois ideologists of the contemporary capitalist world. Every man, as a person, is a member of some family, class, or nation and as such is in constant communication with other members.

It is clear, therefore, that Marx was correct when he said that personality is shaped by the social group. According to Marx there are three stages in the development of personality. The first is the primitive community, when the individual is still "merged" in the clan and his personality has not become distinct from the group. At this stage, there is no problem about rights and obligations and the relations between the person and society or the state.

At the second stage, after private property in the means of production has become established, man's personality detaches itself from the clan and becomes a member of either the exploiting or the oppressed class and a member of public society. In an objective sense this detachment of personality from the clan and attachment to one or the other social class based on property interests is a progressive step. But for the personalities identified with the class of oppressors it is a reactionary moment in which they share extremely selfish interests, opposed to the social interests.

Only the third stage--development of personality in socialist society--is characterized by progressive development of real equality and liberty among the members of society.[1] It is of supreme importance, therefore, to what society and to what class a person belongs. This was expressed by Karl Marx in his famous statement that "... the essence of Man ... is a combination of all social relations."[2]

The most important relations in determining personality are the relations of production. Accordingly, personality develops in one way in a capitalist society and in another way in a communist or socialist society.

However, to correctly understand the Marxist view of personality development, man as the totality of social relations must be supplemented by understanding of man as, simultaneously, a unique individual, having his own personal interests as well as his class interests. As Marx says many times, "Man is a particular individual and it is his particularity that makes him an individual...social being."[3]

The social and the individual in man's personality are dialectically contradictory--and just that ultimately characterizes man's personality in different contemporary societies. There is not and there must not be any metaphysical contradiction between man and society or the state, which would cause, or justify alienation of man from the state. Ortega y Gasset was wrong, therefore, when he said, "Man's life is always a life of the separate man; it is an individual, a personal life ... 'Human,' in the proper and original sense, is only that which I do out of myself with the conscience of my own purposes, or to put it another way, man's action is always a personal action, while society and the collective are not anything of man's."[4] This position of religious existentialism is false. Pretending to take a progressive position and to prove the alienation of man from the state, it aims, in fact, at confusing the workers on this subject.

In capitalist society, the alienation from the state, both for the working class as a whole and for its members, rests on the alienation of their labor in that the fruit of their labor is not destined for the amelioration of their own condition but only for the accumulation of riches by the capitalists. The capitalists and other exploiters are not alienated from the state. They support it and use it in defense of their own selfish interests.

This does not mean, of course, that in capitalist society the worker as a person is alienated from every social group. On the contrary, he lives in collectives--in his class party, his trade union, and his working collective. This is the "natural environment" for the development of the social part of his personality. This is what Marx and Engels meant when they wrote, in THE GERMAN IDEOLOGY, "... it is only in the collective that the individual obtains the means which enable him to develop universally his endowments and therefore it is only in the collective that personal liberty is possible."[5] Further, Marx and Engels revealed the falsehood of "collectives" in capitalist societies as well as the advantages for development of the person of true collectives under socialism. They said:

> In the surrogates of collectivities existing until now in the state, personal liberty has existed only for individuals developing within the limits of the dominating class and only for individuals of that class. The supposed collectivity in which individuals have collected thus far has always been opposed to them as something independent; and it has been a collectivity of one class against another. For the subordinated class it has represented not only an illusory collectivity but also new fetters. In the conditions of the real collectivity individuals will be given liberty for their own organization, and through it.[6]

By these statements, Marx and Engels foretold the liquidation of the alienation of the personality of the working man from the state and the rapprochement to the socialist state and socialist organization in general. At the same time, they showed that the socialist state and the collectivities of the working people are an exceptionally important condition for the realization of the liberty of the person. V.I. Lenin developed this teaching of Marx and Engels, saying that for persons in capitalist society liberty would always be what it was in the ancient Greek republics, namely, "liberty for the slaveholders."[7] Further, Lenin said that "... liberty and equality in the bourgeois system (i.e., as long as private property exists in land and the means of production) and in bourgeois democracy will remain only _formal_, indicating in fact a rented slavery for the workers (formally free, formally equal) and an omnipotence of capital--oppression of labor by capital."[8] This assertion is so convincing and so full that it is not necessary to say anything more about the characteristic of "liberty and equality" in capitalism, and still less in contemporary capitalism. Therefore, we shall only summarize the positive aspects of the Marxist-Leninist conception of the liberty and equality of the person.

Man's personality is a social phenomenon. The personality is the social essence of man expressed by the individual peculiarities of a thinking and acting being. Individuality is a product of historical development.⁹ That is why the liberty and equality of the individual exist and are realized to the extent permitted by one or another society. The government of a society regulates, in one way or another, the liberties and obligations of persons (for the ruling class under capitalism, or for all the working people in general under socialism). Thus arises the question of the legal status of the person as the "frame" for specifying rights, liberties and obligations of the person. It is clear that the nature of the state and law (whether socialist or capitalist, for instance) is decisive of the character of these rights, liberties and obligations. That is why the Marxist theory of state and law pays particular attention to the social aspects of state and law.

II.

We have already indicated that Lenin characterized "liberty," "equality" and "democracy" under capitalism as merely "formal." But Lenin also creatively developed these concepts under socialist society, saying, "... only in socialism does the rapid, real and actual mass movement begin in all fields of social and individual life, with the participation of the majority of the population."¹⁰

This is not demagogy or illusion on the question of liberty, as in the case of bourgeois or petty bourgeois ideologists. Lenin knew very well that "... absolute liberty is a bourgeois or anarchist phrase," and that "it is impossible to live in a society and be free of it."¹¹ He demonstrated this very clearly by founding the state of the dictatorship of the proletariat and socialist law; gradually abolishing private capitalist property, and consolidating the socialist property in the means of production. The socialist revolution destroys all the relations of bourgeois society--of production, property, law, etc., and thus eliminates the system of bourgeois rights and privileges for the exploiting classes, creating instead a new society with the broadest rights, liberties and possibilities for the working people.¹² This is connected with the abolition of every exploitation of man by man, with the relentless suppression of the resistance of the exploiting classes and the establishment of the socialist organization of society.¹³

The homogeneity of socialist society, based on the community of interests of workers, working villagers and the people's intelligentsia, is the basis for building equality in all fields of social and personal life. Thus is built up the system of fundamental rights, liberties, and obligations of the workers. This is the way a new society was built in Bulgaria after the socialist revolution. The new society was firmly set in the Constitution of the Bulgarian Peoples' Republic in 1971. The Preamble stipulates equality and liberty for all the citizens of the country. The Constitution also stipulates the elimination of the exploitation of man by man; the transformation of the means of production to public or cooperative property; the setting of the society upon "the threshold of complete social equality;"[14] the elimination of unemployment and ensuring the right of labor and free choice of a profession for all citizens of Bulgaria who will be brought to a happy, free and equal expression of their capabilities in all fields of material and spiritual production.[15]

Further, having in mind Marx's maxim that "man is a superior being,"[16] the legislature decreed in Art. 3 of the Constitution that "the state serves the people by ... assuring the free development of man, by guaranteeing his rights and by protecting his dignity." Art. 50 provides that, "Every citizen has the right of protection against illegal intervention in his private and familial life and against the violation of his honor and good name." These, of course, are only singled out from among the rights and liberties of the Bulgarian citizen that are legally protected.

The Bulgarian Constitution is a real "great charter" of the rights, liberties and obligations of citizens. It legalizes the right to labor (Arts. 40, 41); the right of rest following a five day working week and during a paid yearly leave (Art. 42); the right of free education and medical treatment; equality of men and women in all fields of social life, with special protection for the mother (Arts. 36, 37); equality before the laws and inviolability of the person, the home, correspondence, etc. (Arts. 48, ff.). It further legalizes the political rights to vote and to be elected (Art. 6) and liberty of organization, of speech, of the press, of association, of meetings, etc. (Art. 54). All three types of liberty and rights--socio-economic, political, and personal--are consolidated and guaranteed by the four possible means inherent in socialist society--material, political, legal, and ideological.[17]

Marx's prognosis about the "new kind of collective" of persons in the socialist state as "a center for securing a free, equal and social manifestation" of these citizens has become a living reality in Bulgaria, as well as in the other socialist countries of the world. Engel's requirement for "the necessity of unity of rights, liberties and obligations in socialism"[18] has been realized. The problems of the state and the problems of the legal theories in the Marxist-Leninist conception of liberty and equality of human personality are a living reality in the practice of the National Bulgarian Republic and that makes it a worthy partner in the family of socialist states, whose representatives discuss with dignity all the aspects of this great conception.

FOOTNOTES

1. K. Marx, F. Engels, ARCHIVES 89, V. 4, Moscow, 1935, (In Russian).

2. K. Marx, F. Engels, WORKS 3, V. 3, (In Russian).

3. K. Marx, F. Engels, FROM THE EARLY WORKS 591, Moscow, 1956, (In Russian).

4. See Jose Ortega y Gasset, EL HOMBRE Y LA GENTE 23 and 27, Madrid, 1957.

5. K. Marx, F. Engels, WORKS 75, V. 3, (In Russian).

6. K. Marx, F. Engels, Op. cit., p. 75, (In Russian).

7. V. I. Lenin, WORKS 432, V. 25, (In Russian).

8. V. I. Lenin, WORKS 350, V. 29, (In Russian).

9. See: D. E. Chesnokov, HISTORICAL MATERIALISM 412, Moscow, 1955 (In Russian).

10. V. I. Lenin, WORKS 99-100, V. 33, (In Russian).

11. V. I. Lenin, COMPLETE WORKS 104, V. 12 (In Russian).

12. See: A. G. Masochina, PERSONAL LIBERTY AND THE FUNDAMENTAL RIGHTS OF THE CITIZENS IN EUROPEAN SOCIALIST COUNTRIES, 62, Moscow, 1965, (In Russian).

13. V. I. Lenin, COMPLETE WORKS 221, V. 35, (In Russian).

14. See: THE PROGRAMME OF THE BULGARIAN COMMUNIST PARTY 66 a.f., Sofia, 1971.

15. See: Boris Spasoff, PROBLEMS OF THE NEW CONSTITUTION 38, 43 a.f., 128 a.f., (In Bulgarian).

16. K. Marx, F. Engels, WORKS 407, V. 1, (In Russian).

17. For details of the classification of these rights, liberties, etc., see: B. A. Patulin, "Universal Expansion and Protection of the Rights of the Citizens--Most Important Problem of the Socialist State," in: COLLECTED ARTICLES 63-86, edited by D. A. Kerimov, Moscow, 1965, (In Russian).

18. K. Marx, F. Engels, WORKS 235, V. 22, (In Russian).

Chapter Two

Participation

Democratic ideals, their historical development and institutional forms: What is the nature of individual participation in the new massive groups, governmental and non-governmental? What place is there for persons participating as functioning parts of an organism with definite common objectives? For persons participating as equal parties to a contractual enterprise? For face-to-face and participatory democracy?

In what ways can new forms of participation be developed or old ones refined to broaden liberties and secure greater equality in participation?

THE CONCEPT OF PARTICIPATION

Kazimierz Opalek

I.

"Participation" is a scientific term of recent date. It does not belong to the traditional language of the study of law (of constitutional law in particular), and is not to be found in the body of classical writings in political theory (or philosophy) and social philosophy. Nowadays, however, this term makes a career, both in the language of practical politics in numerous countries, in programs of manifold political parties, and in political science. It is more and more frequently used in jurisprudence, legal philosophy, and even in legal texts.[1] This development is undoubtedly influenced by sociology, the discipline in which the term was originally coined, particularly in connection with the theory of social roles. Leaving out of account some differences between various sociological conceptions--differences immaterial for the present considerations--participation is understood in general sociology as the "share" of the individual in the process of social interaction, this "share" consisting in some determined course of conduct. Participation so understood varies in its specific traits according to different orientations and expectations linked both with one's own behavior and with that of persons with whom one is interacting. The orientations and expectations in question are conditioned socially, situationally, as well as by the personality factor.[2]

These influences exerted by sociology are, in the present evolution in social sciences, by no means something out of the ordinary. This instance, however, is a peculiar one. The adoption of the term is not only a question of a current fashion (though even that to a certain extent cannot be ruled out); neither is it solely a question of some theoretical requirements. The term "participation" can be said to express adequately social and political demands and strivings one actually meets with on a large scale. The adoption of this term in the above-mentioned domains of practice and science is connected with its getting a new evaluative-ideological tinge of which it was previously devoid. Consequently, some changes in the meaning of "participation" are to be noted. These changes in turn exert some influence on the present uses of the term in sociology itself, particularly in political sociology.

The problem is not to be settled by considerations on the level of the language of social sciences only. Participation has become a motto and a postulate proclaimed by different social forces, in both Eastern and Western social and political systems. At the same time the meaning of participation has been the subject of scientific study.[3] The interest in participation is undoubtedly stimulated by tendencies to be observed in modern States, in their representative and executive agencies, in the new economic-administrative units of government, as well as in mass political organizations (political parties in particular), and interest groups. The tendencies are known under the labels of the predominance of the executive organs over the legislative ones, bureaucracy, technocracy (société structurée, formierte Gesellschaft), and even "refeudalization".[4] While finding acceptance in some circles, and justification in a number of social-political doctrines, these tendencies meet with opposition from others, who consider them a threat to democracy. There have been numerous attempts to counteract them by guaranteeing wide participation in the above-mentioned agencies and organizations, participation in decision-making, in the execution of decisions, and in the supervision of activities (<u>Mitbestimmung</u>).[5]

Thus, the wide use of the new term "participation", parallels some important realities of present political, social, and economic life. Participation appears, therefore, to be a new problem, specific to our time. On the other hand in another sense, and a relevant one, the issue is old and well-known. In its doctrinal foundations, in the processes by which democracy is widened, consolidated, and institutionally guaranteed, it is a problem inherent in every historic stage of the struggle for democracy.

Before the term "participation" was coined, political practice and theory, not without an important contribution by the law and legal thought, considered the problem to be basic, and made notable achievements in ensuring what has come to be called participation. These achievements consist of actually functioning formal (institutional) and informal (legally permitted actions) means of decision making with respect to composition of representative organs, questions of national policy, and influencing and supervising governmental activities.

Especially significant is the extension of civil liberties, including the formation of public opinion, and the gradual elimination of exceptions to the principle of equality, particularly regarding the right to vote.[6]

These matters are pointed out in a sketchy way only, as there is no possibility, or need, to deal here with the complicated features of the machinery of participation in contemporary democracy. It has been our intention merely to emphasize that the problem of participation is by no means a recent one, or a new discovery. In this domain we deal with attainments which are the result of a historical process, but also problems that are a component of present reality. Consequently, all realistic considerations of participation have to take account of these two aspects. One has to examine, firstly, the degree to which the present situation and needs are a "repetition" of former situations and needs, and to what extent they are new. Secondly, according to the answer to the first question, to what extent the existing forms of participation ought to be maintained, refined, and supplemented, and to what extent new forms have to be developed. One can hardly expect, as a result of the considerations to follow, full answers to these questions, or proposals of concrete solutions. The purpose of the present paper is limited to clarifying some aspects of the problems here involved, and thus contributing to the elaboration of a theoretical framework necessary for dealing effectively with questions and demands concerning participation in present time.

One of the most striking problems is the following. On the one hand we observe today intense incitements and strivings toward guaranteeing wide, or massive, participation in government and social-political organizations. On the other hand, however, the results of empirical research conducted by many scholars and teams of social scientists in full mutual agreement demonstrate that people in a mass do not display any vivid interest in participation, that only a very small number of them are active, while passivity, or even apathy, is pervasive. Some authors are disturbed, and not infrequently alarmed by this state of affairs.[7] Many others do not consider it a threat to democracy or an obstacle to its functioning.[8] The seeming paradox may be apparent only. It is possible that people's interests in participation are to be found in other domains than those subjected to the empirical research of political sociology, or the existing forms of participation may somehow fail to induce many people to participate. Anyway, these questions have to be examined further, the vital problem here being that of a realistically possible and desirable "amount of participation".

However, it is beyond the scope of this paper to describe the practice or the institutional forms of participation in a given country, or comparatively in a group of countries. In accordance with the character of the disciplines represented at this Congress, this paper concentrates on general methodological-theoretical problems. The starting-point is that of conceptual-terminological analysis. Such an approach has recurrently met with disapproval on the part of researchers of factual data as "formal" and unpromising for investigations of the social reality. However, the conceptual-terminological analysis will open possibilities for shedding some light on this reality. Further, in the literature on participation, containing a considerable number of works, conceptual-terminological precision is by no means the strongest point. It is not suggested that work on participation be restricted to conceptual-terminological analysis. But it is hoped that a full consideration of conceptual and terminological problems will be helpful in view of the fact that these problems are frequently neglected, or treated but marginally in detailed research.

II.

The problem of participation at present decidedly constitutes a domain of political sociology and political science. We are not going here into the much debated problem of the relation of these two disciplines. Both have attained considerable results in empirical research on participation. This research was initiated, and is still conducted on a very large scale in the United States, and it has gradually gained in importance in a growing number of other countries. Therefore, the research carried out in these two disciplines is the main subject of our attention.

At first glance one can observe the variety of terminology in use. We are dealing not only with "participation" <u>tout court</u> but also with "political participation". It is symptomatic, however, that the latter term also is found to be insufficient. Further qualifications are added by adjectives such as "popular", democratic", "massive", and "active". We also find other nouns substituted for "participation". For instance, we find references to "political activity", "political involvement", etc.[9] It is worth considering.

The term "participation" in its scientific uses and in current political discourse is ambiguous. The concept of "participation" lacks clearness and precision. These deficiences increase as "participation" is adopted in fields in which it was previously not used. (See above, Part I). In the original theoretical usage in general sociology the concept of "participation" was implied in the basic category of "social life" and meant the collective life of people in their interaction. The former is analytically derivable from the latter. The statement, "People participate in social life" is not a synthetic statement; it does not state anything beyond the definition of "social life" which is social in the sense that people are participating in it. Participation thus conceived simply amounts to "people being present together in interaction", this being constitutive of the social life and of its different forms.

Some might object--an objection which can be expected especially from those for whom the term "participation" is associated in meaning with a set of democratic ideals and postulates--that the concept in its general sociology version is useless. They would certainly commit an error, since such a general and "neutral" concept forms a proper basis for distinguishing kinds of participation according to the differing social roles of the participants, and for identifying fields of participation (e.g., politics, economy, education, culture) as well as social structures of participation. This amounts to bringing to light such theoretically important questions as the ways of participation, the subject-matters, and the frameworks (different groups, organizations among them) For this reason the concept of participation, discussed here, is fundamental, and in all considerations on participation proper attention has to be paid to the above three questions in order to avoid current misunderstandings. When speaking in practical politics, political science, and political sociology about "participation" one usually has in mind "political participation"; the latter explicit formulation being also widely used. In this way the concept as understood in general sociology undergoes some restriction as to the field of participation, and seems to be endowed with more concrete content with respect to the ways and frameworks of participation. Some of the implications of qualifying "participation" by "political" need to be considered.

The concepts of "politics", and of "political life", continue to give rise to doubts and to different views in scholarly circles. Thus, there is a marked hesitation as to whether politics should be defined as "... the area of formal government in its widest sense, including processes directly and indirectly associated with it", or as all areas (not only that of "formal government") of "... the potential or actual use of power".[10]

The first definition, when confronted with the problems pointed out above and with the current postulates, is evidently unsatisfactory. The stress is laid actually on problems of participation in various economic organizations, those wholly belonging to the sphere of private autonomy among others, on participation in professional organizations, in local power, and in manifold massive social organizations.[11] Such organizations not infrequently have, to a smaller or greater extent, some connections with, or involvements in, politics (in the sense of the first definition), especially when acting as pressure groups. The question of participation in such organizations is raised, however, in many instances, not as pertaining to the processes associated with the formal government, but explicitly as an independent social-economic problem. There are only some analogies between the latter and the former (related phenomena of bureaucratization, technocracy, etc.). Hence, one speaks by analogy about "economic democracy", <u>Demokratisierung des Betriebes</u>, and the like.[12] The second definition of politics, giving a wider denotation to the term, while well suited also to account for above mentioned phenomena, is nevertheless too broad, as transcending by far their scope. If one were to use such a concept of "politics" in the definition of "political participation", the latter concept would be devoid of the necessary operational value.

A similar significant hesitation can be observed in the reflections of a Polish author, J. J. Wiatr, on the concept of "public affairs". The alternative is either to consider them as an equivalent of "political affairs", in the sense of "... all questions connected with the use of power and the struggle for power" (while the author has evidently in mind the State's power only); or else, to define "public affairs" as those "... which are of importance for the given social group as a whole". The author, pertinently remarks that in the first instance "important economic problems", and many similar ones, would be left out of account, accepts the second definition, while realizing that also this one "... leads to serious difficulties".[13]

Facing these difficulties, the representatives of political sociology are mostly inclined to associate political participation with "formal government" only.[14] This step determines the scope of their detailed research which, consequently, does not pay due respect to some actual problems already mentioned. In some instances again (as, e.g., in the work of R. Batley, H. Parris, and D. Woodhead) a vague and expansible conception of "the political" is used to refer, for instance, to "popular participation" in such fields as housing, education, hospitals, and social security. The fact that all of these problems belong to "political participation" raises doubts about this use of the concept of "politics".[15] When applying a distinction of considerable renown in the literature of the subject, that of the types of political culture (or of roles, and of political behavior), made by G. A. Almond and S. Verba (see also the later work of G. A. Almond and G. G. Powell),[16] it may even appear doubtful if the subject-matter of the work discussed above can be termed "participation". Political culture being defined as "the pattern of individual attitudes towards politics", it is said to have either the character of the "parochial", or that of the "subject", or that of the "participant" political culture. As the range of problems R. Batley, H. Parris, and D. Woodhead are dealing with in their work is typically regional, it would be justified to say that these problems are not those of "participation" but those of "parochial political culture", i.e., one "... in which the inhabitants neither know nor care nor pass judgment on the policies of the government of the entire society".[17] One can refer here to the view that "... participation in local communal affairs ... is not political participation except in so far as the policies taken there are, in some clear way, related to policies propounded for, or administered on behalf of, the public as a whole".[18] Accordingly, one could consider the work in question to be concerned with "participation" but not with "political participation". We do not intend here to criticize the research of R. Batley, H. Parris, and D. Woodhead in itself. We are only pointing out, in this example, the lack of clarity in the concepts and terminology on the subject of participation. The distinction made by G. A. Almond, G. G. Powell, and S. Verba will be discussed below in its other aspects.

On one hand the concept of political participation, limited to affairs of "formal government" today is unsatisfactory. In current research we find this explicitly expressed, or implicitly demonstrated by giving "participation" a wider meaning than it has in traditional usage. On the other hand one can observe that phenomena that are undoubtedly "political" even when that term is restricted to affairs of "formal government", are excluded from the concept of "political participation". Here we have in mind manifold radical political movements (and, within their scope, individual political attitudes and behavior), representing the standpoint of negation of the existing political systems, that is to say, revolutionary movements, among them those which are called extremist, anarchist, terrorist, etc. In some views relating to these problems, so vital in our times, the opinion is expressed that such phenomena are contrary to "sound" political participation ("Gegenpol wirklich sinnvollen politischen Verhaltens")[19] This involves a valuation of some sort, this valuation, however, cannot change facts and must not leave them out of account, as these facts belong to political life.

The concept of participation certainly implies a share in "something common". The problem arises as to how this "common" is to be defined. One of the basic insufficiencies of the concept "political participation" in its common uses is just this lack of precision about what one participates in, or ought to participate in. We have two possibilities. One is that the term "political participation" refers to ways of conduct which do not exceed the limits of the existing political system, i.e., both the ways of conduct of formal (institutional) and informal (legally permitted) character. The second possibility is that the term "political participation" refers to any kind of mass and individual behavior, standing in connection with the affairs of "formal government", that is, including behavior that goes beyond the limits of legality. This would be, in fact, in full accord with the definition of politics cited above--"the area of formal government in its widest sense, including processes directly and indirectly associated with it".

Sociology and political science accept as a rule the first alternative, while only marginally treating phenomena of attitudes and behavior contrary to the political system, as anomalies and deviations.[20] For example, one writer, who uses the term "political involvement", expressly states that it "... does not apply to behavior designed to disrupt the normal operation of democratic political processes or to dislodge a regime from office by violent means; a general strike is an example of the former, and a palace revolt or coup d'etat are examples of the latter".[21] We do not intend to question the value of detailed research on political participation in the usually accepted narrower range (first alternative), according to some canon of problems in more comprehensive or more concise versions such as, "voting at the polls; supporting possible pressure groups by being a member of them; personally communicating directly with legislators; participating in political party activity; engaging in habitual dissemination of political opinion through word-of-mouth communications to other citizens".[22] We have in view only the fact that the picture that emerges from such research is hardly an adequate one, and that the theses on political participation are of limited, or even questionable value. Perhaps some qualifications or corrections are in order even for such commonly accepted statements as that the higher economic status, education and professional position are decisive of greater political activity,[23] or that, "Persons turn to politics only when their basic physical needs, such as food, sex, sleep, safety, affection, have been met".[24] The latter is questionable not only with respect to political behavior contrary to the system but also with respect to that within the limits of the system. (See, for instance, R. E. Lane's considerations on the problem of satisfying unconscious needs by political participation, such needs not infrequently being a consequence of the fact that certain basic physical needs failed to be satisfied.[25] Research on political participation in the narrower scope, furthermore, cannot serve to elucidate problems and situations regarding the very considerable percentage of the population listed in the rubrics of the "politically inactive", or "politically indifferent". Research conducted within the narrow meaning of political participation can deal adequately with instances of political activity, but is not adequate for the areas purportedly covered by current research, or for areas one would wish to have covered by research.

Research on revolutionary, extremist political movements, among others, constitutes one of the separate domains in the works of political sociology, and has some noteworthy results. Such a separation, however, entails certain negative consequences, leading to a deformation of the picture of political participation. There are some traces of dissatisfaction with the traditional scope of research on the latter subject. Here one can refer to the proposal of supplementing the three types of political culture, distinguished by G. A. Almond, G. G. Powell, and S. Verba, by a fourth type, of "oppositional political culture" "Training for revolution" enters here, among others.[26]

Now we intend to examine another aspect of the markedly frequent association, in political science and political sociology, of political participation with the State's affairs. Because the State is not subordinate to any other society, controls the widest range of social relationships, extends its power to all members of this society, and also, in the modern democratic society, endows all of them with political rights, it can be claimed that political participation is something everybody is furnished with. It is by no means a play on words only. In fact, all members of the society participate in the affairs of the State, though in different ways, depending on the various social roles performed. The consequence is a striking extension of the concept of political participation that is awkward for scientific investigations. For example, consider the following statement: "Every person participates at least passively in the political system in which he lives. Mere compliance gives support to the existing regime and, therefore, is a type of political behavior. There are other essentially passive responses to the political system: obeying laws, paying taxes, experiencing order and security".[27] The extension instanced by this statement brings into question the distinction between "participant", "subject", and "parochial", because the last two categories have become modes of participation.

It is just because of this difficulty that one frequently adds the further qualification of "active" to the term political participation (or activity, or involvement). In this way political participation is being contrasted with inactivity, passivity, and political apathy.[28] But this conception of political participation also involved difficulties. It again becomes unclear what should be included in "political activity". "Active participation" is, admittedly, understood currently as participation within the scope of legality but there are no sufficient reasons why this term should not denote as well different types of political activity contrary to the system and illegal. Anyway, as already stated, this sort of activity is a matter worth investigating.

Further, neither the qualification "active" nor even "political" can be attributed to some relevant ways of participation which have been an object of social struggle since the second half of the 19th century, and continue in some countries. This is the area referred to by the term "social democracy", and that of the "Welfare State" in the West. What is involved here is governmental action aimed at correcting the state of social-economic inequality by redistribution of the social product according to principles of the "equalizing justice". While the struggle for such participation can be considered political without any reservations, the participation itself is not political, and not active. Nevertheless, it is an important kind of participation in our time.

Making recourse to the above mentioned theses of a number of political sociologists one could at least assume that participation of this sort, though in se inactive and non-political, would make active political participation more intense and more extensive, as the latter stands supposedly in connection with the satisfying of basic physical needs, attaining influence, education, and a higher social status. Conversely, however, it is pointed out now and again that the increase of social-economic participation leads to the formation of typically "private", non-political consumers' attitudes, and consequently to the diminishing of active political participation.[29] What one is dealing with here is not only the semantic problem of another aspect of ambiguity of the term "participation" but also the considerably more important practical problem of the relation between participation in affluence and active participation in political life. Although the problem calls for further investigation, it seems that a certain dilemma between two values is here involved: that of a democratic meeting of human needs, and that of democratic political activity. Both these values are of immense social importance, and neither can be neglected in favor of the other.

The term "active" (political) participation" is, firstly, not sharp in its denotation, and, secondly, the forms of activity considered under this heading are hardly comparable as to their "participating" quality.

As to the first matter, the contradistinction of active-inactive is only apparently a sharp one. Various forms of inactivity can be decidely active (such as deliberate abstention from certain positive actions in politics). What seems to be lacking here, is the distinction of the concepts of positive and negative action (forbearance), with which the jurists are so well acquainted. In political sociology by "active" one understands only the category of positive actions. By limiting the term "active" in this way one can also get some fairly significant results in empirical research, but the problems get simplified to some extent.

As to the second matter, the term "political activity" refers both to some "politically relevant" activities, and to some scarcely relevant ones, as, e.g., private talks about politics, hearing other people's political opinions, passively taking part in some meetings of political character, superficial interest taken in the looks and in the private life of the politicans, of the candidates to political office among others.[30] Here again the juristic point of view, as it seems, has some advantages, since it expressly distinguishes the legally relevant, institutionally guaranteed forms of political activity. The most important of all here is the proper distinction between political acts (or activities as sequences of such acts) of formalized, or conventionalized character, as set against various kinds of informal behavior and symptoms of political attitudes. The phenomena of the former sort were traditionally rather neglected by sociology.[31]

It is important to distinguish between participation and the right to participate (the latter being admittedly a new term, unknown in the traditional language of law and in that of the jurists.) The concept of active political participation includes both conduct constituting a realization of citizens' rights, expressly instituted, and conduct belonging simply to the sphere of permission.[32] In political sociology the main stress is laid on the latter, whereas the problems of carrying into effect the right to participate, need special treatment.

Still another problem of active political participation arises in connection with the concept of right to participate. As is evident, such participation is a trait common of all political systems, past and present. It is inherent in the political system that a certain number of persons are performing active political roles in it, to a greater or lesser extent (as, e.g., in ancient or in feudal political systems). Almost all research in political sociology has in view only the situation in contemporary democratic political systems. Such an assumption is not satisfactory, because only by introducing the concept of right to participate can we accurately determine the scope of active political participation that is possible in a given political system. This also facilitates comparative research by enabling us to establish the dependence of the range of active political participation in various political systems upon the range of the rights to participate institutionally guaranteed in these systems. The concept of "freedom of participation" is also of considerable importance here, because it permits comparison of the scopes of permitted political activities in different systems.

Last but not least, there is also the vast and complicated problem of the "duty to participate" to be considered. In political sociology and political science we find but vague mention of the impact of "social norms", or "civic norms" on active political participation, while the character of such norms is not made precise, and the concept of "duty" is not specified.[33] We certainly deal, in the domain of participation, with manifold kinds of social norms imposing duties to participate--moral norms, customary rules, norms of social-political organizations, and legal norms which are of special importance within the last-mentioned category. As the problem remains almost untouched, it is hardly possible to treat it here in a detailed manner. One can only underline the need of investigations, in law, political science and political sociology, of the role of various norms instituting the "duty to participate", taking into account both positive and negative effects of such norms as means of ensuring participation (with respect to its "quantity" and "quality").

The qualifications of participation, discussed above, are frequently supplemented by another one, namely that of "popular", "democratic", or "massive" (active political participation). This is undoubtedly due to the already mentioned "neutrality" of the concept of active political participation (in every political system a certain group of people--be it small, or large--is politically active), and, consequently, based on some evaluative assumptions. To such assumptions belong the following: political participation is essential for the full development of every human individual; political participation has in itself a human, moral quality, being an attribute of freedom; political participation is a basic condition of social justice; political participation is decisive of the legitimacy of power according to the doctrine of the sovereignty of the people.[34]

On this point, however, opinions are hardly unanimous. As shown by empirical research, social-political reality does not correspond to the idea of popular participation. Political inactivity, passivity, and even apathy, are characteristic of the majority of citizens, and are increasing. In other words, quite a number of people do not exercise their right to participate, and a still larger number does not make use of their freedom of participation. Some authors are alarmed by this state of affairs and postulate some measures which would effect popular participation. Others--either decidedly or in a more moderate way--raise objections against the idea of popular participation. Opinions are advanced that for many people "... politics is a remote and alien activity"; that politics is rather a necessary evil than a moral good;[35] that "... moderate participation levels are helpful in maintaining a balance between consensus and cleavage in society", while "... high participation levels would actually be detrimental to society if they tended to politicize a large percentage of social relationships", the conclusion being that "... present levels and patterns of participation in politics do not constitute a threat to democracy; they seem, in fact, to be a realistic adjustment to the nature of modern society".[36] These differences in opinion are symptomatic and connected with some general problems which will be discussed below.

III.

We return now to problems raised at the beginning of this paper. The considerations of its second part, while concentrating on conceptual and terminological analysis, constitute a starting point to deal with phenomena and processes of social-political reality. They also lead to some conclusions of theoretical-methodological character.

Participation--although not termed so yet--was one of the basic ideas of the modern democracy since its origins which, admittedly, can be traced back in earlier history, but became manifest in the progressive political thought and revolutionary practice at the end of the 18th century. Thus, participation as a problem was explicitly formulated about 200 years ago. The 19th century was a period of further struggle for, and progress in, the realization of participation by establishing, step by step, equality in the right to participate, and by enlarging the scope of freedom to participate.

The idea of participation as well as the means and forms of its realization cannot be treated now as merely a laudable historical tradition. The idea is essentially the same today, and the gradually developed forms of participation for the most part still are functioning. Under this aspect, therefore, the problem is not a new one, there is a continuation of what was initiated, and then developed, in the historical process. At the same time the stress laid, both in research and in political literature, programs, etc., on the problem of participation is a striking phenomenon of our time. This phenomenon is, in our opinion, symptomatic of important changes within the problem, changes that are running parallel to the aspect of continuation.

These changes are not hard to detect and explain. Here the starting point is the thesis of a functional relation between participation and the political system. Participation, until recently, corresponded in substance to the political system developed in the 19th century, called liberal, or demo-liberal, conforming with the model of the "negative State". There is no need to repeat the well-known characterizations of this system. In accordance with its features participation had a distinct, specific--and from the present point of view, limited--sphere, such as participation reducible to such acts and activities as voting, being a candidate, contributing time or money to political campaigns, taking part, to a greater or smaller extent, in party life, holding public and party offices.

In our time important changes in democratic political systems are to be noted. The "negative State" was replaced--as a result of an evolution which started towards the end of the 19th century--by the "positive State", or the "State with a wide sphere of activity". This State accomplishes many tasks which are not "classically political", first of all the economic and social ones. Accordingly, numerous new governmental agencies came into being, and the character of many of those which previously existed underwent changes. Correspondingly, far-reaching changes in political parties can be observed as well as the development of new mass groups (organizations called interest groups, or pressure groups. In other words, important transformations in hitherto existing political structures are taking place and simultaneously new structures are formed according to the "principle of adjustment" to the requirements of the situation. Running affairs of the political system as a whole became complex and complicated, hardly to be grasped by the layman-citizen, or to be influenced by him in a relevant way. Efficiency and skills of the professionals coming to the fore, the principle of democratic participation has evident drawbacks for the modern bureaucracy and technocracy. These drawbacks have to be dealt with. While lip-service may still be paid to the principle of participation, modern techniques of manipulation are being applied to settle matters conveniently.

This situation is evaluated as a "crisis of participation", consisting in (a) political apathy, (b) "manipulated" participation, and (c) ineffectiveness of participation,[37] (a) being evidently caused to a considerable extent by (b) and (c), as testified to by empirical research (realization of manipulation and of the ineffectiveness of participation among people).[38]

Returning now to our thesis of a functional relation between participation and the political system we can say that while "classically political" participation was adequate to the democratic political system shaped in the 19th century, such participation is scarcely adequate to the system as it is now. The changes in the system were followed by those in participation but in a fragmentary and insignificant way, whereas the situation dictates the need of comprehensive changes, of working out quite a number of new forms of participation both in fields previously not covered but now covered by the activities of the political system, and in "traditional" fields in which these activities underwent changes. This is by no means an easy task because of the obstacles on the part of the professional-executive element of the system, obstacles which have also its deeper, structural aspect. One can add here that there are also obstacles inherent in the up-till-now preserved forms and canons of participation. These forms and canons acquire some degree of independence from social-political development, become petrified and although disfunctional nowadays, are difficult to modify or eliminate.

Political sociology, in spite of its being one of the newly established disciplines, is conservative in its scope of research on participation, and so in correspondence with phenomena mentioned above. In political sociology, stress is decidedly laid on the problems of the "classical" participation. In this field symptoms of growing passivity and apathy are to be detected, and different standpoints concerning the impact of these phenomena on the functioning of democracy are represented. Oddly enough, there is a marked contrast between repeatedly stated and debated political passivity and apathy, and vivid interests in matters of participation shown in different social groups, circles, and strata.

This discrepancy is to a great extent an apparent one, since matters on which attention is focused differ widely. The results of research in political sociology and political science are determined by the scope of the research in question, pertaining to electoral matters (sensu largo) and to those of the individual's involvement in "big politics", his attitudes towards it. On the other hand, the contemporary partisans of the "participatory democracy" generally have in mind partly much wider, and partly different matters--that could be called "permeation by participation" of manifold units (new and hitherto existing) of government and of social-political organizations. The important problem here is that of extending participation to levels lower than that of "big politics" but related directly to citizens' interests. Of vital interest is participation in decision-making in social-economic questions such as housing, education, social security, town and country planning, etc., as well as what could be called participatory democracy in industry. Quite a number of proposals as to other new domains and forms of participation is under discussion.[39] What is postulated in these matters is to create means for eliminating the exclusivity of the executive element with its bureaucratic and technocratic tendencies, nowadays characteristic both of governmental and non-governmental agencies.

Summing up, the interest of the general public in "classically political" participation has in fact diminished, becoming a modest one. Stating and demonstrating it is undoubtedly a valuable contribution of political sociology. At the same time we are dealing with widely shared and intense interests in the "new" participation, characterized above. In the latter, matters close to the citizens are involved, and it is quite realistic to qualify it--in a postulative manner--as "popular" or "massive" participation. The matters involved in the former, matters so complicated at present and not closely related to the interests of the general public, are felt by it to be distant and thence are but superficially treated. Consequently, one can speak here about a discrepancy only as one between two widely differing concepts of participation. It is a merit of political sociology and political science that they have shed light on the problem of participation in "big politics". It is a deficiency of

their research that due attention was not paid to new
phenomena and processes, having in view the descriptive,
explanatory, and diagnostic purposes. One has to stress,
however, that some newer works demonstrate awareness of
the importance of these developments.[40] One has certainly to postulate that attention should be focused in
future research on the new, "non-classical" problems of
participation. This is the necessary condition for establishing, how hitherto existing forms of participation
ought to be developed and refined, and what new forms
ought to be introduced. These matters, however, are not
to be solved in an abstract-general way, as bound to concrete situations and conditions of various countries,
although the basic trend seems to be strikingly similar
in different social-political systems of our time. It
is to be expected that the gradual development of adequate forms of "new" participation will gradually stimu-
late, in an indirect way, also the participation in politics on a bigger scale (the "classical" participation),
and so mostly by creating patterns and conditions enabling the growth of personality, of its political expression,[41] and the growth of political culture in general.

One has to point out that the study of law and legal
philosophy not only "formally" (which can hardly be
blamed) did not make use of the term "participation",
but also (and this has to meet with reproof) did not
show sufficient interest in problems of participation,
or else treated them in a marginal and dispersed way,
when considering matters of separate civic rights. The
scarcity of interests in social-functional aspects of
the problem constituted also a defect of the considerations in question. In this respect the research in political science and political sociology has to be valued
higher than the contribution of the jurists. On the
other hand, two points have to be stressed. Firstly,
jurists--both in their scholarly activities and in prac-
tice--have considerable achievements as far as the for-
mal-institutional aspect of the problem is concerned,
this being easy to understand, as it is inherent in their
profession to concentrate on matters of that sort. These
achievements are notable, when considering the great importance of securing participation by provisions of law.
On the contrary, representatives of political sociology
and political science were almost exclusively interested
in phenomena and processes of informal character, and in
recording facts without evaluating the existing institutional measures and considering those to be adopted in
the domain of participation. Secondly, the legal concept of participation, formerly implicitly assumed in
jurists' considerations, and now (admittedly, under the
influence of sociology) explicitly expressed by the term

"right to participate", is wider and more operational than the concept used in political sociology and political science, one limited to "classical" political participation only. The "right to participate" is a concept applicable to participation of different levels and in different fields, and so, among others, to social-economic participation both in "decisional" and in "consumers'" sense, both within the scope of the "formal government" and within that of other organizations. The importance of the concept of the "freedom of participation" as one differing from the "right to participate" has also to be stressed here, while the concept of the "duty to participate" with complex problems involved in it is evidently in need of a comprehensive study.

In further works on the problem of participation, works corresponding to actual needs, a close cooperation of hitherto isolated domains of the study of law on the one hand, and political sociology with political science on the other, seems indispensable. Such cooperation will lead to combining various relevant aspects of the problem, to enlarging the scope of research, and to enriching its methods.

FOOTNOTES

1. Compare, for instance, HINDERNISSE DER DEMOKRATIE, coll. work, Berlin 1969; "L'idee de participation populaire aux XIXe et XXe siecles" as one of the topics of the VIIth World Congress of International Political Science Association, Brussels 1967; proceedings of Tavola Rotunda su "Potere e partecipazione", organized at the VIIIth Italian National Congress of Philosophy of Law (published in Rivista Internazionale di Filosofia del Diritto, vol. XLVII, 1, 1970); H. I. Kalodner, "The right to participate" in THE RIGHTS OF AMERICANS: WHAT THEY ARE--WHAT THEY SHOULD BE, ed. N. Dorsen, New York 1972.

2. Compare D. Easton, THE POLITICAL SYSTEM: AN ENQUIRY INTO THE STATE OF POLITICAL SCIENCE, New York 1953, p. 23; G. A. Almond, "Comparative political systems", The Journal of Politics, vol. 18, 1956; as to the role of the personality factor, see R. F. Lane, "Political character and political analysis", Psychiatry, vol. 16, 1953.

3. Compare, for instance, "Citizen participation in political life", coll. work under the auspices of UNESCO on participation in England and Wales, A. H. Birch; Finland, F. Allert and P. Pesonen; France, G. Dupeux;

Israel, E. Gutmann; New Zealand, R. S. Milne; Norway and the United States, S. Rokkan and A. Campbell, "International Social Science Journal", vol. 1, 1960. See S. Zawadzki, "Forms of citizens' participation in governing the Socialist State", in Res Publica, vol. XV, 1, 1973.

4. I. Fetscher, "L'idee de participation politique--aujourd'hui", paper presented at the VIIth Congress of IPSA, mimeographed ed., Bruxelles 1967, and the literature cited there.

5. Compare, for instance, F. Vilmar, "Wieweit lasst sich der Betrieb demokratisieren?", and A. Krause, "Wieweit lasst sich die Verwaltung demokratisieren?", in HINDERNISSE DER DEMOKRATIE, op. cit., pp. 58 ff, 64 ff.

6. Compare M. Sobolewski, REPREZENTACJA W USTROJU WSPOLCZESNYCH DEMOKRACJI BURZUAZYJNYCH (REPRESENTATION IN MODERN BOURGEOIS DEMOCRATIC POLITICAL SYSTEMS), Krakow 1962; on developments in U.S.; R. E. Lane, POLITICAL LIFE. WHY AND HOW PEOPLE GET INVOLVED IN POLITICS, New York & London 1959, ch. 2 and the literature cited there.

7. Compare G. Perticone, "Il problema del potere e della partecipazione", in Rivista Internazionale di Filosofia del Diritto, vol. XLVII, 1, 1970, pp. 23-55; S. Rokkan, Introduction to "Citizen participation in political life", p. 7, op. cit.

8. Compare R. A. Dahl, RESEARCH FRONTIERS IN POLITICS AND GOVERNMENT, Washington 1955, pp. 47-66, and a discussion of the problem by B. Berelson, "Democratic theory and public opinion", in Public Opinion Quarterly, vol. 16, 1952.

9. Compare J. L. Woodward and E. Roper, "Political activity of American citizens", in The American Political Science Review, vol. 44, 1950; M. Rosenberg, "Some determinants of political apathy', in Public Opinion Quarterly, vol. 18, 1954. The concept of "political involvement" in L. W. Milbrath, POLITICAL PARTICIPATION. HOW AND WHY DO PEOPLE GET INVOLVED IN POLITICS?, p. 16 ff, Chicago 1965.

10. J. P. Nettl, POLITICAL MOBILIZATION. A SOCIOLOGICAL ANALYSIS OF METHODS AND CONCEPTS, p. 29 f, London 1967.

11. See the discussion by N. Bobbio, "Crisi di partecipazione: in che senso?", in Rivista Internazionale di Filosofia del Diritto, p. 58 ff, vol. XLVII, 1, 1970;

V. Frosini, Due forme di partecipazione", ibid, p. 68 ff.

12. F. Vilmar, op. cit., and W. Strzelewicz, "Burokratisierung der modernen Gesellschaft und die Ohnmacht des Burgers", HINDERNISSE DER DEMOKRATIE, p. 24 f.

13. J. J. Wiatr, "Niektore zagadnienia opinii publicznej w swietle wyborow 1957 i 1958 roku" ("Some problems of public opinion in the light of the 1957 and 1958 elections"), in STUDIA SOCJOLOGICZNO-POLITYCZNE 4, p. 13 f, 1959.

14. J. P. Nettl, op. cit., p. 30. Also L. W. Milbrath, op. cit., p. 2, restricts the scope of politics to "... the process by which decisions about governmental outcomes are made".

15. R. Batley, H. Parris, D. Woodhead, "Popular participation on Tyneside", paper presented at the VIIth Congress of IPSA, p. 2 f, mimeographed ed., Brussels 1967.

16. G. A. Almond and S. Verba, THE CIVIC CULTURE, Princeton 1963; G. A. Almond and G. G. Powell, COMPARATIVE POLITICS. A DEVELOPMENTAL APPROACH, pp. 50 ff, 64 ff, Boston & Toronto, 1966.

17. S. E. Finer, "Groups and political participation", Senior Politics Seminar, 1969-70, p. 1, mimeographed ed.

18. loc. cit.

19. K. Horn, "Politische Unfahigkeit als Hindernis der Demokratie", HINDERNISSE DER DEMOKRATIE, op. cit., p. 6.

20. Compare, for instance, R. E. Lane, POLITICAL LIFE, op. cit., esp. in part IV.

21. L. W. Milbrath, op. cit., p. 18.

22. R. E. Agger and V. Ostom, "Political participation in a small community", in POLITICAL BEHAVIOR, ed. by H. Fulan, S. J. Eldersveld, M. Janowitz, II ed., p. 138 ff, Glencoe 1959.

23. Apart from the results of research in the United States compare quite similar results obtained in research in Western Europe, e.g., J. Steiner, BURGER UND POLITIK. THEORETISCH-EMPIRISCHE BEFUNDE ZUR POLITISCHEN PARTIZIPATION DER BURGER IN DEN WESTLICHEN DEMOKRATIEN MIT BESONDERER BERUCKSICHTIGUNG DER SCHWEIZ UND DER BUNDESREPUBLIK DEUTSCHLAND, Meisenheim 1968, and, by the same author, "Participation politique et statut social",

paper presented at the VIIth Congress of IPSA, mimeographed, Brussels 1967.

24. L. W. Milbrath, op. cit., p. 18.

25. R. E. Lane, POLITICAL LIFE, op. cit., ch. 9.

26. C. H. Dodd, "Participation and education: a comparative essay", Senior Politics Seminar, p. 3 ff, 1969-70, mimeographed ed.

27. L. W. Milbrath, op. cit., p. 9.

28. Compare L. W. Milbrath, op. cit., ch. 1; M. Rosenberg, "Some determinants of political apathy", op. cit. supra, and the literature cited there.

29. As to democracy based on "socio-economic participation" see, e.g., A. Downs, AN ECONOMIC THEORY OF DEMOCRACY, New York, 1957. Contrary views in, e.g., J. Habermas, STUDENT UND POLITIK, Neuwied and Berlin, 1961, p. 18 ff; I. Fetscher, "Funktion und Bedeutung der Politikwissenschaft in der Demokratie", in Gewerkschaftliche Monatshefte 8, pp. 465-473, 1967.

30. Compare, for instance, J. L. Woodward and E. Roper, op. cit., p. 872 ff; R. E. Agger and V. Ostrom, op. cit., p. 138 ff; I. Fetscher, "L'idee de participation politique--aujourd'hui", op. cit. supra, p. 11 f.

31. On these differences between the juristic and the sociological approaches compare K. Opalek and J. Wroblewski, ZAGADNIENIA TEORII PRAWA (PROBLEMS OF LEGAL THEORY), p. 372 f, Warszawa 1969; J. Szczepanski, ELEMENTARNE POJECIA SOCJOLOGII (ELEMENTARY CONCEPTS OF SOCIOLOGY, p. 21 f, Warszawa 1965.

32. About this distinction in recent literature see G. H. von Wright, NORM AND ACTION. A LOGICAL ENQUIRY, p. 86 f, London 1963; A. Ross, DIRECTIVES AND NORMS, p. 120 f, London 1968; K. Opakek and J. Wolenski, "On weak and strong permissions", in RECHTSTHEORIE, vol. 4, 2, 1973.

33. See R. E. Lane, POLITICAL BEHAVIOR, op. cit., p. 350 f; J. Steiner, PARTICIPATION POLITIQUE ET STATUT SOCIAL, op. cit., p. 2 ff.

34. See R. Batley, H. Parris, D. Woodhead, op. cit., p. 7 ff.

35. Compare R. A. Dahl, WHO GOVERNS?, p. 279, New Haven and London 1961, with the differing views of H. Arendt,

THE HUMAN CONDITION, p. 58, Chicago 1958; B. Crick, IN
DEFENSE OF POLITICS, p. 140 f, London 1964.

36. L. W. Milbrath, op. cit., p. 153 ff; M. Rosenberg,
op. cit., p. 349 f.

37. N. Bobbio, op. cit., pp. 55-58.

38. See A. Campbell, G. Gurin, W. F. Miller, THE VOTER
DECIDES, p. 187 ff, Evanston 1954; G. Connelly and H.
Field, THE NON-VOTER: WHO HE IS, WHAT HE THINKS, in
Public Opinion Quarterly, vol. 8, pp. 175-187, 1944;
P. F. Lazarsfeld, B. Berelson, G. Gaudet, THE PEOPLE'S
CHOICE, II ed., pp. 40-51, New York 1948; L. Bean, HOW
TO PREDICT ELECTIONS, ch. 5, New York 1948; R. Centers,
PSYCHOLOGY OF SOCIAL CLASSES, Princeton 1949.

39. Compare, for instance, S. Cotta, "Partecipazione:
a che cosa?", in Rivista Internazionale di Filosofia del
Diritto, vol. XLVII, 1, pp. 61-68, 1970 ("partecipazione
al potere decisionale" and "partecipazione al momento di
liberta della vita sociale").

40. Compare, for instance, research by R. Batley, H.
Parris, D. Woodhead (report in the paper cited above);
H. I. Kalodner, op. cit.

41. Compare R. E. Lane, POLITICAL LIFE, cited above,
p. 357.

PARTICIPATION: A DISCUSSION BASED
ON SCANDINAVIAN EXPERIENCES

Britt-Mari Blegvad

I. INTRODUCTION

The concept "participation" is widely employed in literature on industrial democracy and management. This literature therefore offers a vast material for discussion of the problems relating to the use of this concept. I will mostly use Scandinavian material, but where fruitful will also include material from other countries. As a starting point I shall present four cases taken from experiments with floor level participatory patterns in different industries (II).

Secondly, I will tentatively define participation, which in Scandinavian literature has mostly been regarded as a phenomenon on the group level. Participation by workers at the group level, in decision making and other actions, has presupposed a certain autonomy. Material regarding participation has therefore often been reported under such headings as "autonomous" or "semi-autonomous groups" depending on the degree of autonomy and/or participation in the group. The relation between these two concepts will be briefly discussed and a tentative definition of the latter will be presented (III).

One general assumption has been that the participatory pattern would spread from the group level to other levels of the surrounding organization as well as to society. It is presupposed that the introduction of participation on one level is not enough to reach a stable effect. Therefore, in the further presentation of the Scandinavian material the problem of level will be specially discussed (IV). The following part of the paper will deal with the question under which structural conditions participation may be stabilized and further diffused (V). Finally, I will present a definition of 'participation' for discussion at the congress (VI).[1]

II. THE CASES

The first two cases are derived from a report by
J. Gulowsen on eight autonomous work-groups, set up on
an experimental basis in four large Norwegian enterprises in the middle of the sixties. He also collected
material from twelve small cooperative or semi-independent firms and teams in the fishing, lumber,
and stone industries. The study of the two groups of
lime-workers was originally carried out by T. J. Hegland
and B. Nyhlehn.

Alfa Lime Works: The Oven Group and Quarry Group

"The Alfa Lime Works, which produces calcinated or
burnt lime for industrial purposes, employed a total of
about twenty workers. Within this body of workers there
were two groups, one comprising three quarriers and the
other comprising four workers stationed near the lime-kiln (oven). Each of these groups was autonomous
within certain fields. The members of the two groups,
together with a worker who was in charge of transport
between the groups, were paid according to a joint
piece-rate system. The system included sanctions to
be used against those not conforming to the regular
working hours... The production planning was done during daily morning meetings between the manager and the
workers... The technology was very simple, and the
questions regarding method seldom seemed to cause any
problems. However, there was no doubt that the groups
and the group members had complete control in this
field. They also determined how the different tasks
should be divided among the members... The group members were themselves responsible for the recruitment
of new members."[2]

The following two cases are taken from a report on
experiments with autonomous groups at floor level performed from 1971 to 1973 within the Danish metal industry.* The first experiment took place in a firm with
a highly specialized technology manufacturing advanced
measuring equipment for the determination of protein
content in grain and milk. The work group was composed
of five skilled and two unskilled workers, who did
platework. The job involved measuring, cutting, punching, welding, polishing, and drilling.

The changes in the working situation of the group
during the experimental period were the following: The
group took over certain elements of the production process. The first one was the scheduling of the production process. Previously the foreman had handed out

---
* N. Foss Electric

jobs of a duration of 1-20 hours to each man in the group. Now the group abandoned the practice of clocking in and out at each single operation. At the same time the number of work notes was reduced simplifying the paperwork for managing the production.

After six months the group took over the routine check, i.e. the control of produced pieces. However, on some items the group continued to cooperate with the firm's special control section. During the experimental period the members of the group became involved in the decisions regarding investments in new working equipment. They also took part in construction of new tools for the production process and systematically followed up the results on their proposals.

At the beginning of the experimental period the work drawings were incomplete. Measurements were absent and tolerances impractical. Direct contact with the drawing office was established making quick corrections possible, thereby abolishing the previous slow formalized system for the endorsement of drawings. The members of the group also acquired the right to allocate the work themselves. A welding course augmented the possibilities for job rotation, which nevertheless was not practiced to any large extent. At the end of the experiment the interest for rotation increased due to the acquisition of new machines and an overload of work. It was, however, more difficult to establish a discussion in the group about other changes mentioned.

Wages caused problems as the workers felt that they were entitled to higher pay as they had taken over more advanced work operations than before the experiment. Through complicated negotiations a new and informal wage agreement was reached abolishing adjustments according to productivity and implied regulations every three months in relation to developments elsewhere in the area.

*The second Danish case is taken from the same series of experiments but from a firm specializing in producing truck-cranes and trailers. The firm, which also undertook a substantial amount of repair and service work on trucks and cranes had in fact started as a rather small repair shop. The group in question constituted a naturally demarcated working unit of ten to eleven men, three of whom were unskilled workers. The group assembled, tested, and painted cranes. Work was carried out in accordance with a drawing, but there were possibilities for variation in method. Even though the components varied and the firm manufactured three
_____
* Højbjerg' Toolfactory

types of cranes, the number of operations was few and regarded as very routine.

The workers and the foreman jointly assigned responsibility for the quality and quantity of the production, whereas previous to the experiment these functions were held by a special control unit. The company also started to supply the group with information regarding for example the backlog of work and incoming orders. The group received work notes for a week ahead instead of notes on a day-to-day basis. Furthermore, the group was allowed to have conferences any time during working hours. This possibility was used regularly at the beginning. One met formally once a week, but gradually this pattern was loosened up and the workers had smaller informal meetings in the shop whenever it was regarded as necessary.

The group brought about a great amount of job rotation. Only the younger workers felt that this created variation, as the older ones already knew the different operations before the experiment started. The group's wage system was previously a mixture of individual piece rates and hourly rates, which discouraged the men from talking together, let alone helping each other in their work. In this firm the management was prepared to revise the wage system as part of the experiment. Negotiations resulted in a system of group bonus combined with an individually set basic hourly rate.[3]

III. A TENTATIVE DEFINITION OF PARTICIPATION

Much material regarding participation has been presented as a group level phenomenon. Participation in decision making and actions on this level has often empirically been limited to a certain autonomy. An interesting example of this approach is the previously mentioned work by J. Gulowsen. His studies showed that worker's influence seemed to follow a cumulative step by step pattern, i.e., workers' control in a high level of decision making must be preceded by control on lower levels. If the shop-floor level is not controlled by the group it has little or no means of effective control on long range economic decisions. His criteria could be ordered along a Guttman scale, which suggests that autonomy is a one dimensional property within work groups.[4]

Figure 1.

		LEVEL OF DECISION MAKING
	Distribution of Profits. Financing. Investments.	1
	Choice of products. Quality of products. Quantity of products.	2
	Technology. Planning. Administrative routines.	3
	Hiring new workers. Firing of workers. Distribution of work tasks.	4
TYPES OF DECISION	Decisions regarding individual production methods.	5

Gulowsen hereby defines autonomy in work groups operationally as a lesser or greater possibility for the group to make decisions regarding the group's own tasks and goals. Like many other authors Gulowsen uses autonomy and participation as synonyms. See also, for example, P. Blumberg who in his partial summary of the participation literature includes studies of autonomous groups although he, in his definition of participation, rather stresses the decision making process, whether that decision making is unilateral or not.[5]

While "autonomy" by Gulowsen, and Blumberg in his summary, is used as a concept on the group level, participation in Blumberg's theoretical discussion is regarded as a property useful on the individual level. Blumberg goes from cooperation, where the worker controls decision, to co-determination where the worker controls decision and is responsible for the carrying through of the actions.[6] Likert, on the other hand, defines participation as a continuum of processes, where the degree of information given to the workers represents one side of the scale ("little participation") and action to tackle the problem represents the other side ("much participation").[7]

Other authors, more oriented towards political science, have stressed the societal perspective of participation in work groups and concentrated on the external relations and changes in the power structure. A good representative for this type of approach is C. Pateman who refers to participation as "a process where each individual member of a decision body has equal power to determine the outcome of decisions."[8] The pedagogic element is also very essential for Pateman. She says, "Participation in non-governmental authority structures is regarded as necessary to foster and develop the psychological qualities (the sense of political efficacy) required for participation on the national level. Evidence has (also) been cited to support the argument that industry is the most important sphere for this participation to take place..."[9]

This approach tries to link forces on the group level with those on the societal level. The authors most influenced by social-psychological theories agree on the fact that workers and employers might be regarded as two forces using different means but still having a common goal, an industrial development with good conditions that guarantee a good profit. Both these approaches are based on the assumption that a consensus regarding goal exists between the members of the work group and its social surroundings. But participation may also be regarded as a conflict-solving mechanism in the sense that the members of a work group through participation gain more information about the surrounding social system. The workers' responsibility increases. Fewer misunderstandings arise and thereby fewer conflicts. However, the basic assumptions are still somewhat identical. A consensus exists regarding a common superior goal. Finally, participation may be looked upon by some workers as a cooptation where a long range goal (revolution) gets delayed by short time reforms on the floor level. Here the workers and the employers involved have a different opinion regarding both means and ends.

As most of the available Scandinavian material is obtained through experiments initiated jointly by labor and employers' organizations, I shall in this paper assume a common goal for the systems involved, namely, survival. On the other hand, I find it fruitful to accept that the parties involved in participation do not always accept this means as the best for achieving this goal.

As will be shown, both individuals and certain groups on the floor, for example, female workers, or persons in certain positions, such as foremen, have protested and backed out whenever this was possible. We also find examples of persons on the managerial level who have regarded the introduction of participation as a threat to their previous positions. But even if many of the Scandinavian experiments and attempts to introduce participation have been 'successful' on the group level, it is a fact that the experiences were not necessarily diffused, either within the company or to other firms within the same industry.[10] It is my opinion that many of the discrepancies and unclear points in the discussions of industrial democracy where 'participation' and 'autonomous' work groups are used as basic concepts are due to a limited concern with the level problem.

When dealing with work relations we are not only dealing with individuals but with groups as subsystems to organizations. It has been strongly argued by the British Tavistock group, for example, that the task structure and thus the technology has a strong impact upon work and work structure.[11] This means that the participatory behavior must be linked to formalized rules constituting an organized pattern of social relations and conduct independent of particular human beings. Such rules make a frame of reference for conduct in the groups. Other requirements are that the rules of conduct must be legitimized by traditional and/or accepted rules and, furthermore, enforced by groups rooted in the general power structure.[12]

The problem of levels is therefore also essential for the external relations of the group, i.e. when we relate the work structure in the group with that in the surrounding systems as shown in Figure 2.

Figure 2.

	Levels	Norms and values
Level 1.	Societal level (government)	Statutes, legal practice, political values.
	↕	↕
Level 2.	Organization (factory)	Organizational norms and values, e.g. branch customs, style of management.
	↕	↕
Level 3.	Group level (autonomous working groups)	Group norms and values, e.g. rules on autonomous decisions regarding job circulation, pace of work, amount of work.

I will here focus the discussion on the conditions for participation in an organization such as a factory. The work group will be regarded as a subsystem and, following Pateman, I will emphasize that each individual member of the work group shall have equal power to influence the outcome of the decision. This means that the members of the group must be in possession of the necessary information. The definition also implies that the parties involved influence each other without A subordinating B, or vice versa.[13]

IV. EXPERIMENTS AND EXPERIENCES

The Group Level

Most of the experiments and permanent changes with work structure on the group level are carried out in the ways indicated in the cases presented (see II). Tasks previously taken care of at middle or top management levels are transferred to the floor. The groups take over planning functions and the members participate in decision-making, for example, on buying new tools. The members get opportunities to contact experts within the firm whenever needed. Quality control is also taken over by the groups themselves. The wage systems are changed so that they favor cooperation instead of competition. But the studies show that there are certain technological and social limitations for changes of these kinds.

The general philosophy behind the Scandinavian experiments was developed in Norway as a result of contact between the Institute of Industrial and Social Research and the Tavistock Institute by the end of the fifties. Ideas about the so-called socio-technical system were developed. This system presupposed that certain psychological job requirements had to be fulfilled. Here one is thinking of the fulfillment of needs such as: Opportunities to learn on the job and go on learning; A minimum of variety in the work situation; A minimum of social recognition; The possibility for the individual to relate his social life to that of others, and A minimum of influence over one's own working situation.

These job requirements resemble those formulated by, for example, Hertzberg and others, but what was new was that the Tavistock group tried to link these individual requirements with requirements at group level as well as with technological and social requirements at organizational level. Therefore, a socio-technical system presupposed furthermore that the work group should be organized in a way that facilitated job rotation as well as some possibility of participating in the structuring of the tasks of the group. The technological structure should also be formed so that this was possible.[14]

One of the most recent attempts to link the social design of work rotations with the technological ones has taken place within Volvo, the Swedish car factory. This attempt leaves behind the demonstration phase where participation is introduced in one or two rather small groups. Here we are talking about a new factory planned and built for group work. Technologically this was made possible by the invention of a self-propelling platform on which the cars could be assembled. The factory is divided into independent production units where a certain number of cars are assembled each day. Groups of six to fifteen workers have their own working facilities, restroom, and showers, and even their own entrance. The assembly line is kept only as a transport conveyor. The workers control the flow of material via a data system and also carry out the quality control on their own production. The group is also independent in that its members themselves decide whether the group should engage in job rotation. As long as the group produces the number of units outlined by the management the members themselves can decide, for example, whether the tempo shall be constant during the shift or changed at intervals. The supervision is carried out by an engineer and a foreman.[15]

The groups have little contact with each other, and this has been regarded by some analysts as a way of preventing strikes and other collective actions against management.[16] Although it seems that Volvo has been rather advanced in meeting the technological demands, the groups are not socially autonomous in the sense put forward by Gulowsen. Volvo's measures are in reality more technological than social and at the beginning of the seventies the management has had especially great problems in recruiting men for its highly rationalized and efficient plants.

Although it has not received nearly as much publicity, one of the Norwegian experiments actually fulfills the requirements of full participation to a much higher extent. Here I am thinking of Norsk Hydro where the workers took an active part in both the social and technological planning of a factory and also undertook full responsibility for the fulfillment of a certain production. This is the automatized production of fertilizer implying a highly developed know-how. In the new factory production was divided in three parts where each part has its own control room with three to five workers. These workers were trained to fulfill a large number of maintaining and controlling functions. Job rotation, therefore, entered as a natural element. The workers were paid according to a combined system of wage and bonus, the bonus being related to general production and the workers' qualifications.

However, data from the Danish experiments indicate that the workers were not always positive towards changes that meant more participation in the work situation. "If one gets more influence, it is like giving the company something. They (the workers) were not interested in criticizing anything because that only meant that they gave suggestions that might facilitate the work of management."[17] Problems also arose when discharging members of a work group or hiring new ones. In one of the firms involved in the Danish experiments the steering committee, in which both workers and employers were represented, found it necessary to remove two men from the experimental group because the work load went down. The group wanted to get new types of work instead, but this was refused. After this episode the group did not wish to cooperate in the decision as to which workers should leave the group and the choice was left to the foreman.

The problems of groups with a specific composition have already been mentioned. The results of the Scandinavian participation experiments indicate that women have a more instrumental attitude towards work than men. Work is simply regarded as a means to get certain goods, not as a way to personal success or fulfillment.[18] Female workers were therefore not so interested in participation experiments.

These results may certainly be interpreted in many ways. Participation presupposes efforts that many women who work in factories are not willing to make. They often feel more secure in a supervised work situation. The Scandinavian studies show that female workers are more sceptical towards changes than their male colleagues, yet one would think that they had more to gain as they are often placed in more inferior jobs than men. Participation in forming one's own work or life situation is not a natural element of the female role even in Scandinavia.

The Organizational Level

No experiments with self-steering or autonomous groups can take place without a considerable effort from the top management. A common notion is that a greater autonomy at group level presupposes less power for the management. A certain number of decisions is taken over by other levels in the organization. What could be the motivation for the management to initiate or accept such changes? Most of the Danish and Norwegian experimental results show that productivity is affected positively. In a Swedish report it is claimed that the productivity, measured as value of output per costs involved, increased in the range of 10 to 30 percent. This effect is caused by improved cooperation between workers, by creative suggestions for rationalizations and by effects on workers' motivation. Thus L. E. Karlsson says "the positive effect on economic performance is caused by participation, which allows for a more flexible and rational system of work-organization".[19]

But the results also show that workers have experienced a certain strain as a result of the experiments. Those who take part in the experiments were met by hostility from the other workers who, for example, did not get the same pay as the participants.[20] This was regarded as very unfair as the experimental groups in many ways depended on the assistance from others in the organization. The roles of the supervisors, specialists, and managers in the organization around the experimental group were also re-defined and often in ways which meant that these persons had less influence at floor level without being compensated by greater influence in other areas.

However sometimes this has been possible. In one of the firms involved in the Danish participation experiments (N. FOSS ELECTRIC) the foreman took over some of the managerial tasks. The foreman now spends more time in developing long-term perspectives and in formulating goals for the different sections in connection with the annual drawing up of the budget. The foremen are also, say Fl. Agersnap et al., "...expected by the production management to be the initiators of a continuous definition of the problems in the sections and to come up with suggestions for changes."[21] In order to be able to fulfill these new tasks the foremen have participated in a number of managerial courses. But results from the Norwegian experiments, for example, show that such changes might be difficult to carry through because individuals, such as female workers, felt the change as a threat. The foremen during a long period of experiment regarded themselves as victims of a development and thereby turned to hindering a growth and development that could also have improved their own situation.[22] If the technology used is a process-production this problem can be regarded as especially important.

Another aspect of the technology used is the limits put to change. The type of technology simply limits the extent to which one can use participation. Apparently, changes towards participation are most effective (both from the workers' and the management's points of view) in a production process that is rather simple and easy to see because the results of the changes also are easy to estimate. A complicated production process presupposes a personnel of a high educational standard who are capable of taking personal responsibility. It seems therefore that participation experiments have been most successful in factories where the processes previously had been cut up into rather simple monotonous elements. Probably a change here also meant more, relatively, to the workers' job situation.

On the other hand, all the experiments involve intensive educational efforts in order to prepare the surrounding organization for the changes. Special steering committees were established very early as an essential part of the experiments. In these steering committees the top-management, middle-management, foreman, and other supervisors as well as workers were represented. This pattern was first established as part of the Norwegian experiments and then followed in the other Scandinavian countries.

The Societal Level

In Norway, Denmark, and Sweden participation on the group level has been introduced as results of cooperation between national trade unions, employers' organizations and the governments, which to a certain extent have paid the costs involved. According to E. Thorsrud, an external pressure was established on employers and union officials due to growing international competition as well as internal pressure in the unions from the younger generation.[23] Young people would not accept the old traditional work patterns. Joint acceptance of the need to achieve a greater degree of social justice and the existence of mutual trust between trade union and industrial leaders, together with a perceived need for organizational change, provided the institutional conditions for the research and development program.

In Denmark, too, the experiments with new work forms can be regarded as a result of joint acceptance in government, unions, and employers' organizations of a need for change. There is a long tradition dating back to 1899 of developing and formalizing the relations between employers and employees. Special conflict-solving agents were established. After World War II cooperation councils were set up. In these councils an equal number of representatives of the workers and management participate. During the first years these councils had difficulties in defining relevant tasks within the existing legal constraints. The function of the councils was to be consulted about plans and decisions affecting the workers and to receive information at large about the economic development of the firm. During 1969-70 negotiations regarding a revision of the "co-operation agreement" took place but they were presumably blocked. Experiments with more direct forms of democracy and participation were according to Fl. Agersnap et al., seen as a possible alternative.[24] However, the actual start of the Danish experiment coincided with the entering of a new, revised "co-operation agreement" in October 1970. This agreement increased the authority of the councils and their possibilities to influence the policy of the firms in a number of fields, including personnel policy. It was therefore natural to link the Danish experiments with the activities of the cooperation councils by the provision that they should appoint steering committees for the experiments in the firms involved.

An even more specific background was established during a study tour to the United States in 1969. Various organizations as well as four industrial companies sent representatives. As a result of this tour a committee was established which took the initiative for the Danish experiments reported here.

Also Swedish authorities became interested in experiments involving changes in the worker's possibility to influence his own working situation. As a first step, the Swedish Ministry of Industry, in 1968, established the Delegation for Industrial Democracy. It consisted of seven delegates representing the government, the unions, and the employers within the state-controlled industries, which comprise about seven percent of Swedish industry. The delegation has employed a staff of seven researchers who have actively taken part in five projects covering industries with various types of production such as the Swedish Tobacco Company, two shipyards, and a restaurant chain. The aim of the delegation was to further the establishment of autonomous groups, self-steering departments, and even a self-managed factory.

In 1970 a new delegation was formed in order to carry out experiments within the part of public administration controlled by the central government. The experimental activities comprised collaboration in budget submissions, personnel matters, work environment, and work organization. Various administrative offices took part in the experimental activities ranging from a central post office to the Swedish International Development Authority, a data center, etc. The experiments in these organizations are mentioned here as the implemented changes presuppose another administrative system than the traditional bureaucratic one where responsibility lingers with the head of the office. Decision by joint committees on a parity basis would, for example, call for changes in the legislation regulating decision-making within various administrative agencies. This would probably be the case, too, if management procedures in private enterprises changed considerably.

To a certain extent the experimental activities in Sweden spread from the public to the private sector. Independent activities are performed in many firms that for various reasons prefer to employ their own staff specialists rather than to engage outside researchers. The changes reported here established by Volvo may be regarded as such an independent activity.

V. STRUCTURAL CONDITIONS

In the preceding parts of the paper I have given some examples of participation at floor level, i.e., participation in decision making and in different types of actions that the workers in question have previously not been able, or allowed, to carry out. A common trait has been that participation has only covered limited parts of the work situation. Mostly the experiments have only dealt with internal questions such as the distribution of tasks or questions regarding production methods within the limits set by the management or the supra system. The group has very rarely been able to go further than control of its own products.[25] Choice of production and distribution of profits, questions of investment and financing, etc. have been entirely left to management or second level decision. (See Figure 2.)

Furthermore, only small groups have been affected by the experiments; groups of three to twelve persons. Therefore L. Odegaard has some right to call the Scandinavian experiments "demonstration projects",[26] i.e., they have been used to show that participation in a limited sense can take place without a breakdown of the whole production system. This has been feared especially by managers who view the worker as a person lazy by nature in need of strict supervision. But the diffusion of the experimental results has been rather limited in both Norway and Denmark. The results seem to have been encapsulated.

There might be different explanations of this fact. On the individual level it seems that an essential fact is the previous situation of the individuals engulfed by these experiments. Who will gain by worker's participation? Sometimes the manager; production goes up. But worker's participation may also be regarded as a threat to his power. Sometimes the worker gains. He gets a more satisfying job situation. But some workers prefer to go on working under competition but with a big bonus for the rate buster. Other workers, such as female workers, are not interested in investing efforts in changes. There are also individuals who simply prefer to work by themselves instead of working in a group. I have also mentioned persons in certain positions such as foremen, middle and top management who might have a vested interest in establishing participation at floor level.

It is, therefore, understandable that the effect of the experiments with worker's participation has been limited in Norway and Denmark. Individuals and groups may for different reasons have been ambivalent, or opposed to such changes.

If, however, we look at the results in Sweden, it seems that ideas of participation have spread there. The Swedish government (See Figure 2) has to a greater extent than the governments in Norway and Denmark furthered participation projects by establishing the two delegations mentioned. During the last ten years, while the process of introducing participation has been under way, the general political climate has also been more stable and in favor of this kind of change in Sweden. So, even if the changes at the Volvo factory were taken care of within the management system, they were well suited to the means and efforts carried out in other parts of the Swedish societal systems.

It is a point of this paper that a change at one level can not have any stable effect unless supported at other levels. I have mainly discussed support from the societal level. But the assumption is confirmed by results from some Danish companies where the support over time has been strong on the organizational level. This explains why two of the seven experiments carried out in Denmark had a limited success, measured by diffusion to the rest of the organization. Closer analysis of these results shows that the changes on the group level were supported by previous non-systematic changes towards participation carried out in the firms in question. In one of the companies, which today produces tuner units for TV sets, several joint committees had been established years ago. One, established in the middle of the sixties, took an active part in planning a new factory. Another committee participated in the early seventies, before the experiments discussed here, in solving the firm's problem with personnel turnover. Attempts to assign the work at the long assembly line into group work had already been carried through, etc.

My point is that the social climate in the whole organization had gradually been changed towards a greater degree of participation over a period of twenty years. The personnel had taken part in different courses where the problems of the firms had been discussed directly between representatives from the workers, the middle and the top management. Relatively, both workers and management had gained. They acquired a common goal. The missing support from the societal structure was counterbalanced by educational efforts inside the organization over a long period.

Research on worker's participation in the Western hemisphere has, however, mostly been limited to the effect on the group level. An interesting new approach has, however, been put into practice at the beginning of 1974 in Australia by the Department of Labour and Industry.[27] Here a few companies decided to take part in a change programme where the target was total organization. Every member of the organization, from the shop floor to the board of directors was to be involved. So far no research report has appeared. With this strategy we find optimal conditions for a change programme--support at all levels.

VI. A NEW DEFINITION

Participation has been defined previously in this paper as equal power for the members of the group to determine the outcome of the decisions taken. Based on the material presented here, taken mainly from Scandinavian material, I would like to offer the following amplified definition for discussion: A suprasystem-supported right for members of a social system to participate in the outcome of decisions which might be regarded as relevant for them.

In order to be effective this right should be supported by norms of the system in question as well as by norms of the subsystem and suprasystem, or systems. When discussing work groups at shop level this means that the norms of these groups must be supported by the norms of the individuals as well as be legitimated by the rules of the organization (company) and the society. I have added the word "relevant" although I am aware that by so doing I am opening a discussion which would lead me too far in this context. I only want to underline that the criteria for relevance-- accepted by all the members of the system--is a necessary part of the norms in question. In the legal system these criteria are part of the rules of procedure.

FOOTNOTES

1. I am indebted to my colleagues at the Institute of Organization and Industrial Sociology at the Copenhagen School of Economy, with whom I have discussed the problems raised in this paper. My thanks especially goes to Flemming Agersnap and the team with which he carried out the studies of the Danish participation experiments, as I have been able to draw freely on their reports.

2. J. Gulowsen at 381-382, 1972.

3. Fl. Agersnap et al., 1974, at 37-39, 46-48.

4. J. Gulowsen 1974, at 387; Fig. 1.; simplified by the author.

5. P. Blumberg 1968, at 124-29.

6. P. Blumberg 1968, at 70-72.

7. R. Likert 1961, at 243-244.

8. C. Pateman 1970, at 71.

9. C. Pateman 1970, at 50-51.

10. See, for example, L.A. Ødegaard, 1975.

11. See for example, E. Thorsrud and F. Emery, 1969.

12. See P.M. Blau 1964, at 273-278.

13. See C. Pateman 1970, at 69-71.

14. See, for example, E.J. Trist and F.E. Emery, 1962.

15. See P.H. Engelstad 1974.

16. L.E. Karlsson 1973.

17. Fl. Agersnap et al., 1974, at 41.

18. See J.H. Goldthorpe, et al., 1968, and Fl. Agersnap and F. Junge Jensen, 1974, at 24.

19. Karlsson, 1973 at 51.

20. See Fl. Agersnap and F. Junge Jensen, 1974.

21. Fl. Agersnap et al., 1974 at 41.

22. E. Thorsrud and F. Emery, 1970, at 177.

23. E. Thorsrud, 1969, at 239.

24. Fl. Agersnap et al., 1974, at 34.

25. See L. Gulowsen, 1972, and Figure 1.

26. L. Ødegaard, 1974, at 67.

27. L. Ødegaard, 1974, at 67.

BIBLIOGRAPHY

Fl. Agersnap and F. Junge-Jensen, RAPPORT OM SAMARBEJDSFORSØG I JERNINDUSTRIEN, København, 1974.

Fl. Agersnap, F. Junge-Jensen, Ann Westenholtz, P. Moldrup, L. Brinch, "Danish Experiments with New Forms of Cooperation on the Shopfloor," in 3 Personnel Review 34-50, Summer, 1974.

P. M. Blau, EXCHANGE AND POWER IN SOCIAL LIFE. John Wiley & Son, New York, 1964.

P. Blumberg, INDUSTRIAL DEMOCRACY: THE SOCIOLOGY OF PARTICIPATION. Constable, London, 1968.

P. H. Engelstad, "Bedriftsorganisasjon i utvikling. Volvo-konsernets forsøksvirksomhet i organisasjonsteoretisk perspektiv" in Bedriftsøkonomi, nr. 6, 1974.

J. H. Goldthorpe, D. Lockwood, F. Bechhofer, and J. Platt, THE AFFLUENT WORKER: INDUSTRIAL ATTITUDES AND BEHAVIOUR, Cambridge University Press, 1968.

J. Gulowsen, "A Measure of Work-Group Autonomy" in L. A. Davis and J. C. Taylor, eds., DESIGN OF JOB 374-390, Penguin Books, 1972.

T. J. Hegland and B. Nyhlehn, "Ajustment of World Organizations" in CONTRIBUTIONS TO THE THEORY OF ORGANIZATION, Munksgaard, Copenhagen, 1968.

L. E. Karlsson, "Experiences in Employee Participation in Sweden 1969-1972". Mimeographed. 73 pp.

R. Likert, NEW PATTERNS OF MANAGEMENT, McGraw Hill, New York, 1961.

C. Pateman, PARTICIPATION AND DEMOCRATIC THEORY, Cambridge University Press, 1970.

E. Thorsrud, "Strategi for forskning og social forandring i arbejdslivet" 10 Tidsskrift for Samfunnsforskning 237-263.

E. Thorsrud and F. Emery, MOT EN NY BEDRIFTSORGANISASJON, Tanum, Oslo, 1969.

E. J. Trist and E. F. Emery, "Socio-technical System" in C. R. Walker, ed., MODERN TECHNOLOGY AND CIVILIZATION, McGraw Hill, New York, 1962.

L. A. Ødegaard, "Direct Forms of Participation" paper, 1975.

INDIVIDUAL FREEDOM AND SOCIAL ACTIVITY

D. A. Kerimov

I.

There can hardly be found in the world another more lofty, noble and at the same time so misunderstood a word as "freedom." For many centuries the best men of our planet kept dreaming about a free, just and humane people's community--sang of it as if it were a beautiful, but remote ideal. For a long time, with hardships, at the price of grave sufferings and sacrifices, the working masses searched the truth of life, struggling along towards their cherished dream--towards freedom. That movement was accompanied by theoretical comprehension of the ideal of freedom and fierce ideological struggle.

The outstanding thinker of the 19th century, Hegel, formulated the dictum, impressive in its immensity, "World history is progress in the consciousness of freedom--progress, which we should cognize in its necessity."[1] But alas, the very same Hegel reduced the whole meaning of this progress to a self-realization of the absolute spirit--the mystical demiurge of all that exists.

The genius of Marx was needed in order that the grand idea of freedom would acquire at last its truly scientific basis and development. Resolutely rejecting all forms of alienating history from people and human activity, all spiritualization of the historical process, Marxism understands freedom as cognition and revolutionary transformation of natural and social activities in the interests of the whole of mankind and its members. Thus, Marxist philosophy of freedom provides the world with a prospect of liberating man and mankind from all that suppresses, humiliates and mutilates them. Not only does it inspire hope, it opens up before the people real chances of taking possession of their own destiny, advancing a concrete program of creating a society based on the principles of equality, democracy and progress. In this lies the profound historical optimism of Marxist philosophy; its life-asserting faith in a tremendous potential of the individual, nations and peoples.

II.

The freedom of man depends on the conditions under which he lives, works, and creates. The self-realization of the human personality takes place in a society and because of it. As a member of the society a personality inevitably has numerous and versatile interconnections with other members of the society. Man cannot be isolated from or fully independent of the historically developed relations in the society, the character of which is moulding and conditioning his social entity. At the same time man possesses also his specific individuality, his peculiarities, characteristic features, owing to whose unity he appears to be not just a product of public relations but their creator as well.

Dependency on society and at the same active participation in its transformation are the objective criteria in defining the actual freedom of man. The degree of man's freedom is far from being determined by those good intentions, promises and assurances which are often and rather liberally given by some leaders of different parties, statesmen, ideologists, or which are even legally laid down in the constitutions. First and foremost freedom is determined by the degree that public life itself provides for man in transforming those possibilities into reality. That is why an obligatory precondition of a free self-realization of a personality is public relations that are genuinely democratic, favorable, benevolent, in a word--truly humane.

Free self-realization of personality, social activity of people, acquires full realization only under conditions in which the environment is genuinely humane, that is, the interests of society and of personalities combine harmoniously, where the resulting democracy provides for the real being of freedom. Marxism puts forward scientifically based ways for creating exactly such a society, which has been tested in practice by socialist construction in the USSR and in some other countries.

As a result of developing a public-political system of socialism there came into being a situation under which it is unthinkable to manage social processes without active participation of personalities, the working masses, all the people.

Under socialism there is no private property in the means of production. The public entity no longer appears in the form of disjointed parts working on their own. Here one does not find a number of contradictory crisscrossing forces, aspirations, collisions and struggle of antagonistic social groups, strata and classes. One does not find domination of public relations over people, domination of product over the producer, anarchy of production, crises, unemployment and other natural disasters, engendered by other socio-economic formations. Instead, after the establishment, consolidation, and development of socialist public property, an economic and political entity of people is formed, whose social activity is directed toward attainment of common goals.

The most important consequences of the existence of public property in the means of production are four main objective circumstances:

(1.) Elimination of the possibility of exploitation of man by man, and establishment of a unity of radical interests of all working people in developing and multiplying public wealth, material and intellectual values.

(2.) Transformation of the purpose of public production, the nature of which is the fullest, satisfaction under existing real possibilities, of material and spiritual requirements of the society and the individual.

(3.) Formation of the necessity of planned functioning and developing economy and other social spheres of life of the society.

(4.) Socio-economic development calls for a scientifically based, rational and optimal management on the basis of the principles of democratic centralism.

III.

Up to now the main conditions of freedom of individual and his social activity have been discussed. What is the essence of the freedom of the individual which is necessary for his social activity?

The definition of the notion of freedom, in our opinion, should include three organically interconnected components, namely: (1.) Cognition of the necessity; (2.) Actions in accordance with the cognized necessity, and (3.) Actions in the interests of the society and all of its members. Let us comment briefly on these components of the entity of freedom.

In their existence and activity people depend on objective laws of being, they can not ignore them, they can not fail to consider them. But this does not at all result in fetishization of the objective laws of nature and society. On cognizing these laws, people can use them in their own interests, can manage them, and thus establish their rule over them.

Objective laws express not only relations between current phenomena, but also the attitude towards what should exist; they express the main trend of the development of phenomena. That is why on cognizing the objective regularity of the laws of social development, people acquire a possibility to foresee the results of the action of the objective public forces, to use them consciously and according to plan in their own interests.

In penetrating ever deeper into the nature of phenomena, people can combine actions of different laws or oppose the action of one law through another, thus achieving certain desired results. Consequently, the thesis that mankind has no power to abolish objective laws does not contradict the fact that on cognizing objective laws and on learning how to use them, people are capable of furthering their more effective action, and in case a necessity arises, limiting the sphere of their action through creation of some relevant conditions.

Existing under certain conditions of the objective world, people should inevitably adjust themselves to these conditions, cognize them, master them in order to use objective regularities for satisfying their requirements, for achieving their aspirations, interests and goals. As long as people in their activity are guided only by perceptible appearance, which as a rule differs from the rational nature, they remain slaves of blind necessity, being powerless to cope with it. Their determined actions may per chance coincide with objective necessity and bring about attainment of a desired goal. But in the majority of cases there is no possibility of attaining the results sought. When determined aspirations of people are based on the knowledge of the objective regularities, they do achieve desired results, based on objective necessity. Their activity acquires at the same time a free character.

That is why an actual, not an illusionary, freedom can exist only when people's actions originate from the cognized necessity, when their activity corresponds to the objective regularities of being, when in their practice they lean on the objective laws of nature and society.

If in their existence, activity and development people depend on the objective laws of nature and society, then cognizance of these laws is not only a condition of their existence, but a necessary precondition of the achievement of freedom. The level of people's freedom is determined exactly to the extent that they cognize the objective regularities of social being, conform their actions to these regularities and achieve deliberately assigned goals, and results. If objective necessity did not exist in nature and society, if there were no laws regulating the course of events, then people would be deprived of the opportunity to take decisions as to any actions, including the simplest ones--would be unable to carry them out--because they would not know what to do to arrive at the necessary results.

The purposefulness of thinking and activity is only possible because the objective laws of nature and social development do really exist and these laws are somehow mirrored in human consciousness. The deeper human consciousness takes possession of the cognition regarding the regularities of nature's and society's development, the more successfully it influences the environment. The clearer and deeper the people become aware, from all points of view, of the necessity, the higher their degree of freedom, the freer their will.

So, it is impossible to proclaim, or even to imagine, freedom without necessity--outside nature's and society's objective laws--if we intend to discuss freedom in its strictly scientific and realistic importance. Freedom is based on necessity, it stems from it, representing a product of its regular development. Necessity's regular development involves the evolution of freedom. The level of freedom grows as a result of greater and deeper cognizance of the necessity. Having emerged from necessity and developing alongside with it, freedom in its turn influences necessity's evolution, bearing in mind that the higher the level of freedom is, the more effective its impact on necessity will be.

The knowledge of the character and tendencies in the evolution of the objective realities of the external world, the discovery of the prerequisites and possibilities progressively transforming the reality do not yet make people completely free, because freedom is not only a cognized state of mankind but also its practical change and transformation.

The knowledge of the objective regularities of the evolution of realities is a must, a condition of ruling over that reality, a necessary stage along the path toward freedom of people. But the achievement of freedom calls for human activity on the basis of the cognized necessity. Just as people act more freely if they expand and spread their power over nature, foreseeing optimally precise results of their impact, so their actions become freer as their knowledge gets broader and larger in regard to the regularities of the development of events and processes occurring in society. Thus, human activity must be directed first of all toward attaining socially significant goals, toward obtaining results useful for the entire society because freedom for each individual is only possible under conditions of a free society. Therefore, only socially purposeful human acts, expressing common interests, aimed at society's progressive development as a whole, can be considered as free actions.

IV.

The previous deliberations show that freedom is a socially intense activity based on the cognized necessity and pursuing the interests of the society and all its members. But free implementation of that activity depends on society itself, on the legal rights and guarantees to carry them out, which it grants and ensures to its individuals. In fact, in order that the free cognition of necessity, and the realization of the opportunities emerging as a result of it, can be successfully put into practice, there should be favorable, democratic conditions in social and state life, a genuinely democratic system of social relations, and broad, truly democratic civic rights which will conduce to the utmost cognizance of the objective regularities and their utilization in the interests of society and personality. If an individual is bereft of democratic conditions and the opportunity to make use of everything that society has at its disposal to satisfy his needs and requirements, then it is meaningless to speak about freedom of that individual.

If society, exploiting and oppressing an individual, strips him of the possibility to get an education, to learn spiritual values, to harmoniously develop his diverse abilities, then freedom for that individual becomes only a spectre, an unrealistic ideal or dream. But if all the blessings and achievements of society are put at the complete command of the individual, if it creates all conditions and facilities for his education and cultural development, for bringing forth all his abilities and talents, if that individual lives, works and creates in the conditions of true democracy, enjoying guaranteed rights under law and order then genuine freedom sets in for each individual, all members of the society.

Consequently, freedom is unthinkable without democracy. Democracy is the essence of freedom. Freedom can only be achieved in the realm of democracy where it constantly grows, deepens and broadens. Therefore, the degree of freedom of society and its members is determined by the level of democratization. On the other hand, freedom is not at all a passive mould of democracy. Freedom is an active stimulus in developing democracy, constantly and systematically adapting it to the requirements of freedom. Being a reflection and embodiment of objective necessity, freedom determines the indispensability of democracy, its evolution, development and improvement.

Such a dialectical inter-relationship of freedom and democracy has been attained in socialist society where the social activity of a multi-million collective of people, based on the scientific cognition of the objective regularities, subordinates to its rule in an organized and planned manner the natural and social forces, channeling them along the path most suitable to that society, using them for the well-being of the people, for the satisfaction of their ever growing material and spiritual needs.

The Soviet people's conscientious participation in the construction of communism is a form which manifests individual freedom. In the process of that construction comradely mutual aid and creative cooperation among Soviet people in all walks of their activities contributes to the development, enrichment and improvement of their personal experiences, talents and aptitudes. Mastering theoretical knowledge, applying it in the practical transformation of the objectively existing reality, Soviet people are satisfying their social and personal needs and interests.

People in a socialist society have come to be the creators of their life, have begun to govern their state and regulate the relations among themselves. Thus, they are establishing to a larger degree their rule over reality, achieving a still greater freedom in all spheres of their activity. Not only economic but also political conditions are conducing to the utmost toward that end. All power in our country belongs to the working people of town and countryside vested in the Soviets of the Working People's Deputies. Through the Soviets millions of people are getting involved in the conscientious, unfailing and resolute participation in the political, economic, social and cultural life of the society, in democratically ruling the state.

Ample opportunities for all citizens to participate in governing their state are also ensured by granting them broad, truly democratic socio-economic, political and personal rights and freedoms. The Soviet legislature guarantees the citizens the right to work, rest and leisure and education, the right to maintenance in old age and also in case of sickness or disability, freedom of conscience, speech, the press, freedom of assembly and meetings, street processions and demonstrations, the right to elect and be elected to the organs of state power, the right to unite in public organizations, etc.

Besides the fundamental rights and freedoms fixed by the Constitution of the USSR and the constitutions of the Union and Autonomous Republics, the Soviet citizens enjoy the rights outlined in the existing non-fundamental laws. These supplement or implement the fundamental rights and freedoms of citizens. Together with the duties they constitute the legal status of the Soviet citizen which is characterized by high democratization, humanism, true justice and humaneness. It should be noted, however, that this status is being continuously deepened, expanded and improved.

The characteristic feature of the rights and freedoms accorded to Soviet citizens is that they are not only proclaimed but are really upheld by a system of economic, political and legal guarantees. Soviet democracy rests on the socialist system of national economy and the socialist ownership of the instruments and means of production, full power of the working people, the consistent, steadfast and firm implementation in practice of the requirements of the socialist legality by all state and public organs, all officials. Therefore Soviet democracy ensures real and effective guarantees of the citizen's rights and freedoms.

The entire structure of the organs of the Soviet state in the process of carrying out its functions and first of all in the process of economic, organizational, cultural and educational work guarantees the rights and freedoms of Soviet citizens looking after their practical implementation. Besides, the Soviet state organs are entrusted with the special task of upholding and protecting the citizens' rights and freedoms. Public organizations are also assigned the task of ensuring these rights and freedoms together with state organs.

There is an organic connection among democracy, freedom and legality which stems from everything said above. Just as freedom can exist only in conditions of democracy, so democracy can be genuine and effective only on the basis of legality. In a socialist society legality constitutes a necessary element, a characteristic feature, an inherent property of democracy. It is only due to this that the democratic Soviet social and state order is ensured democratic forms of the activity of state organs and public organizations, the relationships among them and citizens, providing guarantees for a strict observance of the rights of the members of the socialist society.

However, the present conditions of freedom, envisaged

by law, do not suffice so that each individual might become inwardly free. If it is true that genuinely human conditions of life are created by people themselves in the process of the struggle for freedom, then it is obvious that these people should be adequately prepared to struggle for these conditions, for their freedom. If it is also true that genuinely human conditions are required for freedom of the individual, then it is obvious that the individual will only master freedom when he learns to use these conditions in his strenuous public activities. Further, if it is true that there is no final freedom, because it eternally develops together with society's evolution and the whole of mankind, then it is obvious that an individual acquires freedom only in constantly raising his level of knowledge and using it, systematically improving his public activity. That is why given the current conditions of freedom the individual becomes inwardly free, if he, mastering the progressive world outlook, modern scientific knowledge, values of culture and civilization, finds his place in public activity--actively and creatively participating in the production of material and spiritual blessings, in ruling society and state, in the revolutionary transformation of the world. It is only owing to this that an association of people is emerging which is the highest embodiment of humanism in man's community. It is here that the deepest meaning for uniting people should be sought under socialism, ensuring enough room for the evolution of society's freedom and each member of that society.

V.

Public activity in conditions of freedom, democracy and legality, particularly in the epoch of developed socialism, assumes numerous forms and large scale, involving not only individuals but great masses of people, practically spreading among the entire people. Within the framework of this paper there is no possibility to describe even the principal forms, to show their grandiose scale. We shall, therefore, limit ourselves to citing a few examples.

In 1973, 2,193,195 deputies were elected to the local Soviets of the Working People's Deputies, among them: workers, 863,001 deputies; collective farmers, 613,442 deputies; employees, 716,752 deputies. In 1974, 1,517 deputies were elected to the USSR Supreme Soviet of the Ninth convocation, among them: workers, 498 deputies; collective farmers, 271 deputies; employees, 748 deputies. In 1971, 5,879 deputies were elected to the Supreme Soviets of the Union Republics and 2,994 deputies to those of the Autonomous Republics. Among the deputies to the Supreme Soviets of the Union Republics there are 1,742 workers, 1,221 collective farmers, 2,916 employees. Among the deputies to the Supreme Soviets of the Autonomous Republics there are 945 workers, 489 collective farmers, 1,560 employees.

Substantial roles in the organizational work of the Soviets are being played by non-established departments, inspectors and instructors working on a voluntary unpaid basis at the executive committees (organs of state power). In 1973 the executive committees resorted to the services of 7,272 non-staff departments. Among them are 3,153 organizational departments, 2,091 commercial departments, 389 cultural and popular education departments, 615 communal services departments and 1,024 other departments. 63,050 people are employed at these departments on a voluntary unpaid basis. Besides, 396,522 instructors and inspectors are working at staff divisions and departments of the executive committees of the Soviets without pay.

The development of the public activity of the working people, their participation in state affairs finds its expression in a broad, popular discussion of the most important bills. The broad discussion of the draft of the Constitution in 1936, which went on for 6 months, can serve as an example of this. According to incomplete data, the draft Constitution was discussed at 48,189 plenary sessions of the Soviets, 79,294 sittings of the sections and deputy groups, 411,000 meetings of the working people. At these meetings and sittings the draft was discussed by 51.5 million people.

Over the past years broad circles of the population have been intensively, deeply, intelligently discussing such bills of the USSR, Union and Autonomuous Republics as Fundamentals of Legislation On Legal Procedure, Fundamentals of Legislation On Criminal Code Procedure, Fundamentals of Criminal Legislation (1958); Fundamentals of Civilian Legislation Procedure, Fundamentals of Civilian Legislation (1961); Fundamentals of Corrective Labor Legislation (1962); Fundamentals of Land Legislation (1968); Fundamentals of Public Health Legislation (1969); Fundamentals of Water Legislation, Fundamentals of Labor Legislation (1970). For instance, the discussion of the draft Fundamentals of Legislation on Marriage and the Family, published by the newspaper "Izvestia" and Republican newspapers in national languages, went on for nearly two months. During that period over 7,000 proposals and recommendations were received from the population by the Supreme Soviet, "Izvestia" and other newspapers, as well as the magazines which got over 8,000 suggestions and recommendations.

Beside the above-mentioned forms the working people are taking part in resolving national tasks through public organizations. The principal mass public organizations are the Trade Unions, uniting over 99 million members; Consumer Cooperatives of the USSR with a 61.3 million membership; the All-Union Young Communist League numbering about 31 million members.

Mass public activity of people under socialism is by no means restricted by the participation in the activity of state organs and public organizations, but gives birth to many other forms of democracy: emulation, mutual assistance and cooperation, the movement of industrial innovators, voluntary public order squads and comradely courts and so on and so forth. Over the recent years social planning has emerged as a fresh form of democracy following the initiative of the working people, sponsored and actively implemented by them. Its chief meaning lies in the planned management of both economic and social activity of the working collectives, branches of the national economy and regions.

The social development plan is a scientifically founded and materially upheld system of steps aimed at the planned change of the structure and conditions of life of the working people, shaping and more fully satisfying their material and intellectual needs, requirements and interests, toward an all-round and harmonious development of the individual. The realization of the social development plans, judging by the accumulated experience, raises the effectiveness of production on the basis of the growth of qualification, general and specialized education of the workers, introduction of the scientific organization of labor. At the same time the workers' wages are increasing along with the improvement of their housing conditions, medical and sanatorium services, everything is done to organize good rest and physical development, the atmosphere of comradeship and mutual respect grows wider, as well as the activity of people in managing public processes, in the country's cultural life.

The examples that were mentioned, whose number is far from being exhaustive, show that only in the conditions of true democracy can the freedom of the individual be ensured as well as the nationwide social activity in the construction of a new world.

PARTICIPATORY EQUALITY--AN EMERGING MODEL

Elizabeth Flower

The critical role of equality and its intimate relation to rights, liberty and justice in western political thought scarcely need belaboring. But beyond agreement on the importance of equality there has been little unanimity as to what policy is thereby entailed, as to criteria of equal treatment, and indeed as to the meaning of equality. That too needs no belaboring, for versions of equality have flourished like the green bay tree. Nor--at least in this country--has there been agreement about the dimensions of equality as a concrete problem.

Disputes over the appropriate domain of equality are traditional: Does it apply to political, legal, economic, or broadly social activity? There is no accepted standard as to how equality is to be measured. Ability, need, merit, performance, social contribution, or sheer numerical distribution of goods and burdens each has its advocates. And as if this were not sufficiently controversial, there is no unanimity as to the kinds of institutions which are to be trusted to support or encourage equality, whatever form is preferred.

Perhaps the search for a single criterion in each of these dimensions misses the point; there seems rather to be a family of criteria with the provinces of application only roughly marked out. Clearly the kinds of association (of which the political is but one) function differently and interdependently over time. They depend on historical processes, the stage of scientific, social and moral development, and the claims which can be made articulate or pressed at a given time. The following comments explore in outline how the idea of participation may serve as a significant criterion of equality in many contexts and possibly provide a guide to resolve conflicts of principle. The concept of participation here employed connects it with persons (and personality) on the one hand, and with the character of social arrangements (and community) on the other.

A clear contemporary case on which to test our suggestions is that of affirmative action in the United States, in the attempt to secure equality for women, blacks, and ethnics. Every principle invoked to interpret the equality sought seems to run against some other principle of equality (often already entrenched), as if one equality is only to be purchased at the sacrifice of another. If employment opportunities or educational opportunities are limited, any advance to equality for the <u>outs</u> calls for sacrifices on the part of the <u>ins</u>, or of those who have had legitimized traditional expectations of joining the ins. Thus to secure opportunities for blacks as policemen, or women as college teachers, or hispanos as medical and social work personnel, or any of them as lawyers, or indians in the Indian Bureau of the Civil Service, involves in the socio-economic context of the moment curtailing opportunities for those whom the system has previously favored. The deFunis case concerning admission to the University of Washington Law School or the threat to seniority in labor (or the Berkeley decision in university appointment) are notorious present cases. In the present confusion there is often a defensive reaction rising to a counter-attack on the ground that standards of competence will be undermined. On the other hand, attempts at affirmative action through legal means may degenerate into a straight appeal to numbers or quotas without regard for the traditions of applicability or underlying justification.

How can the concept of participation furnish a criterion for the occasions on which affirmative action is warranted? When a group in virtue of social conditions, usually of some historical standing, is not merely at a disadvantage in some field or other, but is subject to a cumulative institutional exclusion, subjection or degradation, which amounts to major alienation from the system and has had or continues to have effects upon the character or self-esteem of the group itself, such a lack of participation can appropriately serve as the criterion for affirmative action. The non-participation is not merely an evil to the group, but a loss to the society as a whole--in the dangers of frustration and destructive disruption, in the weakening of the character of associative relations for all, in the contribution that the group could make. There is also the corruption of the establishment from an inner contradiction in proclaiming an ideal of equality and (forcibly) ignoring it.

Professor Paul Freund has given a paradigm for this in considering why blacks should be given preferential employment as policemen in areas where blacks constitute a sizeable minority. They constitute an insular and identifiable group which seldom forms a part of the shifting coalitions which are effective in making social policy. Their non-participation is due in large part to the workings of the system itself, and the group itself is not responsible for its non-participation. The areas from which they are excluded are fundamental in the society and in the quality of its life, indeed they are often prime targets of legal and law-enforcing institutions. Under these conditions, there is an overriding social interest to bring about the integration and the participation of the group, even by preferential hiring--unless some other constructive method can be found to bring them into the ambit of social participation. Relevant factors will differ from context, but a comparable mode of argument applies in different facets of social life, and to other excluded groups.

Participation may thus serve as a justifying principle. It helps order competing equalities and thus serves as a ground of priority in a given situation. (In somewhat the same way, productive needs in a given period help decide between an egalitarian formula of equal labor time and equal pay for equal output.) In the paradigm above, preferential hiring is intended to break through obstacles which had made for a de facto inequality which used criteria which only the entrenched could meet. The opposing side appealed also to equality, a meritocracy in which equal success rewards equal merit. Almost every egalitarian formula has a prima facie claim; and doubtless there are almost always some contexts in which it should prevail. Hence the role of a principle which provides criteria for the relevance of situations or the grounds of preference is indispensable.

To look at equality in relation to participation or community offers many advantages over the sheer confrontation of competing equalities. It is a fresh approach which may help to unseat the stale deadlock between the standard justifying principles, especially utilitarianism and absolutistic principles of fairness or justice. Like them, the appeal to particpation as an ideal seems to capture a good bit of our pre-systematic notion of a functioning and integrated society, perhaps more than such instrumental conceptions as the Hobbesian basic security or the Benthamite productivity. It frees us from the need to appeal to absolute equalities (often incompatible among themselves) and forged in a vastly different socio-economic setting with aspirations

limited to a narrow group. It helps liberate us too
from a thoroughly competitive or game model that en-
shrines an extreme individualism and turns us toward
the role of community as a necessary feature of personal development. But in so doing it need not rush from
an individualistic extreme to a collectivistic one.
Indeed we have come to realize that between these extremes there is the category of the <u>interpersonal</u>. This
emphasizes neither the individual profit nor the social
group interest but, truer to the workaday relations that
we actually sustain, the immediate character of the manifold of human affiliative association.

The notion of participation would support a formal
statement, starting with something like: No one can
legitimately claim for himself a privilege in respect to
his relations toward others which he would not recognize
for others as a privilege toward himself. It would be
irrational for any man to seek to benefit from association with others in ways which, if institutionalized,
would dispel the possibility that others might similarly benefit from associating with him and thereby destroy the association. But we are carried further by two
supporting premises. One is that the kinds of association relevant to participation constitute a total functional configuration, not the separate functions of
isolated institutions, and particularly not limited to
political. (On this, Professor Opalek's paper is especially illuminating.) The second is that what the associating human being is as person itself depends on the
quality of association itself. (This point is well
developed in Professor Borucka-Arctowa's paper.) The
notion of participation so qualified is open-ended and
is normative in the sense of pointing a direction without foreclosing innovation and freshness in the forms
that it may take. Like the concept of health it enables
us to incorporate the growing knowledge of ourselves and
social relations. At the same time it provides definite
criteria of social ills.

No claim is intended that the appeal to participation
solves all our difficulties. It often raises in its
turn questions analogous to those set forth about equality at the outset. On the other hand, it does seem to
provide an objective measure, to provide for inductive-
ly gained wisdom, to face changing problems in an inno-
vative way, and to provide an empirical measure for social achievement.

This notion of participation is not a novelty on the
American philosophic scene. It is well developed in
the notion of community, for example by Royce. It grew

up in American pragmatic social philosophy in opposition to the view of the isolated individual and of atomic equality. C. I. Lewis, John Dewey, and George Herbert Mead focus clearly on this issue. Lewis works from an enlarged view of person and the attempt to give criteria for authentic association. Dewey works from the notion of community in an effort to unseat the dichotomy between individual and social by understanding of the conditions under which "private" and "public" apply. Mead, of course, is the father of the notion of the interpersonal.

DERECHO Y PARTICIPACION

José Luis Curiel

EL PROBLEMA

"Participación" es un concepto destinado a ocupar un sitio crucial en la esfera de las ciencias sociales. Exige una consideración interdisciplinaria. La Filosofía social y la Filosofía del Derecho tratan de obtener fecundas consecuencias de la participación como hecho y como deber. No puede omitirse su importancia en un Congreso que gira en torno del tema: 'Igualdad y libertad'.

El presente trabajo considera estrechamente ligada la idea científica de participación al supuesto antropo-ontológico del Derecho, ilustrado a continuación mediante tres reflexiones sobre: la norma, el sujeto y la finalidad. Por otra parte los avances de la sociopolítica y de la psicosociología contribuyen indudablemente a esclarecer los caminos microsociales, jurídicos y personales de la participación eficaz.

I. NORMA

Significa orden. Pero como hay varias clases, como el orden natural, el social, el individual, el cultural, el político, el económico, etc., es preciso decir que se trata del ordenamiento de la acción humana. Orden humano que se expresa en el 'debe ser' o en el 'debe hacerse'. En el último caso incluye además una orden, en frase de John Austin.

Ahora bien, el concepto de norma específicamente humano no excluye sino que implica lo genérico del ser humano, contrariamente a lo que ocurre con la serie de objetos que contempla una geometría euclidiana que no distingue en ellos su esencia de sus propiedades esenciales. Conviene aludir aquí a esa segunda clase de evidencia que se da en juicios analíticos en los que el predicado no está incluido en la definición del sujeto, sino que es el sujeto quien necesariamente queda implícito en el predicado en calidad de su propio sujeto. Así ocurre en toda acción o en toda omisión a que se refiera una norma jurídica: surge siempre la exigencia de un sujeto en situación legal concreta, sujeto no solo específica sino también genéricamente humano. El orden jurídico es en este caso un superorden como cuando por ejemplo mi voluntad sirve de un orden físico-natural y al obedecerlo lo domina y lo superordena. Es la vía de la posibilidad entre los dominios de la naturaleza y de la libertad. Tercera vía que en el pensamiento kantiano conduce al concepto de finalidad. La acción humana inicia su partida en la posibilidad de escoger libremente dentro de las limitaciones de la conciencia moral. Acciones humanas no son simple conducta, sino las que aparecen como dotadas de algo aleatorio, futuro, inteligente y querido. Antecedente remoto ofrecen al psicosociólogo y al filósofo social la correspondencia entre Fermat y Pascal o el 'Ars Conjectandi' de Bernouilli. Puede leerse en su cuarta parte: "el arte de conjeturar, arte estocástico, consiste en medir tan exactamente como sea posible las probabilidades de las cosas a fin de que podamos en nuestros juicios y en nuestras acciones escoger siempre lo que habrá de ser reconocido como lo mejor, lo preferible, lo más seguro, o lo mejor deliberado: actitud en la que reside toda la sabiduría del filósofo y toda la prudencia del político".

Pero la vida humana exige por su misma naturaleza un orden todavía más complejo--el orden social que ha sido fuente de términos como ley, deber, obligación, responsabilidad. Porque el concepto de obligación antes que coacción implica solidaridad. La acción humana ordenada racionalmente supone una estructura ontológica generadora, una razón práctica. Su rectitud en ejercicio se llama libertad y entre sus metas se hallan la justicia social y la paz. El camino obvio es la participación. En todo caso la persona humana en última instancia es el sujeto no solamente de toda acción, sino también de toda acción o participación normadas por el Derecho.

II. EL SUJETO

El positivismo jurídico lejos de lograr su loable propósito de precisión ha incurrido en confusiones funestas. Dice que la imputación central nos lleva a una persona, pero nunca a un hombre y al mismo tiempo extiende la imputación central mencionada a sujetos de Derecho que no son hombres como por ejemplo a las llamadas 'personas colectivas'. Expresiones nada positivas pero muy positivistas que no pueden disimular un lenguaje antropomórfico ficticio.

Todos sabemos que los jurisconsultos romanos hablaron de 'collegium', 'corpus' o 'universitas' y no la palabra 'persona' para referirse a propiedades colectivas y a otras entidades que en lenguaje moderno se han denominado 'personas civiles', 'personas jurídicas', 'personas morales'. Ya el jurisconsulto del "Espíritu del Derecho Romano", Ihering, sostiene que el destinatario de todo derecho es el hombre. Afirma que "la llamada persona jurídica es una persona ficticia, artificial, incapaz de gozar, pues no tiene ni interés ni fin. Sujeto aparente que oculta al verdadero. Desde que se pierde de vista esta idea fundamental del Derecho de que el hombre y solamente el hombre es el destinatario de derechos, ya no se detiene en el camino de la personificación. Los verdaderos sujetos de derecho no son las personas jurídicas como tales, sino sus miembros." Es, en palabras de Planiol, el 'mito de la personalidad'. Una de las consecuencias de esta paradójica mitología positivista es el sofisma de la personalidad jurídica de las comunidades y de las comunidades sin personalidad jurídica.

Parece una contradicción interna del positivismo sostener al mismo tiempo que cuando el Derecho personifica ciertas normas y construye una 'persona individual' no por eso ha construido al hombre real, mientras que cuando el Derecho construye una 'persona colectiva' sí produce y otorga existencia a una comunidad. "La comunidad no está frente al orden jurídico sino en él," así brota el razonamiento que pretende fundar jurídicamente el anonimato de las sociedades capitalistas o el anonimato de la masa respecto al tratamiento de la responsabilidad y del cumplimiento de los deberes. El concepto de alienación o enajenación del individuo quien a través de la masa alcanzaría su liberación mal disfraza el desprecio de la persona ante la superioridad del hombre colectivo. Suponen estas diversas posiciones una confusión entre individuo y persona. Se olvida que la persona no es intercambiable, que no es un medio, un número solamente o un auxiliar del pensamiento. Es absoluta frente a las cosas y su sentido profundo permanece imperturbable ante modificaciones individuales, sociales y estatales. Sus posesiones o tenencias pueden ceder ante fines comunitarios, su esencia y su valor no. En un terreno biológico podría afirmarse que hasta Dios mismo respeta su libertad y su intimidad. Deber moral y deber jurídico, carecerían de sentido sin la persona humana individual. Cuanto atrito jurídico se aplica a la 'persona colectiva' como centro de imputación, se traduce en última instancia en deberes o derechos de personas individuales que constituyen sus miembros. La persona colectiva no sufre hambre, ni temor ni anhelos, ni puede contraer matrimonio o procrear hijos. Toda sanción remata en el auténtico titular de obligación o derecho que es la persona individual.

El Derecho penal no puede castigar 'socialmente' sin sancionar a los miembros de un grupo social. La responsabilidad criminal es individual. Cuando el juez puede suspender o hasta disolver una agrupación se da una apariencia de responsabilidad colectiva, pero la verdadera responsabilidad criminal, la exigida sobre la base de la intención o de la culpa que determina la imposición de penas propiamente dichas, es la individual, pues al decir de Cuello Calón, la colectiva más que un carácter penal tiene un sentido de medida puramente preventiva.

Tampoco ha de confundirse persona y personalidad. Decimos certeramente: somos persona, tenemos personalidad. Puede hablarse con propiedad de las 'personalidades de la persona'. Todo ser humano es persona. Esto constituye el fundamento de la igualdad, el punto de apoyo de la libertad, la fuente de la participación. El positivista vuelve su mirada a la Historia y dice: el esclavo en el Derecho romano no era persona, y era hombre. Efectivamente esta consideración ha de tenerse en cuenta. Pero también ha de verse la contradicción que arrastró constantemente al reconocer atributos al esclavo que solamente pueden referirse a una persona en situación jurídica, como por ejemplo se reconoció inteligencia y voluntad bastantes para obligarse civilmente por sus posibles delitos, y si bien se proclamaba que la condición de 'servitus' o de 'servitium' podía obtenerse por el nacimiento, también se admitía que podía adquirirse con posterioridad: "servi aut nascuntur aut fiunt". Lo que implicaba la posibilidad de que una persona libre pudiera llegar a ser esclava mediante el simple cambio de su situación jurídica, como en el caso de la cautividad o por alguna condena judicial. Ha de recordarse igualmente que en algunas épocas mediante procedimientos tan artificiosos como ser registrado ciudadano en los registros del censo, o en la manumisión por la 'vindicta' reivindicaba su libertad. Algunos jurisconsultos romanos protestaron contra esa institución de la esclavitud. Florentinus definió la esclavitud como 'constitutio juris gentium qua quis dominio alieno contra naturam subjicitur'. Cicerón y Séneca, en contraste con algunos pensadores griegos reconocieron la igualdad de naturaleza entre el 'esclavo' y su 'amo'. Y Ulpiano dejó escrito--"omnes homines aequales sunt." El sujeto del Derecho es el hombre como persona. Por esto es también el sujeto de la participación.

III. FINALIDAD

Aun cuando la voluntad psicológica del creador de una norma haya dejado de existir, toda norma sigue expresando un deber ser o una orden de hacer. Sin ello carecería de sentido su interpretación, su continuidad, su duración, su identidad y hasta su eficacia. Aunque un formalismo extremo pretende eliminar del concepto de norma toda nota finalista o teleológica, nadie puede negar que encierra necesariamente un juicio, una afirmación, punto de partida de la interpretación. Montesquieu en su obra tan conocida EL ESPIRITU DE LAS LEYES se refiere a las diversas relaciones que las leyes pueden tener entre sí, con su origen, con el propósito del legislador y con el orden de cosas sobre el que están establecidas.

La regla fundamental del intérprete consiste en el conocimiento de la totalidad del texto de una ley para comprender el pasaje o la disposición más breve. Ya el Digesto decía "Incivile est, nisi tota lege perspecta, una aliqua particula ejus preposita judicare vel respondere". Algunos juristas como por ejemplo Hauriou han distinguido el concepto de fin del concepto de 'idea directriz'. Aquél exterior al ordenamiento jurídico, éste inmanente o intrínseco a él. Para otros juristas como Delos, fin e idea directriz se identifican aunque con doble función. En toda estructura formal se expresa la tendencia a mantener vigente la fuerza del deber emanada de la autoridad moral y jurídica. Si toda norma es expresión de un orden humano social en proceso de fabricación, es evidente que manifiesta una finalidad aunque ésta no determine su esencia. Un pasaje de la Metafísica aristotélica de manera incisiva y sutil enseña que la causa formal indica al mismo tiempo la finalidad específica de un ser. No puedo comprender lo que es un reloj si no entiendo que su estructura sirve para medir el tiempo, aspecto que no se halla afuera sino por así decir dentro de su propia vigencia. Con razón ha dicho Delos que por haber descuidado la inmanencia de la idea directriz en la sociedad y en sus reglas formales, numerosos moralistas han relegado el derecho positivo simplemente al papel de medio al servicio de un fin extrínseco.

En una concepción matemática o en una Etica 'more geometrico demonstrata' no se distingue entre agente y acción, porque reina sin competencia la causa formal, sin agente, ni acción ni fin. Por eso los Antiguos dijeron que las matemáticas no son buenas, ya que las nociones de bien o de fin no intervienen. Pero regular las acciones humanas como lo hacen las normas es dar por supuesto un agente, sus acciones y el orden de sus acciones. Todo agente persigue fines próximos y remotos. El fin más próximo de su propia acción y de acción en acción la meta más remota. Entre lo que es y lo posible, la acción se distingue del agente. En la imputación personal se distingue el sujeto (agente) de lo preceptuado. La relación implica orden con etapas seriadas: conocimiento del precepto, aceptación voluntaria, movimientos ejecutivos en vía de cumplimiento. El enunciado formal de la norma juega en este caso el papel de un fin para el sujeto normado. Se trata de una relación entre acción jurídica y acción moral que mueve a muchas personas a obedecer, obediencia repudiada o desconocida por el positivismo. Rousseau decía "obedecer es hacer la voluntad de otro y por consiguiente renunciar a la suya, a las luces de la razón, a su libertad, a sus derechos, a su personalidad, a su cualidad de hombre."

Pero es que hay al menos tres clases de obediencia como aparece en la clasificación pedagógica de Coussinet: la forzada por la coacción del que manda, la aceptada por conformismo del que obedece y la consentida espontáneamente por el mandato de obedecer que se da el mismo que obedece. "Obedecer no es aceptar la voluntad de otro como individuo humano semejante a mí. Se obedece a la autoridad, no al hombre. La obediencia verdadera es un acto eminentemente razonable y libre, adopta el juicio práctico emitido por la sociedad como necesario." Estas palabras de Soignon significan que la coacción disminuye la pureza de la obediencia. Dice que es preciso una comunión profunda, un acuerdo sobre la finalidad del acto, acuerdo juzgado como indispensable y aceptado por la voluntad libre como otro deber, como el más sagrado. Esta reflexión no se sostendría sin la vigencia de la idea directriz de toda norma que se muestra como un bien dotado de cualidades axiológicas ante las necesidades vitales, sociales culturales y espirituales de la persona humana. Pero esta obediencia es una forma expresa de participación, o si se quiere está condicionada por una participación tácita pero operante.

IV. PARTICIPACION

Igualdad y libertad, normatividad y finalidad, son ingredientes indispensables del concepto de participación. Solamente a la persona humana puede referirse este mundo de relaciones, porque solo ella es hacedora de valores sociales. Igualdad de oportunidades para intervenir en las decisiones de una sociedad, puede ofrecerse como concepto de participación. Sin embargo, conviene advertir que no ha de entenderse por sociedad solamente y ante todo el ente macrosocial, como por ejemplo, el Mundo, la Humanidad u otras abstracciones. La Psicosociología contemporánea no en balde ha mostrado la importancia de la consideración microsociológica ante el enfoque personal. En efecto: las entidades gigantescas operan a través de sus grupos y dentro de éstos funcionan los psicogrupos a que cada uno pertenece a diferentes horas de cada día. Es decir que la idea de participación no ha de quedarse en el concepto sino llegar hasta la existencia concreta. La participación surge hoy en toda teoría política, en todo método pedagógico, en cambio social de nuestro tiempo, en todo afán de dialogar y progresar, en toda consideración de filosofía social.

Pero se refiere necesariamente a la persona humana en situación existencial, al hombre con su implicación no solo específica sino también genérica. Acuñado por la Sociología el término participación es cada vez más de alcance filosófico y jurídico. No es solamente algo entitativo del ser social, del animal político, sino que es una exigencia, un deber y una obligación. Implica una cierta autonomía moral, un cierto respeto al derecho de los conciudadanos, una cierta actividad o iniciativa, una espectativa de presencia fecunda. Pero también una responsabilidad personal. Quizá donde más se ponga de relieve es en una Etica social concebida como urgencia de no omitir el desarrollo de las capacidades que contribuyan al mejoramiento y a la justicia en el seno de la convivencia humana. El Estado se ha hecho para garantizar y acrecentar las realizaciones axiológicas de la persona humana, no tiene sentido sino por y para la persona. El individuo como tal ha de subordinarse al Estado, pero los fines de éste han de coincidir con la comunidad de las personas. Comunidad que no es masa, sino pueblo, no es construcción ficticia como una sociedad anónima, sino vital complejo de relaciones personales. El Estado no es tampoco una persona colectiva dentro de la teoría de la ficción, no tiene como fin principal mediatizar al individuo respecto de su libertad, sino ha de ser viva expresión del orden jurídico al servicio de los altos valores personales.

CONCLUSION

Esta cuádriple reflexión sobre el Derecho considerado como NORMA u orden humano, sobre su SUJETO como persona real y no ficticia, sobre su FINALIDAD como directriz inmanente y sobre la PARTICIPACION dentro de cada grupo social en situación concreta existencial, desemboca en la esperanza de una convivencia social en donde igualdad y libertad no sean antagónicas, sino dimensiones humanas garantizadas y promovidas armónicamente por el Derecho.

Knowledge, duty and hope are based on a fundamental anthropology. The law demands an integral and correct concept about the human person. This paper examines four main aspects in the juridical sphere, the norm, the subject, the finality and the participation.

The normative order has meaning only for human beings, who have realized in their integrity (generic and specific) dimensions like liberty, equality and social solidarity. The theory of juridical personality must not consider a "collective person" identical to the human person.

Legalized slavery constitutes an internal contradiction in Roman Law.

Each norm supposes an intrinsic end as an idea "directriz" expresses the spirit of the Law of Nations. This idea is a good for the human beings and belongs to the axiological world of the person.

Equality and liberty, normativity and finality, are "sine qua non" ingredients of the concept of participation. Participation is equality and liberty of opportunities for producing decisions in a social group. Recognition of the importance of the concept of participation is moving from the sociological sphere to the juridical field.

All the juridical hypotheses turn around the person. Law and ontological anthropology must walk harmoniously together.

ON A THEORY OF PARTICIPATION

Joachim K. H. W. Schmidt

People, who ask for "participation," are concerned with policy making. They want to be in charge, no longer under control. We can observe today an ever increasing number of active selves, often, but not necessarily, rather small social groupings of authentic persons, entering the societal space with the intention to affect societal relevant decisions. Established authorities react nervously, being unable to cope with the growing social awareness in the citizenry, since our Greek-Western tradition focuses on a passive society (contradictious to the "active society" of Amitai Etzioni[1]) stressing legal institutions that prevent any form of participation. The whole framework of our inherited Rechtsdenken is not akin to the set of social institutions to which participation belongs. Thinking about this, I got the idea that it might be useful to distinguish systematically various distinct kinds of Rechtsdenken, assuming that each of these distinct kinds of Rechtsdenken is more or less related to some specific set of social institutions. This proposition is hardly objectionable since we have learned to recognize the way of life of other people(s) as cultured, too, and have to deal increasingly in the present multicultural world even with contradictory kinds of Rechtsdenken[2].

I decided upon a four-dimensional model of Rechtsdenken, labeling these four dimensions (A) curriculares, (B) universalisierendes, (C) didaktisches, and (D) individualisierendes Rechtsdenken. Compare the following illustration:

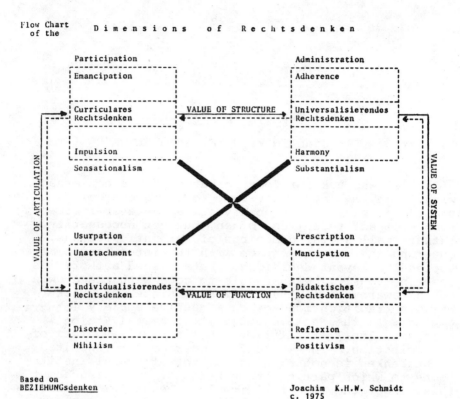

I use this model as a universal order-chart and have tested and applied it several times.[3] According to the model, curriculares and didaktisches Rechtsdenken as well as universalisierendes and individualisierendes Rechtsdenken are contradictory, tending thus to exclude each other, while the four dimensions in other combinations are either contrarious or subcontrarious, seen as inclined to oppose each other, but having something in common that prevents a contradiction in toto. Growing historically out of each other, the four dimensions of Rechtsdenken are related over time, and it is assumed that they are, at least potentially, altogether present in matured societies.[4] However, one is normally dominating, and the dominating dimension tries then to exclude its contradictious dimension, whereby the respective most influencial dimension is functional to a specific state of entropy within the overall societal development.[5] Propelled by the now unfolding value-dimension of articulation[6] the so-called curriculare Rechtsdenken can be predicted to rise in the near future in all the countries previously under the influence of our decayed Greek-Western culture.

Regarding the terminus "curriculares Rechtsdenken" I use an analogy to pedagogics, i.e., I refer willfully to the CURRICULUM. This concept was introduced into the language of pedagogics in the sixteenth and seventeenth century, replacing during the period of baroque style concepts like "studium," "ordo," "formula," and "institutio" as termini technici. "Curriculum" has been a typical reformist term related to its age, loosing its position on the continent already in the eighteenth century, but continuing to be used, however, in the Anglo-Saxon language. Due to recent import from the United States[7] "curriculum" wins its position back, being used now instead of Didaktik. "Didaktik," a Greek term, focuses on the transmission of handed down knowledge. Compare, e.g., the content of the theological treatise DIDACHE, which is past-oriented, referring to some canonical tradition, while "curriculum" is future-oriented, looking for permanent revisions of intentions, contents, methods, and media, whereby it is assumed that these are interrelated. The "curriculare" character of such Rechtsdenken expresses itself in the permanent revisions, signifying high adaptability to changed circumstances. This is exactly what is asked for today, since the impulse-rate of social change is so tremendously high. Focusing on this high impulse-rate of social change, John Platt even speaks of "the greatest and most sudden jump in the whole history of evolution."[8]

To each dimension of Rechtsdenken I have attributed in my model four indicators, referring to (a) data-substratum, (b) intellectual superstructure, (c) psychic manifestation, and (d) disposition to cope with reality. Impulsion, sensationalism, emancipation, and participation are assumed to be the indicators for the curriculare Rechtsdenken, contradicting each the respective one in the dimension of didaktisches. Rechtsdenken, however, are a synthesis of the respective ones in the subcontrarious dimension of individualisierendes Rechtsdenken and in the contrarious dimension of universalisierendes Rechtsdenken. Our present societies are under the dominant influence of the individualisierende Rechtsdenken; while the representatives of the old decayed system try to turn back the wheel of history, ignorant of the deep societal change which has already occurred and will continue to occur due to the still ongoing technological revolution. The advocates of the rising curriculare Rechtsdenken, i.e., people who try to anticipate the working principles of our future social institutions, have to invest most of their energy to eliminate obstacles obstructing their work.

The indicator "impulsion" is up to now best described by Malewitsch[9] and Stockhausen[10]; of course, the number of persons who try to elaborate on this topic, are legion, and their books already numberless. The authors who have discovered the data-substratum of the curriculare Rechtsdenken share high artistic qualities, intense rationality, and are all in favor of constructivism in a general sense, i.e., they consider reality as something that has to be engineered. "Impulsion" focuses on a still not objectivated equality of all constitutive elements, thus contradicting reflection that attributes weights on the basis of arbitrarily chosen viewpoints. Both, impulsion and reflection represent states of order; while the first state of order is oikocratic, the second one is hierocratic. The oikocratic order is unfolded by intervals of universal relations beyond identity among the elements in question, each one has its own time whereby all are simultaneously present; sizes differ due to changing tensions within the field of time without any direction. The hierocratic order is brought forth by various ideologies, each establishing a normative system, territorially bound and exclusive for a given time. For further explanations I have to refer here to other writings.[11]

The indicator "sensationalism" as intellectual superstructure derives from the age of European enlightenment. Etienne Bonnot de Condillac succeeded in working out an adequate theory. According to Condillac the number is constitutive for reality, i.e., things vary only due to the size of the number,[12] assumed already by Philolaos, some 2500 years ago.[13] Condillac emphasized consequently analysis as instrument for dealing with reality, understanding "analysis" as action of decomposition and composition.[14] If we know by exploration how reality works, we are enabled by computation to transform past reality according to our present and future needs. I should add that most writings of Condillac are devoted to education, thus recognizing clearly that material wealth alone is not sufficient to upgrade the human condition. Hopefully nobody expects that I try to give here an introduction to sensationalism. However, it might be important to denote that sensationalism is contradictious to all kinds of positivism, including Popper's rationalism. Positivistic approaches stay within the given, stabilizing established systems, while it is the central concern of sensationalism to transcend the respective given, creating a new world adequate for human beings, based on analysis and with the help of relevant education.

"Emancipation" is here regarded as indicator for the psychic manifestation of the dimension of curriculares

Rechtsdenken. We can measure the frequency of its occurrence like we can measure all other indicators.[15] It is important to know the state, otherwise nobody would be able to govern it.[16] We are, indeed, able to control crises by listening to "facts" about our social reality; but listening to lobbyists cultivates crises. The lobby is a social institution akin to the individualisierende Rechtsdenken that creates disorder and promotes usurpation. Who listens to lobbyists, lacks emancipation, being still an unmatured subject, although he might be not more mancipated but unattached. Like every dimension of Rechtsdenken, curriculares Rechtsdenken, too, penetrates all areas of life and defines the kind of each social interaction, since "Recht" is our program of living, i.e., we live according to the Recht that constitutes society, whereby the kind of the Rechtsdenken is variable. Emancipated citizens "denken" Recht in the manner that active selves project their Recht, i.e., their way of life, themselves, thus being included within the process of defining the conditions of the environment of their living and working places. Cicero has already advocated such kind of Rechtsdenken, i.e., was a representative of the so-called curriculare Rechtsdenken, in emphasizing the participation of all citizens.[17] Cicero was, however, the last prominent advocate of this cause in Rome, he died together with the Roman republic, victim of representatives of the imported Greek culture that was a pure expression of the didaktische Rechtsdenken, growing out of the value-dimension of system.[18]

Finally, in "participation" I see an indicator for the disposition to cope with reality. In my model, participation is only one of four possibilities to deal with reality; the other dispositions are administration, prescription, and usurpation. With "administration" I do not have in mind bureaucratic means, but rendering services in order to help, i.e., I refer to the original sense of the word.[19] "Prescription" envisages the situation that a power center sets up definite rules, participation of the citizen is here considered a crime, everyone has to subordinate. "Usurpation" reflects the common praxis of today, to make use of whatever comes across, disrespecting the rights and dignity of others; normally courts of justice later legitimize these acts, at least in the Federal Republic of Germany. We can gain a real idea of participation only in looking simultaneously at these other indicators. Since the indicators of all dimensions are interrelated, we can look at no one in isolation. It is the very trick of the formal qualities of my flow chart model that every detail is seen in relation to the whole; in contributing to establish the whole. It also derives its meaning exclusively from the integrating whole. My model works like social reality eventually happens: Ecologically; it is impossible to neglect the one or the other part of social reality. The illustration of my model may look conventional; Robert N. Bellah once saw nothing else in it than a Parsonsian chart,[20] since he was not free enough to imagine an alternative to the interpretational approach of his teacher Parsons. Interpretation, however, belongs to the paradigm[21] of the Thought of Being, i.e., it is the old Greek-Western way of representing reality. But there is, indeed, an alternative, provided by the Thinking in Relation. According to this Thinking in Relation we do not interpret and represent reality, but analyze and construct reality.[22] "Reality" itself looks different; it is conceived still as some given, i.e., existing out of data. These data, however, are now considered processable bits of information, thus disposable for our interfering transmutational acts. In other words, reality has lost for us its weight and related immovability, being no more the heavy burden than it once was for our ancestors: Man has finally mastered reality.

Reality looked upon as information, being thus subject to constant change due to the communication of the population,[23] is manageable only by the participation of all citizens, assuming that everyone has internalized society, enabled thus to form active selves, the new centers of society. Formerly society has had only one center, the representative organ, and it did not matter what its nature was: It may be a king or a president; nobility, clergy, professions, or present-day parliaments. Norman O. Brown has worked out the nature of this representative organ in an excellent way.[24] Future societies will have innumerable centers. Each can look at himself as society. Of course, this development has to go hand in hand with the growing responsibility of the societal members who have to regain first their authenticity. It is not to think of any kind of participation without full responsibility of the citizens involved.

Our authorities may be, however, inclined to refuse the transfer of responsibility. The situation is too uncommon; the official might have to supersede himself. Of course, he might have to supersede himself only regarding the specific task in question, but, being honest, who would like it to be superseded? Let me give an example:

> Suppose that you produce widgets; your marketing department makes an error; it does not respond to the needs of the environment; old style widgets are rejected: Result--possible death of your organization.

What is one possible reaction of your governing board? To attempt to impose widgets by controlling markets, by controlling consumer habits, by making it impossible for the state bureaucracy to avoid using widgets. If you look at social reality, you will find many widgets. Many errors are thus imposed.[25]

This example may look out of context, but it reflects exactly our difficulty to accept being superseded. An alternative reaction in the above example would have been to produce something else; and this would be an expected alternative for our officials, since there are so many things to be done, that the participation of everyone is urgently required in order to secure our survival. It is hard to imagine a time when one would be more needed. Even the old and sick ones are needed in our society, since they are a constitutive part of our society. The praxis of the past and still today: Best to forget about them! Humane?

The institutionalization of participation presupposes further the dissolution of the so-called private and public spheres; accordingly, our traditional categories of private and public law are no longer needed. It is, indeed, very hard for me to imagine that any official who is accustomed for many years to distinguish private and public spheres, would feel at ease simply to drop this convention, since it is an essential element of our inherited Greek-Western culture. Moreover, due to this distinction he is somehow legitimized to understand himself as an important person; it is the office that makes him feel great. Participation presupposes that everything belongs to the societal sphere. And this societal sphere is neither public in the traditional sense, nor private in the traditional sense. Consequently, the law will be neither public nor private, but societal law, and as such _ius gentium_. Repeating Cicero: Si enim pecunias aequari non placet, si ingenia omnium paria esse non possunt, _iura certe paria debent esse eorum inter se qui sunt cives in eadem re publica_.[26]

"Ius gentium" is some ius commune, unalienable, the living right, constantly evolving as a consequence of our ongoing communication. In meeting each other and engaging in exchange, admitting responsibility for the other, we author the law and confer to it authority. The "we" refers to active selves, i.e., neither you nor I could bring forth any law, but we could do it together. "Law" is unthinkable on the _individual_ level, it is at all instances _our_ law, i.e., it is something in between, a medium.[27] This character of the law as a "medium" is unknown to the Greek-Western culture, consequently too, "participation." The old Greeks worshiped space, and space can only be separated, it can not be shared. Only time can be shared and shared time enables us to participate in common affairs.

A society that has institutionalized participation, may live without formal officials, since every societal member is potentially himself society. No doubt, things have to be repeatedly done officially--society as a whole could not survive otherwise--but this does not mean that we should be in need of a particular species of human beings, the so-called officials. The old Romans used the fasces in order to indicate the official character of an action (doubtless more congenial than creating a new species) and much more practical. If something needs to be done officially, the official will be the person at hand. Anyway, the officials of today are never around when their advice is needed, i.e., we are already used to being our own officials. If we are able to perfect ourselves, should it be impossible to perfect society? Society is inseparable from the human beings who constitute it, and there is nothing within the limits of our faculties that we cannot reach in common work. We cannot turn ourselves into gods, but we can turn ourselves into human beings. Are we inferior in our faculties to the old Romans? I can hardly believe it, hic Rhodos, hic salta.

FOOTNOTES

1. Amitai Etzioni, THE ACTIVE SOCIETY, New York, N.Y., 1968.

2. Compare e.g., Adda B. Bozeman, THE FUTURE OF LAW IN A MULTICULTURAL WORLD, Princeton, N.J., 1971. Don't miss the added bibliography.

3. See e.g., Joachim K. H. W. Schmidt, ed., PLANVOLLE STEUERUNG GESELLSCHAFTLICHEN HANDELNS--GRUNDLEGENDE BEITRÄGE ZUR GESELLSCHAFTSTECHNIK UND GESELLSCHAFTS-ARCHITEKTUR, p. 315, 319, Opladen 1975. For the quantitative analysis of the model I developed special statistics, so-called _elliptical_ ones with Arabic signs (still unpublished).

4. Regarding the possibility of societies _to mature_, compare Alexander Matejko, Der gesellschaftstechnische Ansatz einer Beeinflussung gesellschaftlicher Systeme und Wertungskriterien für ihre qualitative Entfaltung, in: Joachim K. H. W. Schmidt, ed., PLANVOLLE STEUERUNG, p. 109 ff., see note 3.

5. About "entropy" compare e.g., Hans Reichenbach, THE DIRECTION OF TIME, Berkeley, Ca., 1971.

6. I wrote about this in my paper "On the definition/determination of values within the scope of social engineering," prepared for the Third Meeting of the ISA Research Committee on Sociotechnics, Warszawa, Poland, 1975. Publication soon in English and German.

7. Compare e.g., Herwig Blankertz, THEORIEN UND MODELLE DER DIDAKTIK, p. 117 f., München 1972.

8. John Platt, "The greatest jump in evolution," in: Ekistics 207, p. 57, Feb. 1973.

9. See Kasimir Malewitsch, SUPREMATISMUS--DIE GEGENSTANDSLOSE WELT, Trans. by Hans von Riesen, Köln 1962. Malewitsch, doubtless one of the most genial persons of our age, was co-founder and director of the "Institute for artistic culture" in Petrograd, Russia. I guess there is an urgent need to re-establish this institute somewhere in the world.

10. See Karlheinz Stockhausen, TEXTE ZUR ELEKTRONISCHEN UND INSTRUMENTALEN MUSIK, Bd. I, ed. by Dieter Schnebel, Köln 1963.

11. E.g., Joachim K. H. W. Schmidt, PLÄDOYER FÜR EINEN NEUEN INTERDISZIPLINÄREN FACHBEREICH, in: Joachim K. H. W. Schmidt, ed., PLANVOLLE STEUERUNG, p. 313 ff., see note 3; or books, quoted in note 9 & 10.

12. See my paper "On the computing judge," prepared for the ISA Research Committee on Sociology of Law, Toronto 1974.

13. See Philolaos, in: Hermann Diels, DIE FRAGMENTE DER VORSOKRATIKER, p. 77 f., Hamburg 1957.

14. Etienne Bonnot de Condillac, OEUVRES PHILOSOPHIQUES, ed. by Georges Le Roy, e.g., vol. II, p. 376a (La Logique), Paris 1948. References are, however, to find through all his works; check index in vol. III, p. 592.

15. For the quantitative analysis the up to now common statistics cannot be applied, see second half of note 3. Most social statistics, taken over from economics, imply causal relationships, work with sets of independent and dependent variables; it's a real pity that our reality has not conformed up to today to the blinds of our social scientists..., since so much work and the resources are wasted.

16. Most research in the social sciences is still of a pure academic nature and unsuitable for the praxis; for this reason the ISA Research Committee on Sociotechnics tries to promote a praxis-oriented social science, see e.g., Adam Podgórecki and Albert B. Cherns in the book edited by me, note 3.

17. Marcus Tullius Cicero, DE RE PUBLICA, I, 41 ff., p. 128 ff., Zürich und München 1973, edition by Karl Büchner.

18. See note 6.

19. Compare e.g., K. E. Georges, KLEINES LATEINISCH--DEUTSCHES HANDWÖRTERBUCH, p. 43 f., Leipzig 1864.

20. Personal communication, Berkeley 1971/72.

21. I use the term "paradigm" in the sense of Thomas S. Kuhn, THE STRUCTURE OF SCIENTIFIC REVOLUTION, Chicago, Ill. 1962.

22. See e.g., Joachim K. H. W. Schmidt, "Der Nutzen einer Gesellschaftstechnik für die Gesellschafts-architektur," in my book quoted in note 3, p. 71 ff.; my Toronto paper, note 12; my paper for the ISA symposium on theory in sociology, Tokyo 1975; and unpublished material.

23. Compare my "Paradigma der Kommunikation," published in the book quoted in note 3, p. 75. This paradigm is constitutive for the Thinking in Relation.

24. Norman O. Brown, LOVE'S BODY, e.g., p. 109 ff., 132 ff. (However, the book has to be read as a whole for understanding the theses of Brown). New York 1966.

25. Guy Benveniste, "Die Rolle utopischer Modelle im Bereich der Gesellschaftstechnik," in my book quoted in note 3, p. 59.

26. Marcus Tullius Cicero, book quoted in note 17, I, 49, p. 140.

27. Compare e.g., Joachim K. H. W. Schmidt, "Freirecht-Revisited, Emphasizing its Property-Space," in ARSP 1974, vol. LX/1, p. 113 ff.; my book: DAS "PRINZIPIELLE" IN DER FREIRECHTS-BEWEGUNG, Bonn 1968; my paper: "The meaning of law within the context of social engineering," Tokyo 1975 (prepared for the ISA Research Committee on Sociology of Law).

CLOGS ON PERSONAL PARTICIPATION IN A DEMOCRACY

B. Sivaramayya

Freedom of an individual in a democracy proceeds from the premise of equality. "We hold these truths to be self-evident that all men are created equal, that they are endowed by their Creator with certain unalienable Rights, that among these are Life, Liberty and the pursuit of Happiness."[1] After the American Declaration of Independence, several constitutions of the world and the Universal Declaration of Human Rights[2] used the word "equal" and "equality" with reference to individual liberties. But the concept of equality at a philosophical level and the meaning of equality in practice exhibit divergences. The notions of equality based on merit and on egalitarianism are widely known. Recently Rawls has given us yet another concept, namely, "democratic equality."[3] Some among these variegated concepts of equality are in a measure reflected in democratic constitutions. Political participation is based on the principle of qualified egalitarianism. Generally, the equality of opportunity relating to positions proceeds on the basis of merit.

At lower levels of abstraction the meaning of the word "equality" has shown similar variations. Holmes points out: "A word is not a crystal, transparent and unchanged, it is the skin of a living thought and may vary greatly in color and content according to the circumstances and the time in which it is used."[4] The meaning given to the word equality over the last two centuries tellingly unfolds the truth of the above statement.

At the time of the American Revolution the word equality had a different connotation:

> As we know, the Founding Fathers omitted slaves from the Constitution and equality in consequence means something quite different nowadays from what it meant then.[5]

The Fourteenth and the Fifteenth Amendments to the American Constitution represent yet another milestone. For the first time there was a specific conferment of the right to life, liberty and property, and "the equal protection of laws."[6] But the full measure of the significance of these Amendments was defeated by the decision in Plessy v. Ferguson.[7] It laid down the "separate but equal doctrine;" that equality is compatible with separate facilities for Blacks and Whites.

Fifty-nine years later Brown v. Board of Education[8] overturned this view. It might be that the Supreme Court did not overrule the Plessy's case. But there can be no doubt that a combination of judicial opinions and legislative actions, as expressed in the Brown decisions and the Civil Rights Act categorically reject the notion that separate facilities based on race are compatible with equality.

In contemporary times we find yet another change in meaning and interpretation; that the reach of equality in a democratic constitution encompasses affirmative action to remove gross inequalities of the underprivileged sections in a society. The idea is expressed in different ways and at times in different senses by words like "compensatory treatment," "reservations," "protective discrimination," "quotas," and the like. In 1965 Loren Miller stated:

> At the outset, it is well to recognize that the Fourteenth Amendment was corrective; it was a command to the states not only to drop discriminatory laws and customs supported by law, but to take affirmative action to confer equality on Negroes. The original intent was that where the states failed to act, that obligation would devolve on the Federal Government. When Congress tried to act in the context of its own times through the passage of civil rights laws and other equalitarian statutes, the Supreme Court misread the Amendment to limit the requirement of equal protection to those situations in which the states imposed inequality. A return to the Fourteenth Amendment requires a reinstatement of the original purpose...Negroes must be assisted <u>as Negroes</u>, just as they have been disadvantaged <u>as Negroes</u>, if the Amendment is to serve its equalitarian purpose.[9]

The Constitution of India framed in the mid-Twentieth century, in respect of a similar group that was the victim of social oppression, namely, the Harijans (ex-untouchables), confers such power more specifically.

Article 15 of the Constitution of India lays down: "The State shall not discriminate against any citizen on grounds only of religion, race, caste, sex, place of birth or any of them."[10] But clauses 3 and 4 of that Article expressly provide that nothing in that article shall prevent the State from making any special provision in favor of women and children or any special provision for the advancement of any socially and educationally backward classes of citizens or for the scheduled castes and scheduled tribes.

Social rights and social justice are indispensable to freedom and equality in a democracy. Their absence will result in a real deprivation of individual rights in the underprivileged sections. Consequently, the de facto participation of persons belonging to the underprivileged sections of society will be affected. Marshall observes:

> A property right is not a right to possess property but a right to acquire it, if you can, and to protect it, if you can get it. But, if you use these arguments to explain to a pauper that his property rights are the same as those of a millionaire he will probably accuse you of quibbling. Similarly, the right to freedom of speech has little real substance if, from lack of education, you have nothing to say that is worth saying, and no means of making yourself heard if you say it. But these blatant inequalities are not due to defects in civil rights, but to lack of social rights...[11]

Poverty makes the vulnerable sections of the society succumb to the temptation of money and thereby vitally affects the participation of these groups. The point has been well brought out by Mr. Sen Verma, the former Election Commissioner of India.

> I think that the basic remedy lies in the amelioration of the socio-economic conditions of the electorate. A vast majority of the people still live below the poverty line. Many of them are steeped in brutalizing poverty. With the improvement in the economic lot of the masses of the electorate, many of them will desist from taking bribes, or resorting to undue influence or intimidation or violence with all their attendant risks and dangers.[12]

A. Problems in Enforcement of Social Justice

The acceptance of a concept of equality as including in it a duty to enforce and achieve social justice is bound to bring in its wake certain conflicts, real and apparent, with the preexisting notions of equality. The synthesis between the individual interests and social interests is bound to exhibit differences in approach depending on the conditions in each society. For example, the enforcement of antidiscrimination statutes in the common law countries varies from persuasion and conciliation (carrot and stick) approach in the United Kingdom to penal sanctions in India. Likewise in the United States of America the decision in Shelly v Kraemer[13] holds that racial covenants are unenforceable. But the position is different in the United Kingdom.[14]

A range of difficult problems have been experienced in trying to achieve racial integration. Charles Abrams says:

> The right to dwell where one chooses is pitted against the right to dwell with whom one chooses. Equality under law is confronted by the claim that the long subordination of the Negro's rights demands preferential treatment, which in turn is attacked as "discrimination in reverse." The right of a Negro child to an integrated school is confronted by the right of a white child to attend a school in his own neighborhood.[15]

Further, could a company limit the number of its Negro workers to one-third so that it could maintain its integrated employment policy?[16] Can a housing authority resort to a "phase program" to foster integration?[17] These questions involve difficult dilemmas not easy to resolve.

In developing countries like India, the problem of raising the standards of living and securing social justice to the weaker sections are already stupendous. These problems are compounded immeasurably by the population explosion. Sometimes it is argued that population control can be achieved only by economic development. Even if one accepts the premise unreservedly, one cannot ignore the current vicious circle involved in the statement: Without economic development population explosion cannot be contained and population explosion is nullifying all economic development. Therefore, the adoption of coercive strategies to control population explosion become imperative. The Ehrlichs' say:

> Several coercive proposals deserve serious consideration, mainly because we may ultimately have to resort to them unless current trends in birth rates are rapidly reversed by other means. Some coercive measures are less repressive and discriminatory,

in fact, than some of the socio-economic measures that have been proposed.

One idea that has been seriously proposed is to vasectomize all fathers of three or more children. This was defeated not only on moral grounds but on practical ones as well...But probably India's government will have to resort to some such coercive method sooner or later, unless famine, war or disease takes the problem out of its hands. There is little time left for educational programs and social change, and the population is probably too poor for economic measures (especially penalties) to be effective.[18]

The coercive strategies may involve minor or major coercions, and there should be a progression from the minor to major coercions. By minor coercions we mean deprivation of facilities like restricting hospital facilities of the government to the public as well as its employees, restriction of the travel and educational facilities given to the employees of government and private organizations, and restriction of maternity benefits available to women workers under the law to two children only. Sterilization of the unfit and compulsory sterilization may be categorized as major coercions. These, no doubt, involve restrictions on liberty, as traditionally understood in the developed countries, including the freedom of conscience.[19] But considering the great strides in contraceptive technology the restrictions on freedom involved in coercive strategies would be reasonable.

Apart from the need for population control to achieve social justice, population explosion affects participation in a democracy qualitatively. For functional reasons, the strength of a legislative body cannot be increased beyond a predetermined optimum number. Therefore, with the rise in population the number of voters in a constituency will become unwieldy; the contact with the constituents becomes difficult; the sensitivity to their problems becomes dim; and in an electoral contest the expenses mount making it difficult for people belonging to the lower income groups to contest the elections.

B. Personal Participation

For the vast majority of citizens in democracy, voting is the only mode of participation in government. To be sure, this mode of participation should be based on the principle of universal suffrage subject to limitations like age and citizenship of the voter. The U.S. Commission on Civil Rights recommended that this right to vote

> should not be abridged or interfered with...except for inability to meet reasonable age or length of residence requirements uniformly applied to all persons within a State, failure to complete six grades of formal education or its equivalent, legal confinement at the time of registration or election, judicially determined mental disability, or conviction of a felony...[20]

A significant point is that the Civil Rights Commission prescribed completion of six grades of formal education or its equivalent as a qualification for voting. It is no doubt true that for intelligent participation in the democratic process education is necessary. But the wisdom of excluding citizens from participation on this ground is to be doubted. At any rate in the new democracies in Asia and Africa where illiterates are in a majority, such a limitation on the principle of universal suffrage is unworkable and will undermine the concept of democracy. Mill suggested weightage to literates. But this is undesirable as it will result in an emphasis on the problems of the advanced sections of the community to the neglect of weaker sections. It is now proposed to deal with some of the constraints on effective personal participation.

(1.) Economic Dependence

Economic dependence adversely affects the participation of the weaker sections of a community. The U.S. Commission on Civil Rights observes:

> To be sure, the Voting Rights Act has not resulted in full use of the franchise. Means other than disqualification, such as exploitation of continuing economic dependence of rural Negroes in the South, still constitute deterrents to the exercise of the right to vote.[21]

In India also the Harijans, comprising most of the landless labor, could not take advantage of their political strength because of lack of land and resources. Midha says:

> The Chamars are the second biggest caste in several districts and are the numerically largest caste in the entire State constituting 12.7% of its total population...

Lack of land and lack of education have prevented
the Chamars from leading the Scheduled Castes to an
important position in the State's politics. The
present redistribution of land and the cumulative
effects of efforts to educate them may bring about
a break from the past.[22]

This emphasizes, as noted before, the need for achieving social justice for effective participation.

(2.) Cultural Dependence

The subordination imposed by the social and cultural norms may result in ineffective participation. For
example, in India this kind of dependence is present
in the case of women-voters belonging to rural and lower
middle class background. There are no direct indices
to measure the nature and extent of this dependence.
But some indirect evidence of it is available. When
the Nehru Government wanted to amend the Hindu law to
give better rights to women, the conservative opponents
of the measure wanted a referendum on the subject or a
"clear verdict of the electorate," perhaps confident of
securing a vote against the reforms. Fortunately, owing
to Pandit Nehru's strong determination, they could delay
but not defeat the reforms. Even today a similar position exists in the matter of enforcing minimal essential
reforms in Muslim law. Curiously over the last two
decades the problems of women had never been an election issue and no political party made it an issue. As
Razia Ismail asks:

But what have our women voters won for themselves
in the years since Independence? Have they won any
real independence for themselves?. . . How many
women are still denied the right to decide whether
they would like to work--even at the lowest rung of
the humblest trade--instead of bearing children
from their teenage years?[23]

The family laws and customs of Hindus, Muslims,
Parsis and even Christians in India contain many provisions that discriminate against women. Their continuance is winked at by the Government and is condoned
by the courts.[24] Sometimes it is argued that legal
equality by itself will not achieve social equality.
The implication, therefore, is that an attempt at
achieving legal equality is unnecessary. This argument ignores the point that under any constitution
legal equality is a sina qua non for the enforcement
of social equality. Even if there is no enforcement
it was rightly pointed out "Civil rights laws and

policies by the Federal Government can be of value even when they do not contain strong enforcement mechanisms. [The law] brings about substantial changes in attitudes and behavior."[25] Cultural dependence can be overcome only by big strides in education and social equality.

(3.) Apathy

Apathy as a clog on participation afflicts democracies in developed and developing countries.[26] Some political scientists assign a positive, if not a constructive, role to apathy. It might be so in developed countries. But even in such cases apathy will encourage the maintenance of status quo which is in itself not healthy.

To combat apathy, Australia made voting compulsory by law. Prima facie this policy gives an impression of mistaking quantitative participation for qualitative participation. But Sawer says:

> Opponents of compulsion predicted that it would greatly increase informal voting and lead to electoral irresponsibility at the best and widespread corruption at the worst. Experience has not justified the first prediction...Electoral corruption has certainly not increased--prosecutions for electoral offences of this type are very rare. Whether electoral irresponsibility has increased is a question which none can answer; parties just defeated at the polls think it has and the successful party thinks otherwise.[27]

Compulsory voting has been rejected by many countries as being undemocratic. Compulsory voting will impose considerable administrative burdens in nascent and populous democracies and will be impracticable. Apathy differs in its nature and extent in the developed and developing countries. In the United States of America as to the characteristics of persons who are least likely to vote, "most agree in listing the young, blacks, the poor, the less well educated, women and people in rural areas."[28] Arora, adverting to difficulties in taking voting turn-out as a measure of participation, says, "Illiterate populations of underdeveloped countries, for instance, may exhibit a higher turn out percentage than literate people in developed countries."[29]

But to what extent does participation in a democracy represent a participation qua individual? To what extent is participation free from group, caste or religious prejudices? The ideal of attaining participation free from prejudice is difficult to accomplish but all attempts towards it are worthy of the effort.

FOOTNOTES

1. The American Declaration of Independence.

2. Universal Declaration of Human Rights, art. 1. "All human beings are born free and equal in dignity and rights."

3. Rawls, A THEORY OF JUSTICE 75, 4th ed., 1972.

4. Towne v Eisner, 245 U.S. 418, 425, 1917.

5. Dorothy Kenyon, in EQUALITY 57, 1965.

6. The Constitution of the United States, 14th Amendment.

7. 163 U.S. 537, 1895.

8. 347 U.S. 483, 1954.

9. Loren Miller, in EQUALITY 31-32, 1965.

10. The Constitution of India, art. 15. For purposes of the Constitution scheduled castes and scheduled tribes are the castes and tribes specified by the President by a public notification.

11. T. H. Marshall, SOCIOLOGY AT CROSSROADS AND OTHER ESSAYS 91 (1st ed., 1963). Marshall uses civil rights as rights necessary for individual freedom--liberty of person, freedom of speech, etc. Social right is used by him to indicate the whole range of rights from the right to modicum of economic welfare and security to the right to share to the full in the social heritage and to live the life of a civilized being according to the standards prevailing in the society.

12. S. P. Sen Verma, "A Challenge" in CLEAN ELECTIONS, 159 SEMINAR 23, 1972.

13. 334 U.S. 1, 1948.

14. See generally, Lester & Bindman, RACE AND LAW, 241, 1972.

15. Charles Abrams, Foreword to EQUALITY, supra note 9 at vii.

16. Id. at XI.

17. Id. at XV and XVI.

18. P. R. Ehrlich and A. H. Ehrlich, POPULATION, RESOURCES AND ENVIRONMENT 254, 1970.

19. It may be pointed out that under the Indian Constitution, freedom of conscience is subject to public order, morality and health. Further, under the proviso (b) to article 25, the state can make laws for social welfare and reform.

20. Report of the United States Commission on Civil Rights 26, 1963.

21. Federal Civil Rights Enforcement Effort, A Report of the U.S. Commission on Civil Rights 31, 1970.

22. R. C. Midha, The Future of UP's Caste Pattern in Elections, 4 ELECTION ARCHIVES 30, 1970.

23. Razia Ismail, "Year for Women's Rights," Sunday Standard Magazine, New Delhi, Jan. 19th, 1975, p. 1, col. 6.

24. For the judicial attitudes towards discriminatory provisions against women, see generally, Sivaramayya, WOMEN'S RIGHTS OF INHERITANCE IN INDIA 179-183, 1973.

25. Report of the U.S. Commission on Civil Rights 29, 1970.

26. VOTE POWER, The Official Activist Campaigner's Handbook, 3, 1970. Under the head Apathy it says: "People who are committed to changing the system are not always prepared to act on the basis of that commitment."

27. Sawer, AUSTRALIAN GOVERNMENT TODAY 52, 1961 rev. ed.

28. VOTE POWER, supra note 26 at 16.

29. Arora, "Pre-empted Future ? Notes on theories of political development," BEHAVIOUR SCIENCES AND COMMUNITY DEVELOPMENT 85, 1968.

PEOPLE'S MASSES PARTICIPATION IN THE MANAGEMENT OF SOCIETY IN THE SOCIALIST DEMOCRACY

Ioan Ceterchi

The concept of democracy has been, and still is, one of the most controversial subjects within the complex field of juridical and social philosophy. No political-juridical doctrine, no governing program could be designed without taking a stand and defining its position towards democracy. Perhaps, there is no other concept so intimately and indestructibly linked with all aspects of political-juridical life, with the ideology and political-juridical institutions, with the social relations and psychology. This also explains, among other reasons, why democracy has had a unique place in philosophy and political-juridical practice from the time of Aristotle and ancient city state democracy to the present.

Although it is very clear that democracy in antiquity, under capitalism and socialism is not the same thing, even as travel by mail coach is not the same as travel by train, plane or rocket, there are, nevertheless, certain characteristics of democracy in general or pure democracy. One can, therefore, plead for a concept of democracy irrespective of historical situations, of scientific and technical progress, reduced to a simple form of political regime which is "adaptable to no matter what kind of economic regime." At the same time, the thesis must be rejected that would dissociate political power from economic power, because it ignores the decisive influence of the owners of the means of production over political power, their decisive role in establishing the state policy.

The understanding of the social-historical nature of democracy, of its dependence on the character of the social-economic system, of the importance of people's masses struggle for conquering and defending the democratic institutions, are all crucial points of the dialectical materialistic concept of democracy. For defining democracy it is necessary to specify who, which class, or which classes, possess the economic and political power, which is the way of exerting the power and to what extent the people's masses participate in running the society and state, what is the citizen's political-juridical status, to what extent rights and democratic liberties are ensured for him.

The essence of socialism consists of abolishing private property in the means of production and exploitation of man by man, of ensuring the organization and management of society and social output by the people aiming at the improvement of the living standard of all members of society, the multilateral development of human personality, released from social and national oppressions. Such a transformation is inconceivable without the collective effort of the working people who are no longer the object of power, but the masters, who are consciously building their own history, organizers of the social and political system corresponding to their interests, national specific historical conditions and traditions of the respective people.

Socialism with its new economic, social, political and juridical structures does not emerge spontaneously and it is not introduced by order from above; it is created by the common effort of those who are working, by ensuring the access of all the working people to "the art of state running, the management of the entire state power."[1] Socialism implies a democratic political system set up by the working people, functioning through and for them. "We are convinced that only by ensuring the establishment of a genuine socialist democracy," said the President of Romania, Mr. Nicolae Ceausescu in his Report to the 11th Congress of the Party, in November 1974, "we can successfully implement the Party Programme, its domestic and foreign policy. We unswervingly proceed from the Marxist-Leninist principle that development of socialist democracy, the peoples' conscious making of their own history, is an objective necessity of the building of the new system. It is only together with the entire people that we can ensure the victory of the manysidedly developed socialist society and of communism on the soil of Romania."[2]

Socialist democracy is not achieved automatically, as a consequence of conquering the economic and political power by the working people. The setting up of democracy implies also the necessity of initiating measures and guaranties regarding the way of the organization and implementation of power on the basis of people's masses participation, according to the different stages of socialist construction, the assertion of the democratic rights and liberties within the framework of a new juridical-political status of the citizen. The victory of socialism in the entire economy, the extension of the social basis of power, creates a larger framework for asserting democratism, for making the entire people more active politically. The social-political and moral-ideological unity of the people, achieved as a consequence of eliminating class oppositions from society and raising the civic and cultural general level of the people's masses, offers new and larger possibilities for political participation and manifestation of all citizens.

<u>Under socialism's conditions, the people's masses participation in running the State and society represents the essential aspect of the way socialist democratism is being manifested</u>. This does not mean neglecting the other dimensions of democratism, such as, for example, civil rights and liberties. On the contrary, the participation concentrates in itself, like the focus of a lens, all the dimensions of democratism, for, eventually, the participation is inconceivable, or remains formal without liberty, equality, and other fundamental rights ensuring the full manifestation of the person. On the other hand, democratic rights and liberties would be meaningless and reduced to simple institutional-juridical forms without ensuring the people's masses participation in the government. Of course, democracy, participation mainly, is not an aim in itself, but a means of human emancipation, of free, multilateral development of human personality and of the community he belongs to, a means of using the creative capacity of all the citizens in the work of consciously building a new society.

In the present stage of socialist construction in Romania, intensifying the political activity and participation of the people's masses in running the State and society--essential moment of enlarging socialist democracy--holds a prominent place in the process of improving the organization and management of economic-social life.

According to the Romanian concept and practice, participation as political action and responsible act of a person, is conceived as the real involvement of the working people in political life, in the management of society and State. Participation implies the involvement of the working people in the management of all aspects of the social life--political, juridical, economic, cultural, scientific and civic--for the socialist democracy embraces all these aspects of the social life. The various situations in which a person might be as a <u>citizen</u>, <u>as a member of a working group (wage-winner or collective farmer), as a member of various political and social-civic organizations</u> entitles him to make his own contribution and express his opinion on Party and State's policy concerning the development of the new social system, on the organization of labor and the activity of the group he is working in, on the activity of the political and social-civic organizations he belongs to. The working people, as citizens and holder of the State power, in their double capacity of owners of the means of production and makers of material goods, as voluntary members of various civic masses' and professional organizations, are called to effectively participate in running the State, economic social-cultural and civic-socialist organizations, in the management of the entire society.

Of course, the political decision is not the sum of individual acts, but the general, collective will, expression of the fundamental interests of the nation. That was pointed out even by J. J. Rousseau. But, what really matters is the setting up of the social mechanism enabling the general will to express consensus of the nation. As this mechanism is neither given once and for all, nor the same for all countries, a creative attitude on the part of the responsible factors, mainly the Party, is required to build the most appropriate system and mechanism according to the social, historical and national realities of the respective country, due to permanent perfecting activity in every new stage on the way to socialism and communism, using all that is generally accepted as valid, but giving up imitation, the so-called universal pattern.

Many forms and methods of people's masses participation--at all levels--in the management of the State and society and decision making were institutionalized within the political socialist system of Romania. The socialist constitutional principle of the full and sovereign character of the people's power is implemented in a system of political organization which combines the <u>representative democracy</u> with the <u>direct democracy</u> in complex forms. The organization of the representative institutions of power is based on the new principles which organically combine the eligibility with responsibility and revocability of the deputies, eliminates the Parliamentary professionalism, the representative institutions becoming working bodies with an emphasized control role over the executive. But, at the same time, according to the Constitution and other laws, numerous forms of citizens' direct participation in the implementation of the management and running the State affairs, were settled in the practice of political life especially in the last ten years, after the 9th Congress of the Party (1965). The activity of the representative institutions cannot be separated even for a moment from the people's masses <u>direct participation</u> in solving the most essential problems of managing the State and society.

Thus, the view regarding representation, formulated by Montesquieu and maintained even today by various technocratical and elitist theories stating that the masses, the electors' role and capacity is limited to designating the representatives, is passed over. The democratic principle of organization means that every citizen should be ensured the conditions to participate in the election of his representatives and the discussion and implementation of the laws, as well.

Socialism eliminates the indomitable opposition between political leadership and people's masses, between rulers and the ruled. Having an essential and continuously emphasized role in organizing the economic and social-cultural activity of the socialist construction, in defending the new social order, the socialist one, as well as in the development of the relations with other countries, the socialist political system--which comprises the Party, the State and the social-civic organizations in an organic unity--should find out the most appropriate solutions regarding its internal and external policy. In the development of the socialist political system, the Party, State, and society come closer to one another, a process manifested in the practice of Romania's political life as one of the first steps of the more complex process of integration to be achieved between the political and social aspects of the leading activity in communism, a natural consequence of the complete unification of society, following the evanescence of the classes and the creation of the one working people of the communist society.

But, this process is not unilateral, one-way directed. It is achieved through the development and maximum use of the capacity of all qualities of the representative system as well as through the extension and diversification of the forms of direct democracy. As a matter of fact, the two methods of the socialist democracy, representative and direct, are not decisively separated, but are interconnected in many respects.

In the democratic development of Romania's political system aiming at its continuous deepening and improvement, measures have been taken in order to extend the role of the representative bodies, and improve their activity, and, at the same time, to ensure an intensified people's masses participation at all levels of political social leadership.

Thus, for example, the elective system was improved by ensuring the possibility of designating more candidates for one mandate. In the last elections of March 1975, for instance, in almost 40 per cent out of the total constituencies for the Grand National Assembly, in almost 60 per cent out of the total number of the constituencies for the counties Peoples' Councils, and in almost 80 per cent out of the total local peoples' councils constituencies, the Socialist Unity Front proposed two candidates for one mandate.

The participation form by including some representatives of the social collectivities in the state bodies, both at the central and at the local levels, in the economic and social-cultural organizations, took various aspects. Thus, the President of the Central Council of the Trade-Unions, the President of the National Union of the Agricultural Cooperatives, the President of the National Women's Council, and the First Secretary of the Central Committee of the Communist Youth Union are full members of the Council of Ministers.

Essentially, the economic democracy in enterprises and industrial centrals (industrial combined works) is expressed through the creation of new managing institutions based on a large participation--the working peoples' councils for managing the economic and social activity (in centrals), the working peoples' committees for managing the economic and social activity (in enterprises), and the general assembly of the working people in the centrals and respectively in enterprises.

The direct participation of the citizens in adopting the most important decisions is implemented through <u>public debate</u>, republican or county-level periodical <u>conferences</u> and meetings of the specialists and working people in various sectors of activity, <u>working visits</u> of the high rank Party officials in the counties, towns, villages, enterprises, agricultural cooperatives, institutions, which offers the opportunity to talk with the masses directly on the most urgent problems affecting the general development of society, social-economic organizations or certain localities. The civic organizations and bodies, trade-unions, cooperative unions, youth, women, writers and artists' unions, the Socialist Unity Front, etc., involving millions of people in the direct management of public and civic life, give a strong impulse to the citizens' participation in the management of social life.

The management of contemporary society becomes more complex, requires larger knowledge, needs multilateral educated people. But, contrary to different variants of technocratic theories, socialism is raising the problem of improving the competence and qualification even of the decision-makers and political leaders, the problem of improving the general level of knowledge and culture of the masses. The increasing complexity of the problem related to the management of the society does not determine a reduction but a greater intensification of the people's masses participation, of the citizens', in the management of the society and the process of decision-making. Even if it is not possible for each individual to understand all the specific details in respect to one or another measure, the fundamental problems, the decisive options regarding internal or external policy can be formulated in such a way so as to enable the citizens to speak out their opinion and consciously motivate the option.

The implementation of the socialist democracy, the ensuring of civil rights and liberties have nothing in common with the anarchical views on the so-called absolute liberty. An increased responsibility of each citizen for the collective interests and progress of the nation is a characteristic of the socialist democracy. The fundamental rights and liberties of the citizens can be accomplished only in direct connection with the civic duties.

Being a very complex phenomenon, the socialist democratism develops itself in a continuous process—fighting with the recrudescence of the negative phenomena which might be encountered in the socialist society as well—of gradually shaping the most appropriate forms and methods of its implementation, the best ways for involving the citizens in the management of the society, for the ever full assertion of every individual's personality.

FOOTNOTES

1. Lenin, V.I. "Works" the 24th Volume, E.S.P.L.P., Bucharest, p. 163.

2. Ceausescu, Nicolae, "Report to the 11th Congress of the Romanian Communist Party," Meridiane Publishing House, Bucharest, 1974, p. 71.

PARTICIPATION: AN OVERVALUED, IMPRACTICAL IDEAL

Michael D. Bayles

Demands for participation in social decision-making have erupted at various times and places throughout history. In the last decade, they have again risen to the fore in various countries. Critical analysis of the values of participation and social changes during the last two centuries shows that participation is of relatively small value and generally impractical.[1]

The term 'participation' is unclear and ambiguous which leads to confusion about demands for, and the value of, participation. There are two general concepts of participation--a formal and a substantive one. (1) The formal concept is procedure or process oriented and concerns the methods by which decisions are made. There are at least two variations of the formal concept.
(a) In one formal sense, 'participation' means expressing one's views and being heard as to the merits of alternative policies. This sense presupposes notice of subjects pending decision. Unless one knows a decision is to be made on a topic, one will have little reason to express one's views on it. This notion of participation is most at home in the context of administrative decision-making where an impartial decision-maker will take into account information and arguments presented.
(b) In the second formal sense, 'participation' means that one actively engages in the decision-making, usually by voting as a member of a decision-making group. This sense was involved in struggles for universal suffrage. These two senses of formal participation are logically distinct; one may be heard without being a decision-maker; and one may be a decision-maker but not express one's views. (2) The substantive concept of participation is result oriented; it pertains to actually affecting the decision made. By this concept, one participates when one consciously affects the decision.

The formal and substantive concepts are logically independent. One may express one's views, yet have no effect upon the decision reached; and one may affect a decision (by being one of several decision-makers), yet not express one's views. Similarly, one may have a formal role as a decision-maker, yet never affect the decision, e.g., by consistently being a member of a minority; and one may affect a decision without engaging in the decision-making, e.g., being an informal advisor.

Finally, one must distinguish a right to participate from active participation. A right need not be exercised; it is dispositional--one may do something if one chooses (and perhaps follows prescribed procedures). Thus, a right to participate in sense (1,a) is an opportunity to be heard. However, the value of a right generally depends upon the value of its exercise. If there is no value to active participation, the right to participate is not significant.

The value of participation as an ideal may be considered under the usual categories of value--inherent (contributory) and instrumental. The inherent value of participation derives from its connection with self-respect and the pursuit of happiness. A denial of the right to formal participation suggests a lack of worth. Not to have an opportunity to be heard or vote suggests that one's views and opinions are of no value. As such, it is a denial of equal worth and the dignity of being a reasonable person. A grant of formal but not substantive participation may have the same implication. One is given merely token recognition as a reasonable person. When one substantively participates, one is usually recognized as a reasonable person whose views are worthy of respect. (Substantive participation is not always a sign of respect; it may be mere recognition of power.) The respect of others is important for one's self-respect and dignity. As, for example, the labeling theory of criminality recognizes, it is very difficult to maintain a positive self-image in the face of its denial by others.

Participation may also be inherently valuable as an enjoyable activity. One may simply enjoy participation for its own sake--politics is fun. To be denied participation is to be denied one of the sources of happiness. This value of participation tends to be incompatible with an instrumental evaluation of it. If one takes the results or outcome seriously, losing will tend to take all the joy out of participation. If one participates in a political campaign believing the election of one's candidate to be essential to the welfare of the body politic but the candidate loses, one is unlikely to have found the experience enjoyable. Nonetheless, participation, whether in government, business, or academia, may be valued simply for the enjoyment in the activity.

The instrumental value of participation rests more squarely on the substantive concept than do the inherent values which may sometimes be gained by purely formal participation. If one can affect decisions, then one can promote or protect one's interests or values. However, viewed instrumentally, personal participation becomes less relevant than the substantive participation of someone with similar interests. Thus, groups are organized to press for the common interests of their members. The participation in, perhaps even domination of, the British Labour Party by workers' unions is a striking example of this development. As organized interest groups promote only the common interests of their members, the other interests of members, e.g., laborers' interests in clean air, may be sacrificed to the common ones, e.g., those of full employment. As no one interest group effectively pursues all the interests of its members, an individual must support various organizations to promote and protect all of his interests and values. Moreover, there are no groups representing some interests, e.g., those of tort victims.

When demands for participation erupt, is one of its values predominant? An impressionistic view of history suggests that demands for participation are primarily based on its instrumental value. Claims for participation are usually made by disadvantaged groups in order to promote their interests. Had British policies adequately protected and promoted the interests of American colonials, it is unlikely there would have been an American Revolution. 'No taxation without representation' asserts a claim to participate in decisions affecting one's financial interests. The same analysis seems plausible of the French Revolution. At a theoretical level, the devotion of Bentham and other early utilitarians to universal manhood suffrage was based on a belief that once given the vote, the political influence of workers would enable them to better their lot.

Formal participation has little instrumental value because it need not involve influence over decisions. The Fifteenth and Nineteenth Amendments to the United States Constitution, prohibiting denial of the right to vote because of race or sex, did not directly lead to any significant improvements in the lot of blacks and women. Frequently, a grant of formal participation merely buys off dissident and disadvantaged groups at a very low price. Demands for participation are often presented in moralistic terms of recognition of equal worth. Hence, it is difficult for those making them not to accede to formal participation without admitting that they were merely a disguise for a claim to a larger share of society's goods and thereby discrediting the moral arguments made in favor of them.

At a more personal level, there are reasons for taking, and indications that people do take, the major value of participation in social decisions to be instrumental. Much of the time of the average citizen is occupied in labor, the daily activities of living, and relaxation. Time and energy are not available for much participation in social decision-making. Such participation is for those who have or may arrange the time to be involved (academics, students, farmers, lawyers, etc.) and those whose business it is to be involved (politicians, leading businessmen, etc.). As the widespread reliance on trustees, mutual funds, and business managers attests, people are generally satisfied to place the conduct of many of their affairs in the hands of others so long as the decisions and policies are beneficial to them. Moreover, the percentage of the electorate participating in elections tends to vary inversely with the economic and political well-being of the country. Finally, the inherent values may be obtained from other sources. Pleasure may be found in innumerable other activities. To a large extent self-respect depends upon peer relations among co-workers and acquaintances and family relations, although it may be seriously harmed by membership in a class which is discriminated against.

If the dominant value of participation in social decision-making is instrumental, there are clear limits to its significance. First, participation pertains only to decisions which may affect one's interests or values. Consequently, the relevant public for participation shifts from subject to subject. Second, even when interests are affected, they may be of so little importance that participation in decisions affecting them is insignificant. A universal right to participate supposedly allows for these factors because one may engage in those decisions significantly affecting one while abstaining from others. However, this rationale is undercut by a third factor. When many people are affected, only organized groups have sufficient influence to affect decisions. Those interests not represented by organized groups, e.g., those of consumers and tort victims, are not protected and promoted. Consequently, an individual's right to participate is valueless because its exercise is ineffective.

Valuation of, and demands for, participation are not a recent phenomenon. They have occurred over a large geographical and temporal span. Of course, the nature of the participation sought--suffrage or more substantive participation--has varied. Thus, the ideal of participation has not meant the same thing in the different contexts in which it has been espoused. Changing social conditions require the reevaluation of ideals. In modern, industrially developed, urban societies, the practical possibilities and instrumental value of participation are considerably diminished from what they were two centuries ago. Consequently, as a contemporary ideal, participation may be quite different from what it was at other times and places.

Two major social changes in the last two centuries have significance for the ideal of participation. The first of these changes is industrialization and its concommitant technological development. Feudal society had generally broken down in the West by the end of the eighteenth century. Although industrialization was well under way in England by the last quarter of that century, it developed later in other countries and was not complete in Britain until about the middle of the nineteenth century. The factory system did not reach full flower until the nineteenth century. Moreover, the rapid technological developments of the twentieth century have increased the impact of industrialization. The second major social change is the enormous growth and urbanization of the population. The total population of the thirteen colonies at the time of the American Revolution amounted to only one-third the current population of metropolitan New York City. Only four towns had a population of over 10,000 people. Moreover, the United States' population has shifted from being predominantly rural to being predominantly urban. Similar though sometimes less drastic changes have occurred in most if not all countries.

These changes have significant consequences for the possibility of participation in social decision-making. (1) With large-scale production and distribution of goods, economic (business) decisions affect many more people than before. (2) The interrelations between businesses have become more complex as factories rely upon multiple suppliers and distributors--all of whose activities must be coordinated. (3) Technological complexity results in more people lacking sufficient understanding to make intelligent decisions. (4) The sheer number of people in a geographical area decreases the opportunities for substantive participation in decisions affecting the whole region. (5) In urban areas, one's conduct frequently affects others in ways it does not in rural areas; and there is greater dependence on others for various supplies and essentials. (6) The increased interrelations between people require long and difficult study in order to make intelligent judgments about the long-run impact of alternative choices upon one's interests.

In short, the modern promise of most people intelligently participating in social decisions significantly affecting their lives has not been, and cannot practicably be, fulfilled. The theoretical responses to this situation vary greatly. Some people suggest a technological solution. For example, Robert Paul Wolff has suggested that each person could have a voting machine in his home tied to a large central computer.² A bill could be debated on television for a week followed by a national vote. This proposal overcomes only the difficulties in each person formally participating by vote. It would not provide an opportunity for each to be heard. Nor, substantively, would any individual exercise much influence on a decision; the chance for persistent minorities would increase. Finally, it would not overcome the difficulties in understanding the consequences of decisions. One might reply that the details of decisions are not significant, only the broad policy questions. But most broad policy decisions--to reduce energy dependence on other countries, to lower taxes, etc.--are so general as to be meaningless. Only a few general decisions have a clear significance to the average citizen irrespective of details--to wage war, to reduce highway speed limits.

A second response is to reject the current social organization; to move to small, decentralized living units; to return to the land. But with their current populations, many countries cannot adequately feed, clothe, and house everyone by such techniques. This approach may be a more viable option for developing countries than for fully industrialized ones; indeed, circumstances may force them to adopt it. Yet such an approach would involve renunciation of the interests of citizens in more than basic material welfare. Few citizens believe the gains in the intrinsic values of participation would outweigh the losses in material welfare.

A third response is to turn away from the ideal of participation as impractical and no longer of value. If participation is primarily valued as instrumental to promoting and protecting one's interests and values but is no longer feasible, it has lost its raison d'être as an ideal. Instead, emphasis should be placed upon the substantive aims and values which decisions are to protect and promote. To help insure proper consideration of these substantive interests, decision-makers should be made operationally responsible wherever possible; that is, they should suffer the consequences of wrong decisions. Operational responsibility is more than mere accountability to those adversely affected; it is being adversely affected oneself. One whose decision results in long waits at automobile gas stations should have to wait in line to purchase gasoline. One whose decision results in unemployment for millions should have his salary decreased; at the least, salaries of top governmental officials should fluctuate with real per capita income, not the cost of living. How and to what extent one can organize society so that decision-makers are operationally responsible, have a direct stake in the correctness of their decisions, is the problem which awaits solution. But inasmuch as personal participation is no longer a generally viable ideal, operational responsibility at least offers a possible substitute.

FOOTNOTES

1. One must distinguish between social and personal decisions. Roughly, the former affect a relatively large segment of the community with whom the decision-maker(s) is not personally acquainted. The latter primarily affect an individual or family or small group most of whom know one another, e.g., decisions about one's medical treatment. The argument in the paper pertains only to social decisions. Increased participation in personal decisions may be a counter to the loss of meaningful participation in social decisions.

2. IN DEFENSE OF ANARCHY, p. 34, Harper & Row, New York, 1970.

PRINCIPLE OF COOPERATION IN THE RELATIONSHIP OF THE CITIZEN AND THE SOCIALIST STATE

V. S. Shevtsov

The pre-socialist history of political thought has engendered a kind of a tradition--to regard the relationship of the citizen and the state as an eternal and insoluble contradiction, a ceaseless struggle between the state power, embodying the diktat, various forms of compulsion and constraint, on the one hand and the individual freedom of the particular personality, particular citizen, on the other. The absoluteness of the idea of the individual subordinated to the state, the possibility of their cooperation and forms of such cooperation, all these questions have not lost their topicality in our time, while the answers to them cannot but astound by their diverse and contradictory nature.

One of the aspects of this relationship, which is of basic significance in the entire structure of contracts and indirect ties between the citizen and the state, is that any state comes out as the bearer of the specific state power, while the citizen is the subject subordinated to the political authority, the power of the state.

The individual, and from the legal point of view he is considered as a citizen, cannot be free from state power. The state power in all instances exerts primary influence upon the individual-citizen. Consequently, they are not equal partners. It is only state power that possesses the opportunity to spread its influence upon all relations, existing in society, which makes it all-embracing. The establishment of law and order in society, the fact that the citizens, just like officials and organizations, are endowed with rights and obligations, all this characterizes such a sovereign quality of state power as its <u>supremacy</u>. Supremacy is the manifestation of the determining role of state power in regard to any other relations of power. It is also the embodiment of unity, sovereignty, unlimited state power and its independence.

Consequently, there is a strict regularity in the relationship between the state and the citizen, if one regards this relationship in general. One of the sides is the bearer of power possessing a possibility of exercising power and coercion, the other the subject obligated to subordinate to the authority of the state. The absence in this instance of parity and equality among the relating subjects, a special, prevailing status of the bearer of power cannot be considered as a purely negative phenomenon, irrespective of the concrete socio-historic conditions, which have engendered it and make it necessary.

In this connection first of all it ought be noted the objective nature of establishing such inter-connection between the citizen and the state in a class society. Such a society objectively requires power for its own organization, normal functioning and even self-preservation. On the other hand, state power cannot exist without citizens, without population, consisting of the citizens in the main.

When formulating a number of general provisions characterizing the relationship and association between the citizen and the state, it is necessary to pay attention to the following major circumstance: These relationships differ radically subject to the nature of the social and state order, subject to what they are after in governing social life, subject to the type of state. In exploiter states (slave-holding, feudal, bourgeois) the individual, and consequently the citizen is only the object of state power, whereas in socialist states, he is not only the object but the subject, in the first place, of state governing. Genuine citizenship can only exist where the individual as a citizen belongs to the subject of power. This definition, used by the great French enlightenment scholars of the 18th century, especially Jean Jacques Rousseau, has not lost it scientific and practical value. Moreover, it has assumed its true meaning only in the practice of the socialist society. By its very nature the socialist society and state are called on to satisfy the vital interests of the citizens to an ever larger degree, to ensure, safeguard and guarantee the strict observance of their democratic rights and freedoms.

In a socialist state there is no contradiction between the influence upon a person on the part of the sovereign state power and those rights and duties which he really enjoys. Both aspects of such status of the citizen are reciprocally conditioned. On the one hand, because the person is within the sphere of the sovereign impelling power of the socialist state, the individual enjoys a certain legal status, carrying out a wide range of rights and duties. On the other hand, the legal status of the citizen of the socialist state ensures his effective participation in the formation and implementation of that same state power to which he is subordinated. In this connection of state and citizen finds expression the process of overcoming the alienation between society and the state, the citizen and the state power which were inherited by socialism from previous social formations. This process, not very simple in itself, embraces a relatively long historical period. It will find its completion in the communist public self-governing, where political and state power will cease to exist at all.

The status of a sovereign does not turn the state into a subject absolutely free in its acts and legal judgments toward the citizens. First of all, the ability of the state to have the rights and bear responsibilities is objectively restricted by the material conditions of society. Not taking the risk to deprive itself of support in the society's socio-class structure, to liquidate its own vitality, the state cannot go to such lengths, sufficiently deep and consequential, which run counter to its class and political nature.

In the socialist society the state, concentrating a powerful social force, is the guarantor of the freedom of the individual. Marxism-Leninism has rejected the idea about the absolute freedom of the individual. The freedom of man, all his designs and actions are conditioned in the final analysis by the society's economic, political, ideological and other conditions. It is impossible to live in society and at the same time be free from that society. Marxism understands freedom as cognition and revolutionary transformation of the natural and social environment in the interests of society's progressive development and all its members. The liberation of the individual from exploitation, economic and political oppression, active participation of the individual in the sphere of managing public and state affairs--this is the real basis on which the genuine freedom of the individual rests.

Such interpretation of freedom has never admitted anarchy and anti-social dissoluteness, or "Robinsonianism," when a man becomes sullen, excluding himself from public life and the sphere of collective interests hiding and protecting from the external world. Consequently, the freedom of the individual lies not in being independent of society, but in that constant coordination and inter-connection, in which the acts and doings of the individual stand in relation to society and his vital interests. Neither does it lie in the striving to isolate oneself from society, law and order, but in understanding one's role, its significance for society and concurrently one's responsibility before society, respect for law and order--if it is in complete accord with the ideals and lofty aims of the individual--this is where his genuine freedom reposes.

Socialist democracy is the most important manifestation and simultaneously guarantee of the individual's freedom in society, freedom of the citizen in the state. Without taking due account of this factor one cannot realize and comprehend the very approaches to the problem of cooperation between the citizen and the state.

Socialism revives natural social proportions between the public nature of production and forms of distributing material benefits. Having economically ensured the genuine freedom of the individual, it creates objective prerequisites for the full power of the people, and asserting the full power of the people, it creates objective prerequisites for the freedom of the individual. Thus, socialist democracy, interpreted in its true meaning--namely as the people's power--makes the freedom of the individual directly dependent on the full power, i.e. the sovereignty of the people, and, on the other hand, the full power emanates from the freedom of the individual. Without one, there cannot be the other.

Democracy is a form of the socialist state. This formula finds its concrete embodiment first of all in establishing unity between the state, the society and the people, between the state, the individual and the citizen.

It is well to cite a Marx' postulate that the state is "the mediator between man and his freedom." All that he does within the sphere of the citizen's freedom and first of all his political freedom is evidence of the fact that the state cannot be a neutral mediator. The rights and freedoms of the citizen, prior to being expressed in the judicial acts, must go through the prism of the interests of the ruling class of a given society. Therefore, the state, being the ruling subject of law, may either develop, under given conditions, the political activity of the citizens to the maximum, or considerably infringe the rights and freedoms of its citizens.

The state and the power aspirations of the ruling class are expressed in the legal norms which allow this class to regulate the behavior of society's members and thus exercise state control over society. Owing to that reason the norms of law can be regarded simultaneously as political and legal imperatives.

Law is the principal instrument with the help of which the state conducts its policies, performing the tasks and functions inherent to it, influences public relations. By organizing and classifying them in the interests of the ruling class (in a people's state--of the whole society), the state by means of law imparts to this will a general and compulsory character. The socialist law is absolutely different from the law of the previous social formations by the fact that it has no contradiction between the will expressed in the law and its binding force for the entire society. Under socialism the collective will of the working people, which has found its expression and consolidation in the legal norms, is becoming more and more in accord with the generally binding force of these legal norms.

The political and juridical aspect of the problem of the relationship between the individual and society is most vividly manifested in the constitutional institution of the basic rights and freedoms of the citizen. The rights and duties of the citizens, fixed by the Constitution and the laws of the socialist state, regulate the sphere of the most vital relationships between the individual and society, the state and its citizens. These relationships are based on the socialist public property, being determined by the entire nature of the Soviet social and state order.

The rights and freedoms of the citizens in the socialist society as considered from the viewpoint of their content and principle of cooperation, implemented in the relationships between the citizens and the state, can be generally summarized in the following way: The right of the citizens for the necessary conditions of subsistence provided for the state, for the satisfaction of their socio-cultural and spiritual requirements (socio-economic and cultural rights); The right to participate in managing the affairs of society and state (political rights); The right to be protected from illegal attempts on one's life, freedom, honor and dignity (personal freedoms). This classification of the constitutional rights and freedoms distinguishing socio-economic rights, political rights and freedoms and personal freedoms proceeds from the existence of the three types of social relations in the socialist society: The relationship of the state and the citizens in the socio-economic and cultural field, in the political sphere and the sphere of right protecting activity of the socialist state aimed at defending life, freedom, honor and dignity of Soviet citizens. It is namely these three spheres of the relationship of the citizens with the state that most fully reflect the actual position of the individual, comprising a real socio-economic, political and juridical basis of his legal status in society.

Speaking about a set of rights and freedoms of the citizens such as inviolability of the home, privacy of correspondence, freedom of conscience, one should mention first of all the duty of the state by its actions to protect personal freedoms of the citizens from illegal encroachments on the part of separate organs of the state, officials and citizens. Thus, the rights and freedoms of the citizens in the socialist state are in accord with the duty of the state to ensure the realization of these rights and freedoms, to strictly observe them, and, if necessity arises, to protect them from infringement on the part of separate organs, officials and citizens.

The dual nature of the legal relation between the citizens and the state is expressed not only in the rights and freedoms, but also in those duties which it imposes upon its citizens. In the system of the socialist democracy broad social and political rights of the citizens are inseparable from their obligations toward society. Thus, in conformity with the Constitution of the USSR the citizens are obliged to take care and multiply public property, to work honestly, to maintain public order, to defend the fatherland.

In the socialist society where all relations are based on the trust in man, where the initiative and creativity of the citizens are being encouraged, the citizens are not free of their responsibilities. This is manifested in the spirit of collectivism, morals of the truly comradely community. The consolidation of discipline, the strict implementation of the laws and rules of behavior adopted in the socialist society is a necessary condition of the successful and effective development of the socialist democracy.

Socio-economic rights occupy an important place among the constitutional rights and freedoms: The right to work, to rest, to free education and medical service, to maintenance in old age, in case of sickness and disability. The necessary condition of man's true freedom is the right to work effectively guaranteed. The

LA UTOPIA ROUSSEAUNIANA: DEMOCRACIA Y PARTICIPACION

Andrés Ollero

Doscientos años ofrecen suficiente perspectiva para poder obtener conclusiones de interés en torno a un legado de pensamiento. No es sólo una Constitución la que cumple una destacada efemérides, sino que a la vez todo un núcleo de reflexión política se enfrenta a la verificación histórica, único laboratorio capaz de contrastar su consistencia. Y dentro del pensamiento democrático la figura de Rousseau, cuya muerte será en tres años también doblemente centenaria, puede aportarnos, con el secreto de su renovada actualidad, motivos de reflexión de cara a este balance.

La doctrina de Rousseau dista de poseer la claridad presumible en una teoría capaz de alcanzar tan dilatada relevancia práctica. La discusión sobre lo que dijo o quiso decir ha dado pie a bien diversas interpretaciones y a intentos de explicación de sus contradicciones no exentos de habilidad. Pero lo que ahora interesa es su "sentido" actual, lo que hoy nos dice, o mejor--para no enmascarar con el lenguaje académico pretensiones de interpretación autorizada--lo que dice al autor de estas líneas tras estos doscientos años de teoría y práctica (?) democrática.

EL ROUSSEAU UTOPICO

Tras autoconcedernos este libre examen en torno a los textos rousseaunianos podemos adelantar ya lo que nos parece hoy el núcleo central de su aportación, capaz, no de explicar sus llamativas contradicciones, sino de hacer brotar un sentido fecundo oculto en ellas. La doctrina política de Rousseau, desde su acerado DISCOURS SUR L'ORIGINE DE L'INEGALITE PARMI LES HOMMES hasta su serena y ponderada reflexión DU CONTRAT SOCIAL, nos parece movida por una intención unitaria: la formulación de una utopía democrática, entendiendo el substantivo en su más positiva acepción. Rousseau quiere alcanzar un lugar que no es el de la sociedad de su tiempo. De ahí su insistente afán de no reincidir en el error--para él imperdonable--

de los teóricos del iusnaturalismo moderno: haber hecho aparecer como originarios (y, a fuer de "naturales", como modélicos) elementos de la sociedad actual.

En primer lugar, se enfrentará al espejismo ilustrador que cree haber construído definitivamente, con la sociedad civil, la auténtica comunidad humana. El episodio de Vicennes parece haber puesto en marcha la dimensión utópica de su pensamiento: "si j'avois jamais pu écrire le quart de ce que j'ai vu et senti sous cet arbre, avec quelle clarté j'aurois fait voir toutes les contradictions du système social! avec quelle force j'aurois exposé tous les abus de nos institutions!"[1] La utopía aparece, pues, en una primera dimensión revolucionaria, traicionada aparentemente por el ropaje literario que su autor le presta, cuando--para hacer más drástica su condena del modelo vigente--no duda en considerarlo inferior, no ya a un paraíso futuro, sino a la condición presocial humana. El aparente regreso reaccionario a la vida salvaje está al servicio de la intención revolucionaria. Rousseau disfraza de nostalgia su utopía para que su propuesta no sea fácilmente tachada de visionaria.

Pero lo que concede fecundidad a toda utopía no es la mera decisión de no seguir pisando un suelo condenado, sino la voluntad decidida de encaminarse a una tierra prometida. El Rousseau del CONTRAT se dispone a construir aprovechando la demolición producida por el DISCOURS. A la utopía destructiva que encerraba su nostalgia de la igualdad natural sucede ahora la utopía constructiva de su propuesta de una igualdad civil. La igualdad natural denunciaba la entraña desigual de la sociedad civil de los ilustrados. La nueva igualdad civil propone la democracia como auténtica realización de una comunidad de hombres.

DEL PACTO A LA LEY

Resulta claro que, para pretender que estos textos nos digan hoy algo, es preciso preguntarse con prioridad cuáles eran los objetivos de este doble proceso demolición-construcción. ¿Qué nos destruye el DISCOURS y qué nos propone el CONTRAT? La respuesta nos dibuja una interesante evolución Hobbes-Locke-Rousseau. Locke aceptó el modelo hobbesiano, pero se dispuso a reconstruirlo en profundidad para evitar sus consecuencias políticas: un absolutismo que negaba la libertad y la propiedad humanas, al abortar la sociedad civil que pudiera posibilitarlas. Del pacto hobbesiano, en el que el pueblo vive lo indispensable para dar vida al soberano, pasamos al contrato, que transforma al pueblo en sociedad civil haciéndolo perdurable.[2] Locke ha descubierto cuál era el precio de la construcción del soberano y le ha parecido excesivo. No cabe crearlo a costa de un pueblo.

La audacia de Rousseau ha consistido en adoptar frente a Locke la actitud de éste hacia Hobbes. No ha dudado en preguntarse por el precio de la sociedad civil. La consecuencia puede parecer sorprendente: el contrato le ha parecido leonino. El precio de la sociedad civil ha resultado ser la comunidad humana. La igualdad natural que daba a ésta sentido ha sido destrozada y cubierta púdicamente por el formalismo contractual. El hombre se ha hecho ciudadano para poder ser libre, pero se encuentra con la desagradable sorpresa de tener que contar con las cadenas entre su indispensable equipaje.[3] Rousseau siente la misma repugnancia por el absolutismo que Locke, pero además ha acertado a descubrirlo camuflado en su propuesta liberal. Tenemos, pues, claro qué es lo que el DISCOURS pretende destruir: el contrato y la sociedad política que de él nace legitimando una desigualdad irracional (anti-"natural"). Esa sociedad que no es sino la argucia con la que "le riche, pressé par la necessité, conçut enfin le projet le plus réfléchi qui soit jamais entré dans l'esprit humain: ce fut d'employer en sa faveur les forces mêmes de ceux qui l'attaquoient, de faire ses défenseurs de ses adversaires."[4] La utopía ha comenzado.

Pero la conclusión resulta paradójica. Si Rousseau pergeña su DISCOURS para demoler la sociedad contractualista, ¿qué pretende proponernos en el CONTRAT? Sólo caben dos posibles respuestas: 1, hemos tropezado con las bien conocidas contradicciones rousseaunianas. No hay en él un pensamiento unitario; es una figura romántica de ágil pluma y envidiable temperamento, que doscientos años después sólo cuadraría evocar en el marco exquisito de unos juegos florales y no en el circunspecto ambiente de la reflexión política; 2, podríamos atrevernos a sugerir que lo que Rousseau propone en el CONTRAT no es el contrato. Aunque siga hablando de él y lo erija en título de su portada, Rousseau va a intentar dar paso a la segunda etapa de su utopía proponiéndonos, por exigencias de la clave de su teoría, otra figura jurídica como la capaz de legitimar una sociedad democrática. La opción resulta audaz, pero no por ello evita alguna difícil respuesta: ¿cuál es esa figura jurídica?

Locke ha pasado del pacto al contrato para eludir el absolutismo. Rousseau para eludir el liberalismo ilustrado no halla otro camino que pasar del contrato... a la ley. La utopía cobra así un sesgo acrobático que explicará las peripecias inseguras de la propuesta de Rousseau y los no escasos descalabros prácticos de sus diversos "lectores".

Para Hobbes, que se ha quedado a solas con el soberano —una vez cumplido por el pueblo su obligado suicidio en aras de la conservación—, la ley no puede ser sino un mandato imperativo. Preguntarse por la racionalidad de su contenido sería tratarla como si encerrase un simple consejo,[5] y no es con consejos como cabe evitar la lucha de todos contra todos. Para Locke, que ha hibernado al pueblo como sociedad civil gracias al contrato, la ley aparece como depositaria de un acuerdo consensual. Su contenido racional (muy en línea con la gnoseología lockiana) surgirá de ese grado de probabilidad que es fruto del número.[6] El cómputo cuantitativo de las opiniones ciudadanas engendra la ley de mayor racionalidad a que cabe aspirar. El contrato es, por ello, autoobediencia y la ley juega como vehículo de su dinamismo práctico.[7]

Para Rousseau, por el contrario, el contrato no produce leyes sino cadenas. No es por un cómputo aritmético de la "voluntad de todos" como la autoobediencia se hace realidad. La autoobediencia—núcleo central de la democracia, expresión auténtica de la libertad civil[8]—sólo es posible cuando se impone la "voluntad general". La ley expresión de la voluntad general del pueblo, no mera suma de intereses particulares, es el único posible fundamento de una sociedad democrática que refuerce la sociedad humana sin destruirla. Lo que no pudo conseguir el contrato puede lograrlo ella.

Pero el problema surge en seguida. La ley no es un acto de constitución sino un acto de gobierno. ¿Cómo puede la ley, fruto de una sociedad constituída, constituir la sociedad? La contradicción resulta hiriente y el mismo Rousseau no ceja en superarla. Será preciso remitir la ley a una instancia anterior: el acto por el que el pueblo es pueblo.[9] Y no le importa revestir tal acto de forma contractual. La acrobacia parece haber terminado. Rousseau ha demolido el contrato, ha construído sobre la ley, pero luego no ha podido suspenderla en el aire y ha tenido que volver sobre sus pasos. Locke ha debido contemplar divertido la escena.

La utopía, no obstante, sigue en marcha. Rousseau no duda en intentar el más difícil todavía. Como el acróbata que ha simulado haber establecido su difícil equilibrio sobre un movedizo punto de apoyo, se dispone a eliminarlo coronando su creación, va a demostrar que de hecho ha pasado del contrato a la ley. Y lo hará en base a su peculiar concepción de la soberanía popular. Esta no es sino el continuo "exercice de la volonté générale"[10]: no un simple acto constitutivo sino una tarea de gobierno. La voluntad general es el auténtico centro de gravedad del planteamiento rousseauniano, pues "si on écarte du pacte social ce qui n'est pas de son essence, on trouvera qu'il se reduit aux termes suivants: "chacun de nous met en commun sa personne et toute sa puissance sous la suprême direction de la volonté générale; et nous recevons encore chaque membre comme partie indivisible du tout."[11] Pero Rousseau como veremos, parte de una visión dinámica de la voluntad general (y, con ella, de la soberanía) que hará que lo decisivo sea el gobierno de la sociedad y no su mera constitución. El contrato queda reducido a hipótesis indemostrable, destinada a jugar como condición (lógica!) sine qua non, fundamentadora de la posibilidad de la ley.[12] La voluntad general, que la ley expresa, no es posible sin sociedad. Remitir la constitución de ésta a un contrato no acarreará mayores consecuencias, porque, en cualquier caso, el "acte qui institue le gouvernement n'est point un contrat, mais une loi".[13]

Rousseau no va a conceder al contrato mayor virtualidad práctica de la que Hobbes reconoció al pacto. Hobbes se quedó a solas con el soberano. Rousseau se sumerge en la soberanía popular. Ambas posturas equidistan del liberalismo convencional de corte representativo. Para Locke el contrato seguía siendo el permanente modelo rector de la sociedad civil. El contrato se actualizaba en cada ley al decantarse aritméticamente la voluntad mayoritaria. En Hobbes el soberano se hace más fuerte en cada mandato imperativo. Rousseau, resuelto con el contrato el expediente constitutivo de la sociedad, va a creer posible al fin realizar la utopía democrática: habrá autoobediencia, y con ella igualdad civil, si la ley está por encima de los hombres,[14] si dejan de computarse intereses particulares para lograr sublimarlos en la voluntad general.[15]

Rousseau ha pasado del contrato a la ley esbozando una peculiar utopía democrática: la utopía de la participación. Porque la ley capaz de expresar la voluntad general no es ya simplemente un modelo formal explicativo (como el pacto hobbesiano o el contrato de Locke) sino una exigencia indeclinable de toda auténtica democracia. Sólo cuando el ciudadano comprenda que el sentido de su vida radica en la virtud política, sólo cuando descubra que el ejercicio de la soberanía es condición imprescindible para el mantenimiento de la calidad de ciudadano,[16] sólo cuando renuncie a adquirir el cómodo refugio en su interés particular con el precio exorbitado de su libertad, la democracia será posible. "Dans un pays vraiment libre, les citoyens font tout avec leur bras, et rien avec de l'argent."[17] La vuelta a las cavernas se ve ahora sustituída por la vuelta a la polis. También en este segundo momento ha disfrazado de nostalgia a la utopía.

A LA BUSQUEDA DE LA VOLUNTAD GENERAL

La utopía parece haberse desplegado en todo su alcance. Sin embargo, uno de los pasos del proceso ha sido precipitado y la construcción entera se resiente. Se trata, precisamente, de la remisión al contrato como condición lógica de la ley, que Rousseau concedió con aparente despreocupación. La pregunta sigue en pie: ¿cuál es el fundamento real de la voluntad general que la ley expresa? Si del cómputo aritmético (encerrado en el contractualismo de Locke) sólo podemos obtener la voluntad de todos, ¿de dónde obtendremos la voluntad general? El interrogante pone de manifiesto hasta qué punto el paso del contrato a la ley ha sido absoluto en Rousseau. El contrato se ha limitado a camuflar el movimiento menos airoso de su ejercicio, la clave más radicalmente metafísica de su utopía. Esta sólo conservará su fuerza destructiva si podemos contestar a la pregunta ¿qué es lo que sublima al interés particular elevándolo a interés general?

La respuesta no es ociosa, ya que de ella depende la auténtica soberanía popular, la "conversion subite de la souveraineté en démocratie"[18] que haría "que le peuple et le souverain"..."soient une même personne".[19] Y, como consecuencia, la posibilidad de llegar, si fuese necesario, a "obligar a ser libre"[20] a todo aquél que anteponga su voluntad particular a la general.

Ley y voluntad general componen, pues, un círculo vicioso que queda ahora al descubierto. La ley expresa la voluntad general. La voluntad es general cuando actúa bajo la ley. La utopía rousseauniana se agota dejando inexplorado el camino de la voluntad general, y privando con ello de su apoyo indispensable a la figura jurídica sobre la que bascula su construcción: la ley.

Las peripecias de las distintas interpretaciones de Rousseau, de las distintas "lecturas" que sin pretenderlo ha acabado suscitando, no encierran sino la búsqueda reiterada de ese último eslabón de su utopía. Tres actitudes nos parecen especialmente significativas al respecto.

La que entronca con más aparente facilidad con la doctrina rousseauniana es la mística anarquista de la comuna. El mismo Rousseau insistió repetidamente en que no pretendía esbozar un modelo político para grandes Estados, y cuando se vió en el brete de hacer propuestas concretas para ellos no dudó en olvidar muchos aspectos de su teoría. Su utopía democrática tiene en sus últimos escritos menos de propuesta que de pasión inútil. La democracia sólo sería soñable en una sociedad de dioses.[21] La nostalgia del DISCOURS reaparece. El hombre que aspire a la igualdad y a conservar su libertad en la autoobediencia habría de retrotraerse a épocas pasadas de la civilización. Quizá tendría que dinamitar una construcción inhumana, si no para volver a convivir con los osos, sí al menos para intentar salvarse del apocalipsis del gigantismo.

Más rigurosa en el entronque con el desarrollo de su pensamiento, aunque radicalmente alejada de su opción vital, sería la interpretación totalitaria. Esta sí que intenta derechamente proponer el secreto de la voluntad general. El Estado ético de la derecha hegeliana ofrece una fácil traducción. El círculo vicioso ley-voluntad general-ley se rompe a favor de la ley. Esta expresa la voluntad general en la medida en que expresa la voluntad del Estado que la encarna. Hobbes ha resucitado, pese a las protestas de Rousseau.[22]

Idéntico sesgo siguen las propuestas recientes de descubrir en Rousseau el precursor de la temática marxista. El eslabón habría quedado cerrado: Hobbes-Locke-Rousseau-Marx. Este habría repetido con Rousseau lo que él hizo con Locke y éste con Hobbes. La ley será, en efecto, la clave radical del modelo político; pero no una ley cualquiera sino la expresiva de la única "voluntad general" imaginable: la de la historia. La ley se ha hecho "legalidad socialista" y el proceso que arrancaba del pacto y pasaba por el contrato ve ahora cumplido históricamente su sentido. Pero las protestas de Rousseau contra el totalitarismo siguen en pie. Resignarse con el hobbesianismo sería poner fin a la utopía y esta, aun falta de su último eslabón, conserva no obstante su virtualidad negativa.

Una última posibilidad es devolver a Rousseau al museo. Celebrar con Locke el fracaso del desprecio al contrato. No pocos modernos planteamientos funcionalistas o sistémicos parecen animar a ello. El paso del contrato a la ley implicaría un retroceso en la evolución hacia la racionalización de la vida política que con la Modernidad había comenzado. La ley sólo tiene sentido como consecuencia del contrato. Es éste el que tiene que continuar guiando los derroteros del ejercicio de la soberanía. La participación generalizada sería algo peor que una utopía, sería una irracionalidad. La Ilustración no se ha hecho en vano y la técnica engendrada por ella ofrece una inaudita multiplicabilidad de los intereses particulares. Sacrificarlos en aras de un ficticio interés general ideológico sería algo más que un error, sería un derroche. Una sociedad despolitizada permitiría la compatibilización técnica de los intereses particulares.

Desde esta perspectiva el modelo de Locke sigue siendo

válido. Sólo en un extremo ha sido preciso rectificarlo: no cabe ya gobernar con un mínimo de leyes. La actualización del contrato exige hoy día una continua legislación. Pero también el temor que le llevaba a desconfiar de ellas era infundado. Las leyes no son ya amenazas para la libertad sino que expresan la auténtica "voluntad general": el bienestar garantizado por la "legalidad técnica". Dejemos incluso que ella misma construya el consenso. Al fin y al cabo ya el primer Rousseau había dicho que "l'homme qui médite est un animal dépravé..."[23]

Doscientos años después el pensamiento de Rousseau nos aparece como una gigantesca utopía. Por su estructura, sin duda, que ofrece aún su doble proceso destructivo-constructivo. Pero quizá pueda serlo también por su relevancia práctica. Utopía productiva que mueva aún a buscar el eslabón perdido de su trabajoso pensamiento, recordándonos todavía que el hombre para vivir necesita sentirse tal; que "en politique comme en morale, s'est un grand mal que de ne point faire de bien";[24] que "sitôt que quelqu'un dit des affaires de l'Etat: que m'importe? on doit compter que l'Etat est perdu";[25] que el supremo ideal de la democracia está condicionado por la existencia de "ciudadanos" en el sentido más ético del término: expertos en la virtud política; que el divorcio entre intereses privados y cosa pública no es una exigencia de la "racionalización" de la vida política sino el sello de su destrucción.

De lo contrario sólo nos quedará la utopía frustrada que se resigna pensando que al fin y al cabo el hombre ha nacido para ser puesto bajo las leyes. No tendría, pues, sentido empecinarse en resolver "la quadrature du cercle", en encontrar aquella "forme de gouvernement qui mette la loi au-dessus de l'homme". "Si malheureusement cette forme n'est pas trouvable, et j'avoue ingénument que je crois qu'elle ne l'est pas, mon avis est qu'il faut passer à l'autre extrêmité, et mettre tout d'un coup l'homme autant au-dessus de la loi qu'il peut l'être."[26] Preguntarse desde esta perspectiva de frustración si el hombre acabará bajo la ley de la técnica o bajo la ley de la historia es una pregunta ociosa. El Rousseau frustrado nos invitaría a esperar otros doscientos años.

FOOTNOTES

1. Carta a M. de Malesherbes de 12-1-1762 (en OEUVRES COMPLETES, tomo X 301, Hachette, Paris, 1910. En adelante añadimos entre paréntesis esta referencia a continuación de la relativa a la obra correspondiente).

2. Cfr. al respecto T. Hobbes, DE CIVE VII, 7 y 12, LEVIATHAN II, 18 /89 y 90/; J. Locke, AN ESSAY CONCERNING THE TRUE ORIGINAL EXTENT AND END OF CIVIL GOVERNMENT VIII, 95 y 99. Rousseau utilizará "pacte" y "contrat" como sinónimos. marginando la distinción apuntada por T. Hobbes en DE CIVE II, 9.

3. "L'homme est né libre, et partout il est dans les fers." La frase no está arrancada del DISCOURS sino del CONTRAT--I, 1 (III, p. 306)--donde, según algunos, Rousseau aceptaría la sociedad que en aquél rechazara. Por el contrario, cuando se propone ahora resolver la cuestión de la posible legitimación de esta situación de esclavitud, no intenta anular aquella crítica a la sociedad de su tiempo sino delimitar la utopía de la sociedad futura.

4. DISCOURS seconde partie (I, p. 114).

5. T. Hobbes, DE CIVE XIV, 1.

6. J. Locke, AN ESSAY... VII, 89 y XI, 134. Cfr. AN ESSAY CONCERNING HUMAN UNDERSTANDING 4º, XVI, 6.

7. J. Locke, AN ESSAY... VI, 57.

8. Ella es realmente "le problème fondamental dont le CONTRAT SOCIAL donne la solution", I, 6 (III, p. 313).

9. CONTRAT I, 5 (III, p. 312).

10. CONTRAT II, 1 (III, p. 318).

11. CONTRAT I, 6 (III, p. 313).

12. "La difficulté est d'entendre comment on peut avoir un acte de gouvernement avant que le gouvernement existe", CONTRAT III, 17 (III, p. 364).

13. CONTRAT III, 18 (III, p. 364).

14. "Le grand problème en politique" es "trouver une forme de gouvernment qui mette la loi au-dessus de l'homme", carta al Marqués de Mirabeau de 26.7.1767 (XII, p. 25).

15. "Il y a souvent bien de la différence entre la volonté de tous et la volonté général; celle-ci ne regarde qu'à l'intérêt commun; l'autre regarde à l'intérêt privé, et n'est qu'une somme de volontés particulières", CONTRAT II, 3 (III, p. 320). La validez de la mayoría de sufragios está siempre subordinada a su conexión con la voluntad general. Cfr. CONTRAT IV, 2 y I, 5 (III, pp. 368, 312).

16. Rousseau exacerba este aspecto hasta llegar a la negación del sistema representativo: "la souveraineté ne peut être représentée, par la même raison qu'elle ne peut être aliénée", cuando "un peuple se donne des représentants, il n'est plus libre; il n'est plus", CONTRAT III, 15 (III, pp. 361, 362).

17. "Sitôt que le service public cesse d'être la principale affaire des citoyens, et qu'ils aiment mieux servir de leur bourse que de leur personne, l'Etat est déjà près de sa ruine", CONTRAT III, 15 (III, p. 360).

18. CONTRAT III, 17 (III, p. 364).

19. Propuesta utópica presente ya en la dedicatoria del DISCOURS (I, p. 72).

20. "...quiconque refusera d'obéir à la volonté générale, y sera contraint par tout le corps: ce qui ne signifie autre chose sinon qu'on le forcera d'être libre", CONTRAT I, 7 (III, p. 315).

21. Ya en el CONTRAT, tras señalar entre las condiciones de la democracia que se trate de un Estado pequeño con un pueblo fácil de reunir, apunta: "s'il y avoit un peuple de dieux, il se gouverneroit démocratiquement. Un gouvernement si parfait ne convient pas à des hommes", III, 4 (III, p. 344).

22. "Les peuples se sont donné des chefs pour défendre leur liberté et non pour les asservir. Si nous avons un prince, disoit Pline à Trajan, c'est afin qu'il nous préserve d'avoir un maître", DISCOURS seconde partie (I, p. 117). "On dira que le despote assure à ses sujets la tranquillité civile" "qu'y gagnent-ils, si cette tranquillité même est une de leurs misères? On vit tranquille aussi dans les cachots", CONTRAT I, 4 (III, p. 309). Rousseau personifica este rechazo en Hobbes, al que no duda en comparar con Calígula, CONTRAT I, 2 (III, p. 307).

23. DISCOURS première partie (I, p. 87).

24. DISCOURS SUR LES SCIENCES ET LES ARTS seconde partie (I, p. 11).

25. CONTRAT III, 15 (III, p. 361).

26. Carta al Marqués de Mirabeau de 26.7.1767 (XII, p. 25).

Rousseau's works, although apparently inconsistent,

can be unified if we concentrate on the author's utopian intention. In this way DISCOURSES (which destroy the Enlightenment social model paradigmatically fixed in Locke's contractualism) and SOCIAL CONTRACT (where the social life of a future society, different from the bourgeois model, is legitimized) can be harmonized.

Paradoxically, SOCIAL CONTRACT does not establish the contract as the model for legitimizing political power, but leaves this function to law. In the same way that Locke substitutes the concept of contract for the Hobbesian covenant, Rousseau replaces the Lockean contract with the concept of law as General Will. This General Will is not a theoretical or formal model, but the material condition of a true democracy in the future.

The "will of everyone," resulting in the contract model of an arithmetical computation is to be replaced by the General, which is able to give rise to law. The key to substituting the common interest for particular interests is the real participation of the citizens--that is the material condition of Rousseau's democratic utopia.

Aside from romantic treatment of anarchic communes, there have been three attempts to replace real participation as a condition of effective democracy. The Hegelians replace participation with obedient adhesion to a State that has its own personality. Marxists replace participation with docility to the dictate of History. The technocrats replace participation with rational determination of the General Welfare. Rousseau whould not be read as giving support to any of these versions of a "general will."

PARTICIPATION FROM SOME ASPECTS OF
SOCIALIST LEGAL THEORY

Mihaly Samu

A. Some General Brief Reflections

From the socialist viewpoint the problem of participation is an essential social, political, economic, cultural and legal question. So it seems to be natural to approach this problem from the aspect of philosophy of law and it is on the agenda of the present World Congress.

According to socialist social theory, democratic thinkers of mankind have set up the requirement of participation in the public life, in the solution of public affairs, in the direction of society as a political and legal idea. This idea can be followed with attention from different aspects, and concrete solutions can be pointed out: The people's decision on war and peace, their participation in legislation and jurisdiction, the equal distribution of the produced goods by the decision of the people, the direct election and recall of leaders, the solution of everyday problems by the participation of the given communities, the solution of disputes in the community, the judgment by the people, etc.

Some forms of participation can be found in primitive societies. They are the result of the instinctive development of communities: The solution of public affairs with the participation of every member of the given community was the exigency of the whole community. This basic interest dictated that the primitive communities created special forms for safeguarding the participation in public life. These forms are living and developing in the modern world as well, especially in the functioning of direct democracy in socialism.

Marxist history stated that historical development resulted in the gradual dissolution of participation of communities in settling public affairs in the consequence of changing the social power. The ruling class determined the forms and possibilities of participation. So democratic participation depended on the interest of the political power, i.e., the interest of sustaining and strengthening class power sometimes made some limited participation possible.

The connection between social power and participation is influenced by social alienation. It can be generalized by historical experiences on the field of social power: The ruling classes expropriated the political power and ended common participation in the exercise of social power. The alienation of social power influenced the character of law and forms of participation in law as well.

In our age the socialist requirement claims the dissolution of alienation of social power so that the people and the communities have to exercise social power. Consequently social power has to be common to all and can function by the participation of people. Moreover in socialism conscious social development results in new solutions in the field of people's participation. Social science works out the proper forms and institutions for efficacious participation in the different areas of society.

In the marxist social theory participation is a special notion and it has special approaches from different aspects. We speak of the people's participation in abolishing capitalist society and building up socialist society. The people's participation indicates an active influence in the direction of society and in building of a new social order; it refers especially to the people's activity in the political, economic, cultural and juridical life. The people's participation is the basis of the development towards socialism and communism.

Some doubts might emerge in Western social sciences regarding the socialist concept of people's participation. Nevertheless the tendency of historical development and the results of socialist building prove the following consequence: The activity of the people in the direction of society, in public affairs, in the solution of joint problems, consequently the people's participation is a realistic historical aim and fact and not only a democratic desire or doctrinaire dream.

But it is important to point out that the requirement of people's participation depends on time and environment. This requirement can be realized in the life of a nation considering the historical tradition, peculiarity and circumstances. It is not a scientifically based concept which proposes to realize the requirement of people's participation independently from the national conditions and circumstances. Therefore we have to emphasize that the people's participation has a lot of forms and implementations and the general requirement can be realized in a lot of special institutions and mass movements depending on concrete situation.

It is well known that in the modern world many discussions are going on regarding the forms, institutions and rules of the worker's participation in the direction of industry, enterprises and institutions. Moreover there are discussions on the problem of the participation in the cultural life as well, e.g., in the work and governance of schools or universities. In principle we have some generally accepted requirements but in the reality the development of concrete solutions requires further investigation. The socialist development indicates that new ideas concepts, and institutions are expected.

It is necessary to remark that the people's participation has more aspects: It can be found in the political, economic, cultural life and as a specific problem in the legal life. I should like to point out some peculiarities in the field of law.

B. Participation In Law

With respect to law the problem of participation is expressed first of all in law making and in jurisdiction.

The participation in law making is generally regulated and investigated in constitutional law. This question is connected with the election system, with the possibility of the control of representation, with the relation between representatives and citizens. From the side of participation the problem of democratic law making and the effect of direct democracy in this respect are of great importance.

In socialist law there is a trend to enlarge the democratic character of law making. From this, the principal and practical problems come forward: The people's initiative in the law making, the people's discussion of drafts and the acceptance of basic rules by direct people's decision.

The people's initiative expresses the direct interest of the people in law making; the different communities have the right to initiate new rules and amendments. This question is connected with the democratic consciousness of the people, with their experiences in public life. The people's initiative could be realized in a mass movement demanding some social changes including new rules or annihilation of old ones, in different organizations and communities. The forms and solutions depend on the national peculiarities, on the political, economic and legal system.

The second form of the people's participation in law making is the discussion of drafts. In the course of discussion the diverse opinions of the communities, social or state organizations can be expressed. It is obvious that public opinion plays an important role in the development of special and controversial alternatives regarding the drafts. But it is necessary in modern societies that the lawyers play a special role in the formulation of different alternatives and in the acceptance of drafts.

The next step of democratic law making is the people's decision on the acceptance or rejection of a draft. In this case the people decide by direct suffrage on the suggested bill. The people's referendum is generally fixed in the socialist constitutions, and its practical application has come forward lately, as when the new constitution was accepted by a people's referendum in the German Democratic Republic and in Bulgaria.

I must mention that the development of democracy in the socialist countries can create new forms and solutions of the people's participation in law making.

It is a general requirement that the people have the right to participate in decisions of common interest. One expression of this is the role of lay judges in the judicial system. In the socialist countries a special characteristic of the people's participation is the participation of lay judges in the work of the courts. We can say that the participation of lay judges is a generally accepted solution in the socialist countries. But it is not the only form and is not required; the most important thing is the people's representation in the judicial system. It can safeguard the will of the people and give an opportunity for expressing common sense in the work of courts and for the control of professional judges by lay judges.

The importance of the lay judges is often denied, because of their sometimes passive role at the trials. Nevertheless the lay judges incorporate the common sense of the people and even their passive role or behavior influences the practice of courts as a whole. Further, the lay judges make possible the development of collective opinion and assure its influence on the higher level of legal culture. Lay judges provide a legal guarantee in the observance of human rights.

The people's participation in law also raises another question, namely, the problem of election of judges. In principle we are for the election of judges, but we have a lot of practical problems. It is to be expected that the further theoretical and practical investigations will give clear-cut suggestions and solutions. Finally the people's participation in the judicial system emerges not only in the work of professional courts but of the social courts as well. Direct administration of justice by the people is appropriate to the high level of democratic common life; its realization requires further investigation. Socialist development will give some new solutions and forms in the direct adjudication by the communities.

C. Participation In Public Affairs Is A Human Right.

In the development of law an essential qualitative change was the introduction of human rights as a legal institution. At the beginning of the capitalist society the right to life and the right to liberty were fixed in the legal systems and in some constitutions. Now, economic, social and cultural rights have become important. In international life there has been some development of human rights, in international documents, such as in the Universal Declaration of Human Rights, the International Covenant on Economic, Social and Cultural Rights and the International Covenant on Civil and Political Rights.

From the socialist viewpoint human rights are the fundamental rights of all citizens and they are guaranteed rights for all members of society instead of being merely formal. Basic human rights are not only proclaimed, but are guaranteed and consistently implemented. The general requirements of people's participation are expressed in legal institutions. These requirements are fixed in the socialist constitutions and in other documents, and are included in the catalogue of human rights. The democratic development of socialism claims new democratic solutions in the field of human rights as well. We could acknowledge that its new form is the participation in public affairs as a human right.

The right of participation in public affairs is realized in the institutions of democracy first of all in the form of workplace democracy. It extends into the field of industry, agriculture, trade and cultural life. So participation builds an organic part of economic, political and cultural life, it permeates the everyday work of common institutions and activity. Furthermore, participation is fixed as a human right in constitutions and is safeguarded by legal guarantees. For instance, in the Hungarian Constitution it is formulated in Section 2. "Citizens directly participate in the settling of public affairs at their workplace and their residence," and it is fixed in Section 68; "Every citizen has the right to take part in settling public affairs; he is obliged to carry out his public commission conscientiously."

The realization of the right to participate in public affairs is influenced by the social structure and by the political system. The acknowledgement and implementation of this right depends first of all on power relationships. Political, economic and ideological power includes the realization of human rights as well, especially the right to participate in public affairs. In this respect the main problem is to participate in the decision in the different fields of social life. Therefore, we can understand the realization of this right by studying the decision system. The realization of the right of participation in public affairs expresses itself in the measures and solutions for participation in decision making. As O. Friedman has said, "Participation is intended to provide equal access to certain parts of the decision making process for all members."

The realization of human rights in decision system is not only a legal problem, it is an important social correlation. The implementation of the right of participation in public affairs ensures the general realization of people's participation in political, economic, cultural and legal life.

In closing I should like to emphasize that participation in public affairs has an interesting essential relation to humanism. The dignity of the human being demands the acknowledgement of rights for activity and participation in the society and communities. So the right of participation in public life is not only a legal question but an essential social claim for the development and perfection of man. The personal development of man results in building up a collective personality and in this process the realization of human rights contributes to the strengthening of the collective features of man.

BÜRGER-MITBESTIMMUNG AN DEUTSCHEN GERICHTEN: DEFIZITE UND CHANCEN IN DER LAIEN-AUSWAHL UND DER EINSTELLUNG DER BERUFSRICHTER

Ekkehard Klausa

I.

Bürger als ehrenamtliche Richter an den meisten deutschen Gerichten--ist diese Form der Mitbestimmung[1] ein demokratischer Fortschritt oder ein Einfallstor für Willkür, Emotion und Unverstand auf der Richterbank? Oder gibt es diese "Mitbestimmung" am Ende gar nicht, sondern sind die Laien nur demokratische Attrappe? Sind sie Nachkommen der "stummen Schöppen" im Frühabsolutismus, die nur noch neben dem gelehrten Richter sassen, um dessen Würde und Autorität zu unterstreichen?

Meine Überlegungen haben zwei Ziele, ein inhaltliches und ein methodisches:

(1) Ich will aufgrund empirischer Befunde[2] induktiv die These vertreten, dass die Partizipation ihr Problem nicht nur in der Qualifikation der mitbestimmenden Bürger hat, sondern zu allererst in der Einstellung der Fachleute, die ihre Macht mit Laien teilen sollen.

(2) Dabei soll der mögliche Beitrag einer empirischen Rechtssoziologie zu rechts- und sozialphilosophisch wertenden Diskursen erprobt werden, hier am Beispiel der Diskussion über den Wert des Laienrichtertums.

Unser Ausgangspunkt ist der mit Partizipation angestrebte Wert. Mitbestimmung vermindert ihrem Anspruch nach Herrschaft und wird damit Voraussetzung und Mittel zur Entfaltung der Persönlichkeit mitbestimmender Menschen, die im kommunikativen Prozess gemeinsamer Willensbildung und Entscheidung ihre Fähigkeiten anwenden und vervollkommnen (HARTFIEL 1972: 440).

Ob man diesen Wert anstreben soll, ist empirisch nicht diskutierbar. Wer das staatliche Führerprinzip der kom-

munikativen Willensbildung vorzieht, der ist weder mit
logischen Ableitungen noch mit empirischen Ergebnissen
zur Umkehr zu zwingen (KLAUSA 1974: 76). Zwar gibt es
auf jeder historischen Sinnstufe gewisse suggestive
Wertformeln, denen keiner gern widerspricht. Kaum
jemand wird z. B. sagen, er sei gegen die Gerechtigkeit.
Nur sind solche Wertformeln oft leere Worthülsen, die
selbst gegenteilige Inhalte beherbergen können (die "Gerechtigkeit" etwa das ständische "suum cuique" ebenso
wie das egalitäre "allen das Gleiche"). Aber selbst
eine seit kurzem so suggestive Wertformel wie "Partizipation" ist nicht evident, sonst könnte nicht Michael D.
Bayles seinen Diskussionsbeitrag unter den Titel "Participation: An Overvalued, Impractical Ideal" gestellt
haben.

Über letzte Wertprämissen lässt sich zwar mit "guten
Gründen", aber nicht empirisch streiten. Anders schon,
wenn sich ein Werturteil ausdrücklich oder implizit auf
bestimmte Tatsachenannahmen stützt, etwa auf die Behauptung, die Verwaltung sei im Führerstaat in dieser oder
jener Hinsicht effektiver. Solche empirischen Behauptungen sind häufig, ja sogar in der Regel in Werturteilen mitenthalten (RYFFEL 1969: 157; OPP 1973, 33 ff.).

Hier geht es aber nicht um die Begründung eines Werturteils zugunsten der Mitbestimmung als Prinzip, sondern um den nächsten Schritt. Befürworter des Laienrichtertums behaupten etwa, diese Einrichtung wirke sich
im Sinne der Mitbestimmungs-Werte aus, fördere also die
Entfaltung von Persönlichkeiten durch gemeinsame Willensbildung in einem kommunikativen Prozess zwischen
Berufs- und Laienrichtern.

Der erste Schritt, die Entscheidung für den Mitbestimmungs-Wert, ist rechts- und sozialphilosophischer Natur.
Dies selbst dann, wenn dieser Wert zum Grundstein
rechtssoziologischer und allgemeinsoziologisch-
empirischer Denkgebäude wird. (Rechts-) Soziologie ist
ohne philosophische Grundentscheidung nicht möglich.
Sie muss von einer übergreifenden "Philosophie des Politischen" ausgehen und in sie einmünden (RYFFEL 1974:
147).

Der zweite Schritt jedoch ist rechtssoziologisch. Die
Behauptung, dass eine bestimmte gesellschaftliche Institution, das Laienrichtertum, bestimmte Wirkungen im
Sinne des Mitbestimmungs-Ideals hervorbringe, ist mit
soziologischer Theorie und soziologischer Empirie diskutierbar.

In der deutschen Diskussion um den Wert des Laienrichtertums finden wir zahlreiche Argumente, die der Empirie

zugänglich sind. Ein Beispiel: Der Altmeister des
deutschen Gerichtsverfassungsrechts, Eduard Kern, ein
Befürworter des Laienrichtertums, erhebt Bedenken gegen
ehrenamtliche Arbeitsrichter, die von Arbeitgebern und
Arbeitnehmern gestellt werden. Dies verstosse gegen das
Gebot objektiver Rechtsprechung, weil solche "Vertreter
gegensätzlicher wirtschaftlicher Gruppen . . . geneigt
sind, Interessenstandpunkte zur Geltung zu bringen"
(KERN 1932: 487). Meine empirische Untersuchung ergab
aber, dass sich die Voten der Arbeitsrichter kaum oder
gar nicht nach Gruppenzugehörigkeit unterscheiden. Die
ehrenamtlichen Arbeitsrichter sind grösstenteils so
peinlich um "richterliche Objektivität" bemüht, dass
selbst ein "Rollentausch" vorkommt: Ein Arbeitsrichter
aus der Gewerkschaft beurteilt die Verfehlung eines Ar-
beiters härter ("Der ruht sich auf den Knochen seiner
Kollegen aus--die Kündigung ist in Ordnung!) als der Ar-
beitgeber-Beisitzer ("Halb so schlimm, solche Leute habe
ich auch im Betrieb und kündige nicht gleich") (KLAUSA
1972: 140 ff.).

II.

Vor der inhaltlichen Diskussion der Partizipations-
These (oben 1) noch eine Warnung: Es ist naheliegend,
aber irrtümlich zu glauben, jeder Demokrat sei automatisch für Laienrichter und jeder Gegner sei automatisch
ein Reaktionär--es gehe also nur noch darum, zum Lobe
des Laienrichtertums der Rechts- und Sozialphilosophie
die passenden Orgeltöne zu entlocken und der Empirie
ein paar dienliche Belegstücke. Denn einerseits sind
die Laienrichter auch im Dritten Reich und in der DDR
mit fast den gleichen Formeln offiziell gefeiert worden
wie in der offiziellen Bundesrepublik (BAUR 1968:
49f.). Andererseits hat ein Verfassungsrechtler wie
Helmut Ridder, der schon Radikaldemokrat war, bevor dies
kurzfristig Mode wurde, das Laienrichtertum abgelehnt:
Das Recht sei so kompliziert, dass nur der rechtsgelehr-
te Richter den "gesunden Menschenverstand des Laien" an-
gemessen sublimieren könne (RIDDER 1953: 115 f.).
Andere Autoren halten den ehrenamtlichen Richter für
überfordert. Er habe praktisch keinerlei Einfluss auf
die Entscheidung (GÖRLITZ 1970: 304).

Schliesslich haben auch die Befürworter des Laienrichtertums durchaus unterschiedliche Motive. Die Interes-
senverbände etwa, die am Arbeits-, Sozial- und Finanz-
gericht ehrenamtliche Richter stellen, erhoffen sich
Einfluss auf die Rechtsprechung. Ein Grossteil der Berufsrichter, so erschien es bei der Analyse meiner Interviews, bejaht das Laienrichtertum eher aus träger Ge-
wohnheit, weil es eben so ist und immer so war. Vor

allem aber, das sagte die Mehrzahl der Strafrichter ganz
deutlich, trauen sie dem Laien keine Verbesserung der
Rechtsprechung zu, sondern nur eine Propagandawirkung
für Richter und Rechtsprechung im Volk (KLAUSA 1972:
66). Das bedeutet offenbar nicht, dass man gedenkt,
diesem Volk eine Einfluss im Sinne der Mitbestimmung
einzuräumen.

Uns geht es jetzt also um einen empirischen Beitrag zu
der Frage, ob die Laienbeteiligung eine Wirkung im Sinne
des Mitbestimmungs-Ideals ausübt (Deskriptionsfrage)
oder, wenn nicht, unter bestimmten Bedingungen ausüben
könnte (Prognosefrage).

Der Aussage von Ridder, Schöffen seien durch die "Kompliziertheit des Rechts" überfordert, steht ein Ausspruch gegenüber, der dem Verfassungsrechtler Leibholz
zugeschrieben wird: "Im Strafverfahren gibt es nur zwei
interessante Fragen--erstens: ist er's gewesen?, und
zweitens: was kostet das?" Beide Fragen haben mit
fachjuristischer Subsumtion nichts oder fast nichts zu
tun und sind intelligenten Laien durchaus zugänglich.
In der ersten Frage, wo es etwa um die Glaubwürdigkeit
von Zeugen geht, kann der Berufsrichter sich ruhig die
Meinung anderer Laien anhören, denn was Zeugenpsychologie betrifft, ist er selbst einer (OPP 1973: 109). Ob
allerdings Ridders oder Leibholz' empirische Prämisse
stimmt, ob nämlich im Strafprozess überwiegend komplexe
Rechtsfragen oder die beiden anderen Fragen auftauchen,
wäre wiederum nur empirisch zu entscheiden. Zugunsten
der Laienrichter würde schon sprechen, wenn die beiden
Fragen von Leibholz neben den Rechtsfragen nicht völlig
zurückträten.

Zahlreiche Autoren sehen die wichtigste Aufgabe der
ehrenamtlichen Richter (besonders an Straf- und Verwaltungsgerichten, wo sie nicht nach spezifischen Kenntnissen oder Erfahrungen ausgewählt werden) in der "Plausibilitätskontrolle": Die ehrenamtlichen Richter
werden jeweils den Bevölkerungskreisen entnommen, die
von der Jurisdiktion des Gerichts unmittelbar betroffen
sind. Bei den Straf- und Verwaltungsgerichten ist das
jedermann, bei den Arbeitsgerichten sind es Arbeitgeber
und Arbeitnehmer, bei den Kammern für Heilberufe Ärzte
und Apotheker usw. Diese "formale Repräsentation" (RÜGGEBERG 1970: 197) kann in erster Linie den Sinn haben,
durch kritische Teilnahme von Artgenossen des Betroffenen den Berufsrichter v o r Erlass seines Urteils zu
zwingen, die Verständlichkeit und Überzeugungskraft
seiner Argumente zu überprüfen.

Diese Aufgabe können ehrenamtliche Richter nur dann
erfüllen, wenn ihre Anwesenheit den Prozess deutlich be-

einflusst (und nicht bloss demokratischer Zierrat ist).
Nicht nötig ist dabei, dass sie die Berufsrichter hin
und wieder überstimmen. Es genügt, wenn sie den Berufsrichter dazu bringen, bewusst oder unbewusst seine Argumentation auf die Bedürfnisse und den Verständnishorizont von Nichtjuristen einzustellen.

Wertlos wäre es allerdings, wenn der Berufsrichter die
Laien rein manipulativ berücksichtigte, etwa indem er
die eigentlich entscheidungsbedürftigen Zweifelsfragen
des Falles verschweigt und die ehrenamtlichen Richter
durch einseitige Darstellung zum Kopfnicken anhält. Das
kommt sicherlich häufig vor. Meine Untersuchung ergab,
dass an den meisten Gerichten die Berufsrichter sich
schon vor der mündlichen Verhandlung absprechen. Abweichende Meinungen werden dann in der Beratung mit den
ehrenamtlichen Richtern sehr häufig gar nicht mehr erwähnt.

Ein besonders krasses Beispiel versuchter Manipulation der Laien berichtete mir ein befragter Berufsrichter am Landessozialgericht: In einer Vorbesprechung
waren er und der zweite beisitzende Richter anderer Ansicht als der Senatspräsident. In der Hauptverhandlung
nahm ihm dieser daraufhin die Berichterstattung ab und
trug in der Beratung den Fall einseitig aus eigener
Sicht vor. Die Gegenargumente seiner Kollegen erwähnte
er nicht einmal. Dann fragte er die beiden ehrenamtlichen Richter: "Sie sind doch auch dieser Meinung?" Die
nickten. Darauf der Senatspräsident: "Gut, dann ist ja
eine Abstimmung gar nicht mehr nötig." Daraufhin erhob
sich der zitierte Richter, zog seine Robe aus mit der
Bemerkung, er werde hier offenbar nicht benötigt, und
ging in sein Arbeitszimmer. Es gab einen grossen Krach
und danach eine neue Beratung. Als der Berichterstatter
sein Votum vorgetragen hatte, wurde der Senatspräsident
überstimmt (KLAUSA 1972: 170).

Hier wurden die ehrenamtlichen Richter erst durch
einen offenen Krach zwischen den Berufsrichtern in ihre
Entscheidungsrechte eingesetzt. Wer das hierarchische
Bewusstsein und den Corpsgeist von Berufsrichtern kennt,
wird nicht zweifeln, dass dergleichen ein Ausnahmefall
ist. Diese "Manipulationsanfälligkeit" darf nun nicht
ohne weiteres auf dem Negativkonto der Laienrichter verbucht werden. Auch Volljuristen hätten ohne Vorbereitung und Kenntnis der Spezialmaterie nach dem einseitigen Vortrag des Senatspräsidenten keine eigene Stellung
beziehen können.

Daraus ergibt sich die Hypothese: Partizipation der
Bürger am Gericht ist nur möglich, wenn die Berufsrichter Mitbestimmung wollen und ermöglichen. Ob ehrenamt-

liche Richter in der Lage sind, die Aufgabe der Plausibilitätskontrolle zu erfüllen, ist kein objektives "Datum" (= Gegebenes). Vielmehr sind es die Berufsrichter, die jene "Realität konstruieren" (i. S. v. BERGER/LUCKMANN 1969), welche die Laien zur Mitbestimmung befähigt oder zu "stummen Schöppen" herabwürdigt.

So gesehen sprechen sich die Berufsrichter in gewisser Weise ihr eigenes Urteil, wenn sie angeben, die Laien trügen zur Verhandlung nichts bei. Görlitz (1970: 304) plädiert aus diesem Grunde für die Abschaffung des Laienrichtertums. Görlitz war zwar der erste, der dieser Frage in der Verwaltungsgerichtsbarkeit empirisch nachging, aber er befragte allein die Berufsrichter. Schiffmann (1974: 222), der auch ehrenamtliche Richter und andere Prozessbeteiligte befragt hat, widerspricht Görlitz. Meine eigene Untersuchung vermittelte den Eindruck, dass der Einfluss tatsächlich gering ist--und dass dies nicht zuletzt an der konservativen Einstellung zahlreicher Verwaltungsrichter liegt.

Ein Test auf die Hypothese, dass der Berufsrichter selbst den Laien zum fähigen oder nutzlosen Mitarbeiter macht, war mir am Berliner Kriminalgericht möglich. Am Anfang stand ein subjektiver Eindruck: Gerade diejenigen unter den befragten Berufsrichtern, die den Laienbeitrag ernst nahmen und nicht Jasager, sondern Partner einer Gewissensgemeinschaft suchten, wirkten auf mich im Interview aufgeschlossener und liberal-progressiver als die meisten Verächter laienhafter Ignoranz. Darin liegt natürlich noch kein empirischer Beleg. Vielmehr könnte es ein Vorurteil zugunsten der Laien sein, das ihre Ablehnung kurzerhand mit ideologischer Rückständigkeit gleichsetzt. Möglich wäre allerdings, die Begriffe "liberal-progressiv" und "konservativ" zu operationalisieren und der Korrelation dieser Merkmale mit der Einstellung zum Laien empirisch nachzugehen.

Einen wichtigen empirischen Beleg kann ich aber unabhängig von diesem möglichen Vorurteil anführen. Er ist um so glaubwürdiger, als er überraschend kam und nicht eine vorgefasste These nachträglich belegte, sondern mich überhaupt erst auf diese These brachte. Auf die Frage nach dem Wert der Laienbeteiligung antworteten die Berufsrichter am Kriminalgericht extrem verschieden. Von zwei Amtsrichtern an Wirtschaftsabteilungen wählte der eine unter fünf Antwortalternativen die beste ("grosser Vorteil"), der andere die schlechteste ("sollte abgeschafft werden"). Der eine hält Schöffen speziell in Wirtschaftsstrafsachen für "absolut notwendig", der andere gerade in Wirschaftsstrafsachen für "nutzlos". Der eine hält die Schöffen für überfordert, der andere lobt ihren Sachverstand. Der eine

ist mit seinen Schöffen derart zufrieden, dass er
sogar--unzutreffend--an eine Sonderauswahl von wirtschaftlich besonders bewanderten Schöffen für Wirtschaftsabteilungen glaubt (KLAUSA 1972: 66).

Wenn zwei Berufsrichter derart unterschiedlich urteilen, so kann das mehrere Gründe haben: Entweder ist ihr Erfahrungsmaterial verschieden. Es ist aber unwahrscheinlich, dass in vielen Jahren das Los dem einen Kammervorsitzenden immer die schlechten, dem anderen immer die guten Schöffen zuspielt. Zweitens wäre denkbar, dass beide Richter das gleiche Erfahrungsmaterial nach unterschiedlichen Wertmaßstäben beurteilen. Wenn beide erlebt haben, dass von zehn Schöffen einer intellektuell unfähig ist, so könnte der eine dies schwerer nehmen als der andere. Doch kann auch dieser Unterschied der Wertmaßstäbe nicht erklären, wieso der eine den Fachverstand und die Auswahl der Schöffen über den grünen Klee lobt, der andere ihnen jede Ahnung abspricht. Deshalb ist eine dritte Erklärung die wahrscheinlichste. Die Berufsrichter machen mit den gleichen Schöffen unterschiedliche Erfahrungen. Das liegt nicht an der Qualität der Schöffen, sondern an der Einstellung der Berufsrichter dem Laien gegenüber.

In gewisser Hinsicht formt sich jeder Richter seine Schöffen selber. Will er die Schöffen "dumm" halten--und das liegt nahe, wenn er sie von vornherein für dumm hält--so hat er es meist nicht schwer. Erheblich anstrengender ist es für ihn, wenn er Verfahren und Beratung für Laien durchsichtig machen will. Erst dadurch kommt die Mehrzahl der Schöffen in die Lage, einen eigenen Beitrag zu leisten. Begreift er nichts, so muss das nicht immer an seinem "juristischen Unverstand" oder an der schlechten Auswahl liegen. Wahrscheinlich ist vielmehr, dass der Berufsrichter ihn aus Absicht oder Unvermögen nicht genügend eingewiesen hat.

Die Einstellung, aus der heraus der Berufsrichter seine Erfahrungen mit Schöffen nicht nur (rezeptiv) "macht", sondern (aktiv) "schafft", ist letztlich politisch. Eine autoritäre Persönlichkeit mit Vorbehalten gegen demokratische Mitentscheidung von Nichtfachleuten wird die Schöffen zu unmündigen Jasagern machen oder in die Opposition treiben. Im ersten Fall empfindet er sie als nutzlos, im zweiten als störend. Für diese These spricht auch die Erfahrung mehrerer befragter Schöffen, die an verschiedenen Spruchkörpern verschiedene Berufsrichter erlebt hatten. Bei dem einen Richter fühlten sie sich entmündigt und nutzlos, beim anderen als geachtete und brauchbare Mitarbeiter. Über eine bestimmte Jugendkammer hörte ich immer wieder, dass sie Laien geringschätze, eine andere wurde immer wieder von den Ju-

gendschöffen gelobt. Allgemein scheint es, dass die
Amtsrichter die Schöffen stärker beteiligen, weil sie--
zahlenmässig--mehr auf sie angewiesen sind. Daran mag
es liegen, dass ich am Amtsgericht nicht so viele ge-
ringschätzige Urteile hörte wie am Landgericht (KLAUSA
1972: 79).

III.

Nur um den Anschein zu vermeiden, dass ein so komple-
xer Sachverhalt wie die "Funktionsfähigkeit des Laienrichtertums" monokausal zu erklären wäre, deute ich
neben der Einstellung der Berufsrichter noch weitere
Einflüsse an. Es gibt viele, die sich zum Teil nach der
Rechtsart unterscheiden. So sind z. B. die ehrenamtlichen Richter in der Revisionsinstanz am Bundesarbeitsge-
richt einflussreich und gern gesehen, am Bundessozialge-
richt einflusslos und geringgeschätzt, obwohl sich ihre
Auswahl nicht nennenswert unterscheidet (KLAUSA 1972:
197 ff.). Damit ist einmal die Ansicht zahlreicher Berufsrichter widerlegt, Laien seien, wenn überhaupt, nur
in der Tatsacheninstanz zu brauchen. Zum anderen sieht
es so aus, dass der Gegensatz zwischen dem formalisti-
schen sozialgerichtlichen Prozess und dem auf Interes-
senausgleich bedachten arbeitsgerichtlichen am ehesten
diesen Unterschied erklären kann.

Zuletzt verdient aber noch Erwähnung, was für Wert-
schätzung und Funktionsfähigkeit der Laien äusserst
wichtig und bis heute sehr schädlich ist: Die Auswahl
der Schöffen. Keinen Pförtner würde man einstellen,
ohne ihn vorher in Augenschein genommen zu haben;
dagegen werden die Schöffen grösstenteils wahllos und
unbesehen aus der Einwohnerkartei gezogen. Bei solcher
Zufallswahl kann es nicht ausbleiben, dass jeder Berufsrichter gelegentlich unfähige Schöffen bekommt. Die
Zufallsstichprobe meiner Interviews brachte mich z. B.
an einen Schöffen, der mir auf den ersten Blick als
schwer neurotischer Paranoiker erschien. Bei allen Gerichten ausser den Strafgerichten ist es selbstverständ-
lich, dass die ehrenamtlichen Richter von Verbänden oder
Parteien vorgeschlagen werden, die sie als geeignet be-
zeichnen. Soll Mitbestimmung keine Farce oder Fassade
sein, so muss für Schöffen die Zufallswahl abgeschafft
werden. Sonst entsteht der böse Schein, die auswählen-
den Behörden nähmen die Bestellung von Laienrichtern
weniger ernst als die Anstellung eines Pförtners.

Ausserdem haben wir gesehen, dass der Laienbeitrag
weitgehend vom guten Willen der Berufsrichter abhängt.
Von manchem Richter ist dieser gute Wille nicht zu er-
warten, solange er den Eindruck hat, ihm werde der erste

Beste von der Strasse als Kollege oktroyiert.

Vielleicht liegt an dieser Stelle, wo sich heute ein doppeltes Defizit des Laienrichtertums zeigt, eine doppelte Chance für die Zukunft der Partizipation in Deutschland.

FOOTNOTES

1. Ehrenamtliche Richter wirken mit an fast allen strafrechtlichen Spruchkörpern ausser am Oberlandesgericht und Bundesgerichtshof, in der Zivilgerichtsbarkeit nur an den Kammern für Handelssachen am Landgericht und in Sonderfällen (z. B. Landwirtschaftsgericht), in der Arbeitsgerichtsbarkeit und der Sozialgerichtsbarkeit in allen drei Instanzen, in der Verwaltungsgerichtsbarkeit in erster und zweiter Instanz, in der Finanzgerichtsbarkeit in erster Instanz, in der Disziplinargerichtsbarkeit, der Ehrengerichtsbarkeit der Rechtsanwälte und Notare und der Berufsgerichtsbarkeit (z. B. Heilberufe, Architekten, Steuerberater usw.); vgl. Klausa 1972: 216 ff.

2. In einer Untersuchung (KLAUSA 1972) mit ca. 400 Befragungen von Berufs- und Laienrichtern sowie Auswahlberechtigten.

Literatur:

Baur, Fritz	"Laienrichter--heute?," in FESTSCHRIFT FÜR EDUARD KERN, Tübingen, S. 49 ff. 1968.
Berger, Peter/ Luckmann, Thomas	DIE GESELLSCHAFTLICHE KONSTRUKTION DER WIRKLICHKEIT, Stuttgart, 1969.
Görlitz, Axel	VERWALTUNGSGERICHTSBARKEIT IN DEUTSCHLAND, Neuwied-Berlin, 1970
Hartfiel, Günter	WÖRTERBUCH DER SOZIOLOGIE, Stuttgart, 1972.
Kern, Eduard	"Rechtspflege: Grundsätzliches und Übersicht," in Gerhard Anschütz/Richard Thoma, HANDBUCH DES DEUTSCHEN STAATSRECHTS, Bd. 2, Tübingen, S. 475 ff., 1932.
Klausa, Ekkehard	EHRENAMTLICHE RICHTER: IHRE AUSWAHL UND FUNKTION--EMPIRISCH UNTERSUCHT, Frankfurt, 1972.
————	SOZIOLOGISCHE WAHRHEIT ZWISCHEN SUBJEKTIVER TATSACHE UND WIS-

	SENSCHAFTLICHEM WERTURTEIL, Berlin, 1974.
Opp, Karl-Dieter	SOZIOLOGIE IM RECHT, Reinbek, 1973.
Ridder, Helmut	"Empfiehlt es sich, die vollständige Selbstverwaltung aller Gerichte im Rahmen des Grundgesetzes gesetzlich einzuführen?, Gutachten," in VERHANDLUNGEN DES 40. DEUTSCHEN JURISTENTAGES, Bd. 1, Tübingen, S. 91 ff., 1953.
Ryffel, Hans	GRUNDPROBLEME DER RECHTS- UND STAATSPHILOSOPHIE: PHILOSOPHISCHE ANTHROPOLOGIE DES POLITISCHEN, Neuwied-Berlin, 1969.
————	RECHTSSOZIOLOGIE: EINE SYSTEMATISCHE ORIENTIERUNG, Neuwied-Berlin, 1974.
Rüggeberg, Jörg	"Zur Funktion der ehrenamtlichen Richter in den öffentlichrechtlichen Gerichtsbarkeiten," in VERWALTUNGSARCHIV, S. 189 ff., 1970.
Schiffmann, Gerfried	DIE BEDEUTUNG DER EHRENAMTLICHEN RICHTER BEI GERICHTEN DER ALLGEMEINEN VERWALTUNGSGERICHTSBARKEIT, Berlin, 1974.

This paper has two aims; the first concerns the substance of citizen participation, the second concerns methodology:

(1) Previous empirical research (400 interviews with professional and lay-judges and those selecting them) has led the author to the broad hypothesis that effective citizen participation may be hindered not only by lack of qualification but also by the attitudes of professionals who are reluctant to share their power and to accept non-experts as their equals.

(2) The methodological aim of this paper is to show how an empirical sociology of law can make a contribution to a philosophical discourse on values (in this case, the discussion on the value of lay-judges). Ultimately, a commitment to a value such as citizen participation cannot derive from empirical evidence. However, value judgements in general and particularly those concerning laymen's participation in legal process often imply empirical assumptions which can and should be put to the test.

The author has found that professional judges dealing with the same subject matter in the same court with lay-judges selected in the same manner had widely differing views on the value of their lay colleagues. The most plausible hypothesis to explain this is that the professionals did indeed have very different experiences with lay-judges--but that they themselves were in the last resort responsible for this experience, as they "constructed their reality" themselves.

Liberal professionals willing to accept the lay-judges as colleagues contribute significantly to their positive performance. Conservatives who take a dim view of "incompetent" laymen easily reinforce their incompetence. This thesis is supported by evidence from lay-judges who had very different experiences with different professional judges. Depending on the attitude of the professionals, they felt respectively incompetent and useless or helpful.